NEAR-RINGS AND NEAR-FIELDS

NORTH-HOLLAND MATHEMATICS STUDIES 137

NORTH-HOLLAND – AMSTERDAM • NEW YORK • OXFORD • TOKYO

NEAR-RINGS AND NEAR-FIELDS

Proceedings of a Conference held at the University of Tübingen, F.R.G. 4-10 August, 1985

edited by

Gerhard BETSCH

Mathematical Institute
University of Tübingen
Federal Republic of Germany

1987

NORTH-HOLLAND – AMSTERDAM • NEW YORK • OXFORD • TOKYO

ISBN: 0 444 70191 5

Publishers:
ELSEVIER SCIENCE PUBLISHERS B.V.
P.O. BOX 1991
1000 BZ AMSTERDAM
THE NETHERLANDS

Sole distributors for the U.S.A. and Canada:
ELSEVIER SCIENCE PUBLISHING COMPANY, INC.
52 VANDERBILT AVENUE
NEW YORK, N.Y. 10017
U.S.A.

Library of Congress Cataloging-in-Publication Data

Near-rings and near-fields.

(North-Holland mathematics studies ; 137)
Papers from an international Conference on Near-rings and Near-fields, held at the Mathematical Institute of the University of Tübingen, Aug. 4-10, 1985.
Includes bibliographies.
1. Near-rings--Congresses. 2. Near-fields--Congresses. I. Betsch, Gerhard, 1934- . II. Conference on Near-rings and Near-fields (1985 : Mathematical Institute of the University of Tübingen) III. Series.
QA251.5.N43 1987 512'.4 86-32931
ISBN 0-444-70191-5 (U.S.)

PRINTED IN THE NETHERLANDS

PREFACE

This volume contains the substance of two invited lectures and 30 communications delivered at an international

Conference on Near-rings and Near-fields,

which was held in the Mathematical Institute of the University of Tübingen, 4th - 10th August 1985.

All papers in this volume have been refereed. The editor expresses his gratitude to all colleagues who helped in examining the manuscripts.

Special thanks are due to Günter Pilz for his unfailing help and advice, and to his secretary Waltraud Eidljörg for a lot of work on the way to a camera-ready manuscript.

The participants of the conference greatly appreciated the hospitality of the Tübingen Mathematical Faculty and Mathematical Institute. The editor (and organiser) expresses his thanks to Dean Prof. W. Kaup and the staff of the Institute for their valuable assistance.

Several participants received financial support. In this context the editor gratefully acknowledges generous support by the

- Deutsche Akademische Austauschdienst, Bonn

and by the

- Vereinigung der Freunde der Universität Tübingen e.V. (Universitätsbund).

Finally, the editor would like to thank the North-Holland Publishing Company for including this volume into the Mathematics Studies series and for excellent cooperation.

G. Betsch

LIST OF REGISTERED PARTICIPANTS

Johannes André, Saarbrücken, F.R. Germany
Howard E. Bell, St. Catherines, Ontario, Canada
Gerhard Betsch, Tübingen, F.R. Germany
Donald W. Blackett, West Newton, MA, USA
Anne Buys, Port Elizabeth, South Africa
James R. Clay, Tucson, AZ, USA
Simonetta Di Sieno, Milano, Italy
Ulrich Felgner, Tübingen, F.R. Germany
Celestina Ferrero Cotti, Parma, Italy
Giovanni Ferrero, Parma, Italy
Yuen Fong, Tainan, Taiwan
Peter Fuchs, Attnang-Puchheim, Austria
Gert K. Gerber, Port Elizabeth, South Africa
Artur Grigori, Northridge, Calif., USA
N.J. Groenewald, Port Elizabeth, South Africa
Theo Grundhöfer, Tübingen, F.R. Germany
Werner Heise, München, F.R. Germany
Gerhard Hofer, Linz, Austria
Helmut Karzel, München, F.R. Germany
Hermann Kautschitsch, Klagenfurt, Austria
William E. Kerby, Hamburg, F.R. Germany
Suraiya J. Mahmood, Riyadh, Saudi Arabia
Carlton J. Maxson, College Station, Texas, USA
John D.P. Meldrum, Edinburgh, U.K.
J.H. Meyer, Stellenbosch, South Africa
Rainer Mlitz, Wien, Austria
Hans Ney, Saarbrücken, F.R. Germany
Dorota Niewieczerzał, Warsaw, Poland
Alan Oswald, Middlesbrough, U.K.
Silvia Pellegrini Manara, Brescia, Italy
Günter Pilz, Linz, Austria
D. Ramakotaiah, Nagarjunanagar, India
Maic Sasso-Sant, Saarlouis, F.R. Germany
Raffaele Scapellato, Parma, Italy
Mirela Stefanescu, Iasi, Romania
Alberta Suppa, Parma, Italy
V. Tharmaratnam, Thirunelvely, Sri Lanka
S. Veldsman, Port Elizabeth, South Africa
Andries P.J. van der Walt, Stellenbosch, South Africa
Heinrich Wefelscheid, Duisburg, F.R. Germany
Hanns Joachim Weinert, Clausthal-Zellerfeld, F.R. Germany
Richard Wiegandt, Budapest, Hungary

INTRODUCTION

By Gerhard Betsch

The papers in this volume provide a fairly complete picture of current trends and problems in the theory of near-rings and near-fields. In the sequel we shall describe a few condensation points. But first we shall mention some

Basic definitions

A right near-ring is a triple (N,+,·), where N is a set and + and · are binary operations on N such that

(i) (N,+) is a group (not necessarily commutative);

(ii) multiplication is associative;

(iii) $(a+b)c=ac+bc$ for all $a,b,c \in N$.

A left near-ring is defined in an analogous manner: (iii) has to be replaced by the other distributive law

(iii') $c(a+b)=ca+cb$ for all $a,b,c \in N$.

In the sequel we shall - with few exceptions - simply speak of "near-rings", because frequently it is inessential for the purpose of this introduction, which type of near-rings (right or left) is investigated in the paper under discussion.

Examples of near-rings are abundant: They arise in a natural way whenever one deals with systems of "non-linear" mappings. E.g., let $(\Gamma,+)$ be a not necessarily commutative group. Then all mappings of Γ into itself (written as left operators of Γ) form a (right) near-ring $M(\Gamma)$ with repect to pointwise addition and multiplication by composition. In most cases, $M(\Gamma)$ has many subnear-rings. Conversely, it can be proved, that any (right) near-ring N may be embedded into a suitable near-ring $M(\Gamma)$.

A near-ring N is called zero-symmetric, if $n0=0n=0$ holds for all $n \in N$ (0 is the neutral element of the additive group (N,+,)).

A near-field is a near-ring N with the property that $(N \setminus \{0\},\cdot)$ is a group.
All near-fields with more than 2 elements are zero-symmetric. The additive group of a near-field is commutative [11, Prop. 8.1 and Theorem 8.11].

(Numbers in square brackets refer to the short bibliography at the end of this introduction, while names in capital letters refer to the paper of the quoted author(s) in this volume.)

The ideals of a near-ring N may be defined as the kernels of (near-ring) homomorphisms of N. If N is a right near-ring, then an N-group is the near-ring analogue of a left module over a ring. (N,+) is an N-group by left multiplication. The left ideals of N are precisely the kernels of N-group

homomorphisms of (N,+).
The concept of a primitive ring, the various radicals, etc. can be generalized to near-rings, but in such a way, that e.g. different Jacobson type radicals J_ν (ν=0,1,2,3) of near-rings correspond to the Jacobson radical J of rings. If N is a ring, then $J_0(N)=J_1(N)=J_2(N)=J_3(N)=J(N)$.
For further definitions and details we refer to the monograph [11].
Now we turn to the subjects treated in this book.

Near-fields

The investigation of near-fields is the oldest branch of the theory. In 1905 L.E. Dickson gave the first known example of a near-field which is not a field, and proved a series of basic results [5], [6]. One of the main questions in near-field theory is, whether a given abstract near-field is a Dickson near-field, i.e. can be constructed by a method going back essentially to Dickson. Zassenhaus [13] determined all finite near-fields. There are exactly seven isomorphism types which are not Dickson. Until now all known infinite near-fields are Dickson near-fields. GRUNDHÖFER proves, that every non-discrete locally compact (topological) near-field of characteristic zero is a Dickson near-field. KARZEL in his paper provides a far-reaching generalization of Dickson's construction to "structure groups". These are multiplicative groups with an additional structure such that left multiplication by group elements induces an automorphism of the structure. "Dickson structure groups" are then obtained from structure groups by means of coupling maps. The referee of Karzel's paper suggested, that it might be advisable now to look for a general theory of coupling maps, which covers all known applications.
In studying infinite near-fields, suitable finiteness conditions proved very useful. FELGNER presents a series of results on near-fields which are pseudo-finite (= infinite, and elementarily equivalent to an ultraproduct of finite near-fields). Felgner's results give an example of the impressive power of model theoretic methods.

In this context we also mention the paper by FUCHS, in which ultraproducts of affine near-rings are investigated. The vitality of near-field theory will also be shown in a forthcoming monograph by H. Wähling, to appear in 1987.

Links with Geometry

As early as 1907 near-fields have been applied to geometry [12]. In the 1960's Karzel and his students investigated near-fields in connections with incidence groups [10]. KARZEL's paper in the present volume is still related to this line of research. G. Ferrero, the editor, and J.R. Clay established the interplay between planar near-rings and Frobenius groups with combinatorial structures, like Balanced Incomplete Block Designs (BIBD's), PBIBD's, etc. [7], [3].
In this volume, ANDRÉ presents a survey of non-commutative geometry, which he initiated in 1975 [2]. The paper by SASSO-SANT deals with relations between near-rings, non-commutative geometry, and combinatorial structures, thus combining non-commutative geometry with the problems and methods of [7] and [3].

Also, SCAPELLATO's contribution fits into this context.
Finally, in MAXSON's paper near-rings are associated to generalized translation spaces, and the influence of the geometry on the algebraic structures (near-rings under consideration) is investigated.

Radicals

A lot of work has been done to define and investigate concrete near-ring radicals corresponding to the various known ring radicals. Most results, which were obtained until 1983 in this context, may be found in the monograph by G. Pilz [11]. The paper by GROENEWALD "On the completely prime radical in near-rings" follows this line a bit further, applying methods from general radical theory. Since zero-symmetric near-rings are Ω-groups in the sense of Higgins [8], the paper by GERBER on "Radicals of Ω-groups defined by means of elements" bears immediate relevance for near-rings.
Having studied various concrete radicals it was then fairly natural to turn to abstract Kurosh-Amitsur radicals (KA-radicals) of near-rings. In [4], the editor and R. Wiegandt proved, that KA-radicals of associative near-rings differ from KA-radicals of rings in a remarkable way: There are plenty of KA-radicals of (associative) near-rings with non-hereditary semisimple classes. Hence the famous theorem of Anderson-Divinsky-Sulinsky [1] does not carry over to KA-radicals of near-rings. But while in the class of associative near-rings we still have both hereditary and non-hereditary semisimple classes, the situation may be rather bad, if one drops associativity: VELDSMAN shows, that in the class of all not necessarily associative zero-symmetric near-rings with commutative addition the only KA-radicals with hereditary semisimple class are the two trivial radical classes.

The "classical" radicals of near-rings are not necessarily KA-radicals. In fact, among the Jacobson type radicals J_v the radicals J_2 and J_3 are KA-radicals, while the radicals J_0 and J_1 are not KA-radicals. (See [11; p. 143/144] for some details and references.)
Recently (1986) Kaarli and Kriis proved, that the prime radical of near-rings is also not a KA-radical [9]. Hence one might ask, whether the KA-radicals are an appropriate type of abstract radicals in the case of near-rings. This is the background of the paper by MLITZ. He investigates, whether the Jacobson type radicals of near-rings might fit into the more general concept of M-radicals.

Matrix near-rings

Matrix near-rings over arbitrary near-rings were introduced recently by Meldrum and van der Walt. The purpose of van der WALT's paper is, to give a detailed account of the correspondence between the two-sided ideals of the base near-ring and those in the matrix near-ring.
MEYER and van der WALT construct a matrix near-ring over an infinite near-field, which is a 2-primitive non-ring with identiy and with a certain additional property. The existence of such a non-ring solves a problem posed by the editor in 1971.

One may expect that the matrix near-rings of Meldrum and van der Walt will provide quite an effective machinery for constructing crucial examples or counterexamples.

Of course, a substantial portion of the papers in this volume may not be subsumed under a precise headline, but have to be listed under

Miscellaneous topics

The papers by FONG and MELDRUM and by MAHMOOD and MELDRUM are devoted to certain distributively generated near-rings - since many years an attractive subject in near-ring theory.
CLAY also deals with d.g. near-rings. He proves, that certain near-rings are in fact commutative rings with identity; the structure of these rings and the structure of their group of units is determined. Two notes by DE STEFANO and DI SIENO deal with distributive near-rings. The paper by MELDRUM and van der WALT is devoted to the important subject of Krull dimension and tame near-rings. (For the definition of a tame near-ring see [11; § 9g].)
The medial near-rings studied in the contribution of PELLEGRINI MANARA have a subdirect structure which is very close to that of strongly IFP-near-rings (see [11; § 9a]).
Reduced near-rings (= near-rings without non-zero nilpotent elements) are investigated by RAMAKOTAIAH and SAMBASIVARAO. Result: The set of all idempotents of a reduced near-ring with identity forms a Boolean algebra under a specific partial ordering.
The paper by FERRERO COTTI is devoted to near-rings with the property, that all endomorphisms commute with right (or left) translations. In many cases near-rings of this type have only the identical automorphism.
WIEGANDT proves: Let A be a subdirectly irreducible right near-ring. If a left invariant subset of A satisfies a certain permutation identity, then A is a field. The main tool for the proof is a lemma, which is purely semigroup theoretical.
BELL and MASON present several commutativity theorems for near-rings admitting suitably constrained derivations.
KAUTSCHITSCH studies maximal ideals in certain near-rings, the structrue of which is not too far from polynomial and power series near-rings.
BETSCH presents a class of near-rings, which cannot be one-sided ideals in a near-ring with identity. Also, he characterizes those near-rings, which do have an embedding as a one-sided ideal into a near-ring with identity.

Applications: System theory and automata

The subject of the paper by PILZ are "separable" (non-linear) dynamical systems (Q,A,B,F,G), where Q,A,B are groups (consisting of the states, inputs, outputs, respectively) and the mappings F,G can be "separated" in a certain way.
Separable systems with identical input- and output groups form a near-ring with respect to parallel and series connections. System-theoretic questions like feedbacks, reachability questions, invertibility,... are studied within this framework. Concerning the paper by HOFER: If the set R of

states of an automaton is furnished with the structure of a module, the mappings of R into itself form a near-ring, called the syntactic near-ring of the automaton. The present paper shows how the knowledge of the ideals of the zero-symmetric part of the syntactic near-ring may be applied to determine reachability in the automaton.

Generalizations of near-rings

Partially and fully ordered seminear-rings and near-rings are studied by WEINERT, while HEBISCH and WEINERT deal with euclidean seminear-rings and near-rings.

STEFANESCU presents a ternary interpretation of infra-near-rings - to be more precise: of the addition in infra-near-rings. This interpretation explains the relationship between infra-near-rings and prerings. Furthermore, in this way left infra-distributivity may be understood as distributivity of the multiplication with respect to a ternary composition.

Bibliography

[1] Anderson, T. - Divinsky, N. - Sulinski, A., Hereditary radicals in associative and alternative rings. Canad. J. Math. **17** (1965), 594-603.

[2] André, J., Affine Geometrien über Fastkörpern. Mitt. Math. Sem. Gießen **114** (1975), 1-99.

[3] Betsch, G. - Clay, J.R., Block designs from Frobenius groups and planar near-rings. Proc. Conf. finite groups (Park City, Utah 1973), p. 473-502. Acad. Press 1976.

[4] Betsch, G. - Wiegandt, R., Non hereditary semisimple classes of near-rings. Studia Math. Hungar. **17** (1982), 69-75.

[5] Dickson, L.E., Definitions of a group and a field by independent postulates. Trans. Amer. Math. Soc. **6** (1905), 198-204.

[6] Dickson, L.E., On finite algebras. Ges. d. Wiss. zu Göttingen, Nachr. math.-phys. Klasse (1905), p. 358-393.

[7] Ferrero, G., Stems planari e BIB-disegni. Riv. Mat. Univ. Parma (2) **11** (1970), 79-96.

[8] Higgins, P.J., Groups with multiple operators. Proc. London Math. Soc. (3) **6** (1956), 366-416.

[9] Kaarli, K.- Kriis, T., Prime radical of near-rings (1986). To appear.

[10] Karzel, H., Bericht über projektive Inzidenzgruppen. Jahresber. Dt. Math. Ver. **67** (1965), 58-92.

[11] Pilz, G., Near-rings. Revised Ed. Amsterdam: North-Holland 1983.

[12] Veblen, O. - Maclagan - Wedderburn, J.H., Non desarguesian and nonpascalian geometries. Trans. Amer. Math. Soc. **8** (1907), 379-388.

[13] Zassenhaus, H., Über endliche Fastkörper. Abh. Math. Sem. Univ. Hamburg **11** (1936), 187-220.

CONTENTS

Preface v
List of Registered Participants vi
Introduction vii

INVITED LECTURES

J. André, Non-commutative geometry, near-rings and near-fields 1
U. Felgner, Pseudo-finite near-fields 15

COMMUNICATIONS

H.E. Bell and G. Mason, On derivations in near-rings 31
G. Betsch, Embedding of a near-ring into a near-ring with identity 37
J.R. Clay, The near-ring of some one-dimensional noncommutative formal group laws 41
S. De Stefano and S. Di Sieno, On the existence of nil ideals in distributive near-rings 53
S. De Stefano and S. Di Sieno, Distributive near-rings with minimal square 59
C. Cotti Ferrero, Near-rings with E-permutable translations 63
Y. Fong and J.D.P. Meldrum, Endomorphism near-rings of a direct sum of isomorphic finite simple non-abelian groups 73
P. Fuchs, On the ideal structure in ultraproducts of affine near-rings 79
G.K. Gerber, Radicals of Ω-groups by means of elements 87
N.J. Groenewald, Note on the completely prime radical in near-rings 97
Th. Grundhöfer, On p-adic near-fields 101
U. Hebisch and H.J. Weinert, Euclidean seminear-rings and near-rings 105
G. Hofer, Ideals and reachability in machines 123
H. Karzel, Couplings and derived structures 133
H. Kautschitsch, Maximal ideals in near-rings 145
S.J. Mahmood and J.D.P. Meldrum, D.g. near-rings on the infinite dihedral group 151
C.J. Maxson, Near-rings associated with covered groups 167
J.D.P. Meldrum and A.P.J. van der Walt, Krull dimension and tame near-rings 175
J.H. Meyer and A.P.J. van der Walt, Solution of an open problem concerning 2-primitive near-rings 185
R. Mlitz, Are the Jacobson-radicals of near-rings M-radicals? 193
S. Pellegrini Manara, On medial near-rings 199
G. Pilz, Near-rings and non-linear dynamical systems 211
D. Ramakotaiah and V. Sambasivarao, Reduced near-rings 233

M. Sasso-Sant, Non-commutative spaces and near-rings including PBIBD's planar near-rings and non-commutative geometry 245
R. Scapellato, On geometric near-rings 253
M. Stefănescu, A ternary interpretation of the infra-near rings. 255
A.P.J. van der Walt, On two-sided ideals in matrix near-rings. 267
S. Veldsman, Some pathology for radicals in non-associative near-rings 273
H.J. Weinert, Partially and fully ordered seminear-rings and near-rings 277
R. Wiegandt, On subdirectly irreducible near-rings which are fields. 295

Near-rings and Near-fields, G. Betsch (editor)

NON-COMMUTATIVE GEOMETRY, NEAR-RINGS AND NEAR-FIELDS

Johannes ANDRÉ

Fachbereich Mathematik
der Universität des Saarlandes
D-6600 Saarbrücken, Federal Republic of Germany

Some relations between near-rings (especially near-fields) and spaces with a non-commutative join are considered.

0. INTRODUCTION

Eighty years ago L.E.Dickson (1905) discovered the nearfields thus proving that the both distributive laws are independent of each other. Veblen, Wedderburn 1907 gave the first geometric application: They constructed projective planes using nearfields. Zassenhaus 1935 rediscovered the nearfields when he considered the structure of finite sharply twofold and threefold transitive permutation groups (see also e.g. Wielandt 1964 and Kerby 1974). In the sixties further geometrical applications of nearfields follow (see e.g. Arnold 1968, Karzel 1968). Not only plane geometries but also higher dimensional spaces were constructed by nearfields. Arnold 1968 e.g. considered the following spaces: Let F be a nearfield (with the distributive law $(a+b)c = ac+bc$) and let F^n be the point-set of this space. The lines are defined by

$$(0.1) \qquad \overline{xy} := F(y-x) + x \qquad (x,y \in F^n) \quad ,$$

where, as usual, addition and multiplication are defined componentwise. Two different points are incident with exactly one line (but the general Veblen-condition only holds if F is a skewfield). The spaces thus defined are special Sperner-spaces (cf.Sperner 1960) which Arnold was able to characterize purely geometrically.

Consider now spaces with the same point-sets; the lines, however, are defined by

$$(0.2) \qquad x \sqcup y := (y-x)F + x \quad .$$

Here the operation $\sqcup$ is not commutative in general. Historically we thus obtained the first examples of non-commutative spaces (cf. e.g. André 1975). But there are simpler types of such spaces: Consider e.g. the circles of the euclidean plane together with their centres. In this case $x \sqcup y$ means the circle through y with the centre x (circle space). Besides to nearrings and nearfields the non-commutative geometry can be applied (among other things) to groups, especially permutation groups, and to graphs and other combinatorial structures (see e.g. André 1984).

In the first section the non-commutative spaces are introduced in a way stated by Pfalzgraf 1984,1985b (which is more general than that originally given by the author). Section 2 gives some geometrical closure conditions (Schließungsaussagen): The well-known conditions of Veblen and Desargues can be embedded into an infinite hierarchy of simplex-conditions (Pfalzgraf 1985a). The next section brings the first contact to nearfields. The nearfield spaces already indicated in (0.2) will be defined in more detail. Special interests seem to have the improper structures of such spaces. They lead to special rank-two-nearfields the so-called Biliotti-nearfields considered in section 4. The following section gives an other way to introduce noncommutative spaces over nearfields closely related to sharply twofold transitive groups (Rößler 1985). The last two sections concern spaces over special types of nearrings (the one and the higher dimensional case resp.) which have been considered especially by E.Theobald (1981) and H.Ney (1983).

1. GEOMETRIC SPACES

Let X and R be non void sets whose elements are called <u>proper</u> and <u>improper points</u> resp. A <u>geometric space</u> or more briefly <u>space</u> (Pfalzgraf 1984,1985b) is a collection

(1.1) $\underline{S} = (X,R,\langle,\rangle)$

where

(1.2) $\langle,\rangle : X^2 \setminus i_X \twoheadrightarrow R, \quad (x,y) \mapsto \langle x,y\rangle \in R \quad (x,y \in X, x \neq y)$

is a surjective mapping*. The <u>union-line</u> or more briefly <u>line</u> (Linie) $x \square y$ of the two different points x,y (<u>in this order</u>) is defined by

(1.3) $x \square y := (x \sqcup y) \cup \{\langle x,y\rangle\}$

where

(1.3') $x \sqcup y := \{x\} \cup \{z \in X \mid \langle x,y\rangle = \langle x,z\rangle\}$.

Here $x \sqcup y$ is the <u>proper part</u> of the line $x \square y$ and $\langle x,y\rangle$ its

*Sometimes it is advisable to extend this mapping to X^2 (instead $X^2 \setminus i_X$), where $\langle x,x\rangle$ does not depend on x and $\langle x,x\rangle = \langle x,y\rangle$ implies $x=y$. (See also Pfalzgraf 1984,1985b).

improper point. The point x is called a *basepoint* (Aufpunkt) of $x \sqcup y$ (or $x \square y$), denoted by

(1.4) $$x \prec x \sqcup y \quad \text{or} \quad x \prec x \square y \quad .$$

Obviously $x \prec L$ implies $x \in L$ but not vice versa.

The following two properties on lines are easy to check.

(L1) $x, y \in x \sqcup y$ (*incidence condition*) ,

(L2) $z \in (x \sqcup y) \setminus \{x\}$ *implies* $x \sqcup y = x \sqcup z$ (*exchange condition*) .

Two lines $x \square y$ and $x' \square y'$ are called *parallel*, denoted by

(1.5) $$x \square y \parallel x' \square y' \quad ,$$

iff they have the same improper point, i.e. iff $\langle x,y \rangle = \langle x',y' \rangle$. Clearly $\parallel$ is an equivalence relation. Moreover the following condition holds

(P1') *Given an* $x \in X$ *and a line* L *there exists at most one line* L' *with a basepoint* x (i.e. of the form $L' = x \square y$) *such that* $L \parallel L'$. *This line* (if it exists) *is denoted by*

(1.6) $$(x \parallel L) \quad .$$

Obviously

(1.7) $$(x \parallel L) = L \quad \text{if} \quad x \prec L \quad \text{and} \quad (x \parallel (y \parallel L)) = (x \parallel L)$$

provided that these lines do exist. A refinement of (P1') is the *Euclidean parallel-condition*

(P1) *To any line* L *and any* $x \in X$ *there exists exactly one line* $L' \parallel L$ *with* $x \prec L'$. In this case $(x \parallel L)$ thus always exists.

EXAMPLES OF GEOMETRIC SPACES

(i) *Group spaces*. Let G be a group acting on X . Define

(1.8) $$\langle x,y \rangle := G(x,y) := \{(gx,gy) \mid g \in G\} \quad .$$

Then (1.3') yields

(1.8') $$x \sqcup y = G_x\{x,y\} = \{x\} \cup G_x y \quad .$$

In this case $\underline{S} = \underline{V}(G)$ is called a *group-space* with respect to G.

(ii) *Distance-spaces*. Let X be a metric space and d its distance. Define $\langle x,y \rangle := d(x,y)$. An example of a distance-space is the circle space stated in the introduction. Also any connected graph leads to a distance-space in a natural way.

(iii) Let be $X = R = \mathbb{R}$ and put $\langle x,y \rangle := y-x$. In this case we have $x \sqcup y = \{x,y\} = y \sqcup x$.

DEFINITION 1.1. The geometric space $\underline{S} = (X,R,\langle,\rangle)$ is called <u>properly determined</u>* (<u>p.d.</u>) iff

(1.9) $x \sqcup y = x' \sqcup y'$ <u>implies</u> $\langle x,y\rangle = \langle x',y'\rangle$,

i.e. iff the lines are uniquely determined by their proper parts (c.f. Pfalzgraf 1985b). Inthis case it makes sense to say that two proper parts of lines are parallel, i.e.

(1.9') $x \sqcup y \parallel x' \sqcup y'$ means $\langle x,y\rangle = \langle x',y'\rangle$.

DEFINITION 1.2. A space is called <u>self-adjoint</u> or <u>symmetric</u> iff

(1.10) $\langle x,y\rangle = \langle y,x\rangle$

holds for all $x,y \in X$.

REMARK. There exist spaces not being p.d. (Pfalzgraf 1985b), see e.g. example (iii). However

PROPOSITION 1.1. <u>Any self-adjoint space is properly determined</u>.
PROOF. Assume $x \sqcup y = x' \sqcup y'$. Then $\langle x,y\rangle = \langle x',y'\rangle$ is obviously true if $x = x'$. Assume now $x \neq x'$. Applying (L2) twice we obtain $x \sqcup x' = x \sqcup y = x' \sqcup y' = x' \sqcup x$. Hence $\langle x,y\rangle = \langle x,x'\rangle = \langle x',x\rangle = \langle x',y'\rangle$ by (1.3') and because the space is selfadjoint. □

DEFINITION 1.3. Let be $\underline{S} = (X,R,\langle,\rangle)$ a space. Then $U \subseteq X$ is called a <u>subspace</u> if $x,y,z \in U$ imply $(x \parallel y \sqcup z) \subseteq U$. The space $\underline{S}$ is called <u>primitive</u> if it contains only the trivial subspaces $\emptyset$, $\{x\}$ $(x \in X)$ and X , otherwise <u>imprimitive</u>.

2. PFALZGRAF'S HIERARCHY (Pfalzgraf 1985a)

The geometric spaces defined above are still too general to obtain interesting properties. It may be reasonable to require a hierarchy of geometrical closure properties (Schließungsaussagen).

The following <u>q-simplex-condition</u> Sim_q $(q \in \mathbb{N})$ is due to J.Pfalzgraf (1985a): <u>Given</u> $x_0,x_1,\ldots,x_q,x_0',x_1' \in X$ <u>with</u> $\langle x_0,x_1\rangle = \langle x_0',x_1'\rangle$. <u>Then there exist</u> $x_2',\ldots,x_q' \in X$ <u>such that</u> $\langle x_i',x_j'\rangle = \langle x_i,x_j\rangle$ <u>for</u> $i,j \in \{0,1,\ldots,q\}$, $i \neq j$.

If additionally $x_0' = x_0$ holds we obtain the <u>diminished</u> (verjüngte) <u>q-simplex-condition</u> Sim_q^b .

REMARKS. (i) Sim_1 holds in any space. Sim_2 is the <u>Tamaschke-condition</u> (see e.g. André 1981). Spaces with Sim_2 are called <u>skew-affine</u>. A skewaffine space additionally with $x \square y = y \square x$ for all $x,y \in X$ is an affine space in the usual sense. The <u>Veblen</u> and the

* In German: <u>Eigentlich bestimmt</u>.

Desargues condition are exactly Sim_2^b and Sim_3^b resp.

(ii) For any $q \in \mathbb{N}$ there exist spaces with Sim_q (and hence Sim_r for $r \leq q$) but not with Sim_{q+1} (Pfalzgraf 1985a).

The following proposition gives a purely geometric characterization of finite group-spaces.

THEOREM 2.1. A finite geometric space is a group-space (cf. example (i) in Section 1) iff Sim_q hold for all $q \in \mathbb{N}$.

PROOF. Obviously all Sim_q hold in any group-space (not necessarily finite). Conversely let $\underline{S}(X,R,<,>)$ be a finite space with Sim_q . Assume $X = \{x_0, x_1, \ldots, x_q\}$ and $x'_0, x'_1 \in X$ with $\langle x_0, x_1 \rangle = \langle x'_0, x'_1 \rangle$. Then Sim_q implies the existence of $x'_2, \ldots, x'_q$ with $\langle x_i, x_j \rangle = \langle x'_i, x'_j \rangle$. As a consequence we have $x'_i \neq x'_j$ for $i \neq j$, hence $X = \{x'_0, \ldots, x'_q\}$. Let g be the permutation on X with $gx_i = x'_i$ for all $i \in \{0,1,\ldots,q\}$. The permutation group G generated by all these g is transitve on all pairs (x,x') with a given $\langle x,x' \rangle \in R$. This implies $\underline{S} = \underline{V}(G)$. □

REMARK. Theorem 2.1 is not true for infinite spaces (cf. Pfalzgraf 1985a). The problem of obtaining a purely geoemtric characterization for arbitrary group-spaces thus remains unsolved.

3. SKEWAFFINE, NEARAFFINE AND NEARFIELD-SPACES

DEFINITION 3.1. A line in a geometric space is called a straight line (Gerade) if all of its proper points are basepoints. Hence a line L is straight iff for any two of its proper points x,y with $x \neq y$ we have $L = x \sqcup y$.

REMARK. If the space $\underline{S}$ is skewaffine (cf. sect.2, remark (i)) then a line L is already straight if it possesses two different basepoints. Moreover, if L is straight and $L' \| L$ then also L' is straight (André 1981, Satz 1.1).

PROPOSITION 3.1. Let $\underline{S}$ be a skewaffine space in which any line has at least three proper points. Then $\underline{S}$ is p.d.

PROOF. Let be $L = x \sqcup y = x' \sqcup y'$. The case $x = x'$ is clear. Assume $x \neq x'$. Due to the last remark $L = x \sqcup x' = x' \sqcup x$ is straight. Because of $|L| \geq 3$ there exists a $z \in L \setminus \{x,y\}$. Repeated application of (L2) yield $\langle x,y \rangle = \langle x,x' \rangle = \langle x,z \rangle = \langle x',z \rangle = \langle x',y' \rangle$. □

DEFINITION 3.2. A selfadjoint skewaffine space $\underline{S} = (X,R,<,>)$ is called nearaffine if

(i) any straight line meets any line different from it in at most one proper point,

(ii) any two proper points can be connected by a finite chain of straight lines.

It is called regular if given two straight lines L and L' with $L \cap L' = \{x\}$, then there exist $y \in L \setminus \{x\}$ and $y' \in L' \setminus \{x\}$ such that $y \sqcup y'$ is straight (André 1975).

Examples of regular nearaffine spaces are the nearfield-spaces: Given a nearfield F (with $(a+b)c = ac+bc$) and an index-set I define

(3.1) $X := F^{(I)} := \{(x_i)_{i\in I} \mid x_i \in F, x_i \neq 0$ for only finitely many $i \in I\}$,

(3.2) $R := \{uF \mid u \in X \setminus \{0\}\}$,

(3.3) $<x,y> := (y-x)F$ for all $x,y \in F$.

(Here addition and multiplication are defined in the usual way.) Easy computation yields

(3.4) $x \sqcup y = (y-x)F + x$.

By definition $\underline{S} = (X,R,<,>)$ is a nearfield-space. Such a space is nearaffine, desarguesian (i.e. Sim_3^b holds) and regular. Conversely any regular and desarguesian nearaffine space $\underline{S}$ is a nearfield-space $\underline{S}(F^{(I)})$ and $|I|$ and F are uniquely determined by $\underline{S}$ up to isomorphism (André 1975*).

Moreover $\underline{S}$ is a group-space with respect to the group of all transformations of the form $x \mapsto xa + b$ $(x,b \in X,\ a \in F \setminus [0])$.

Let $\underline{S}(F^{(I)})$ be a nearfield-space over F. Then $0 \sqcup u$ is straight iff for all $a,b \in F$ there is a $c \in F$ such that $ua+ub=uc$ (André 1975,II,Satz 1.2). An improper point uF with such a $u \neq 0$ is called internal, otherwise external (see also Dembowski 1971).

Let be $p := uF$ an internal and $q := vF \neq p$ an arbitrary improper point. The set of all improper points wF such that $w = ua+vb$ for suitable $a,b \in F$ with $(a,b) \neq (0,0)$ is an improper-line. It is called internal if q can be chosen as an internal point otherwise external.

The set R together with the improper lines is called the improper space to $\underline{S}(F^{(I)})$ and is denoted by $\underline{J}(F^{(I)})$ (see also André 1984, Chap.II,sect.3).

4. BILIOTTI-NEARFIELDS

It would be desirable to get further informations about the structure of $\underline{J}(F^{(I)})$. If $|I| = 3$ (in this case we write F^3 instead

*In that paper the other distributive law is assumed for a nearfield.

of $F^{(I)}$) then analogous to the desarguesian case $\underline{J}(F^3)$ should possess a plane geometric structure. Indeed, one can show that $\underline{J}(F^3)$ becomes a projective plane and, more precisely, a generalized Hughes-plane iff the follwowing configuration holds in $\underline{S}(F^3)$ (cf. André 1981, Satz 5.2):

Given two different lines L and L', both not being straight, with the common basepoint x, then there exist $y \in L \setminus \{x\}$ and $y' \in L' \setminus \{x\}$ such that $y \sqcup y'$ is straight (Biliotti-configuration)*.

We are now looking for an algebraic equivalent for F. Let K be the kernel of F defined by

(4.1) $K := K(F) := \{c \in F \mid c(x+y) = cx+cy \text{ for all } x,y \in F\}$.

Obviously K is a skewfield and F a left K-vectorspace. The dimension of F over K is called the rank of F. If F is a nearfield of rank 2 then any set $\{1,e\}$ with $e \in F \setminus K$ is a basis of F over K, hence

(4.2) $F = K + Ke$.

DEFINITION 4.1. A rank-2-nearfield $F = K+Ke$ is called a Biliotti-nearfield (cf. Biliotti 1979) iff

(4.3) $F = K + eK$

for $e \in F \setminus K$ holds.

REMARK. If (4.3) holds for a special $e \in F \setminus K$ then it is easy to see that it holds for all such e (Biliotti 1979).

PROPOSITION 4.1. Let F be a nearfield. Then the Biliotti-condition holds in $\underline{S}(F^3)$ iff F is a Biliotti-nearfield.

PROOF. André 1981, Satz 5.3**). □

REMARK. H.Zassenhaus (1986) recently pointed out that there do exist rank-2-nearfields not being Biliotti.

The following two propositions, however, show that many rank-2-nearfields usually ocurring are Biliotti.

PROPOSITION 4.2. A rank-2-nearfield F is Biliotti if one of the following conditions hold in F:

(i) F is finite.

(ii) Centre and kernel of F coincide.

PROOF. (i) by counting arguments, (ii) is trivial (see also André 1981, Satz 4.1). □

---

* In André 1981,p.458, called DrL . It is analogous to the regularity condition noted in sect.3.

**A Biliotti-nearfield is there called "Fastkörper mit Umkehrbed."

Let $\mathbb{H}$ be the skewfield of quaternions (with the standard-basis 1,i,j,k) and $t \in \mathbb{R}$. Define a new multiplication o on $\mathbb{H}$ by

$$(4.4) \qquad a \circ b := \begin{cases} |b|^{it} a |b|^{-it} b & b \neq 0 , \\ 0 & b = 0 . \end{cases}$$

Then $\mathbb{H}_t := (\mathbb{H},+,\circ)$ becomes a rank-2-nearfield (with kernel $\mathbb{C} = \mathbb{R} + \mathbb{R}i$ if $t \neq 0$), the Kalscheuer-nearfield (cf. Kalscheuer 1940, where he proved that any nearfield of finite rank over $\mathbb{R}$ with a continuous addition and multiplication is isomorphic to an $\mathbb{H}_t$ with a uniquely determined t).

PROPOSITION 4.3. Any Kalscheuer-nearfield is Biliotti.
PROOF. We must show $\mathbb{C} + \mathbb{C} \circ j = \mathbb{C} + j \circ \mathbb{C}$. Because $\mathbb{C} + j \circ \mathbb{C} \subseteq \mathbb{C} + \mathbb{C} \circ j$ is trivial it suffices to prove that for a given $c \in \mathbb{C} \setminus [0]$ the equation

$$(4.5) \qquad c \circ j = j \circ w$$

has a solution $w \in \mathbb{C}$. Due to $jc = \bar{c}j$ for all $c \in \mathbb{C}$ and (4.4) the equation (4.5) takes the form

$$(4.5') \qquad c = |w|^{2it} \bar{w} .$$

This yields $|c| = |w|$, consequently $c = |c|^{2it} \bar{w}$, hence $w = \bar{c}|c|^{2it}$ is a solution of (4.5). □

THEOREM 4.1. If F is a Biliotti-nearfield then the improper space $\underline{J}(F^3)$ of the nearfield-space $\underline{S}(F^3)$ is a generalized Hughes-plane and vice versa. If K is the kernel of F then $\underline{J}(K^3)$ is the (uniquely determined) desarguesian Baer-subplane of $\underline{J}(F^3)$.
PROOF. André 1981, §5. □

REMARK. Recently E.Scheid (1984) considered the improper space $\underline{J}(F^4)$ if F is a Biliotti-nearfield. Besides the (internal and external) improper points and lines also internal and external improper planes occur. The internal planes are (pairwise isomorphic) generalized Hughes-planes. Every external plane contains exactly one internal line and only external points outside this line. Only little is known about the structure of external planes. All internal points, lines and planes form a (desarguesian) three-dimensional projective space over K(F) analogous to the desarguesian Baer-subplane of $\underline{J}(F^3)$. It might be fruitful to search for further geometric properties of $\underline{J}(F^4)$ or, more generally, of $\underline{J}(F^n)$.

5. OTHER SPACES OVER NEARFIELDS

The following construction is due to Rößler 1985. Given a nearfield F (it is possible to extend this construction to neardomains, cf. e.g. Karzel 1968) and a subgroup U^* of $(F\setminus[0],\cdot)$, put $U := U^*\cup\{0\}$. Then define a space

$$(5.1) \qquad \underline{S} = \underline{S}(F,U) = (F,R,<,>)$$

with

$$(5.1') \qquad R := \{Ux \mid x\in F\setminus\{0\}\} \;, \quad <x,y> := U(y-x) \;.$$

A simple computation gives

$$(5.1'') \qquad x \sqcup y = U(y-x) + x \;.$$

The space $\underline{S}(F,U)$ is closely connected to the sharply twofold transitive groups of operations $x \mapsto xa + b$ $(x,a,b\in F,\ a\neq 0)$ on F. It is, however, no group-space (in the sense of sect.1, example (i)) in general. Rößler 1985 stated some sufficient conditions for this property.

PROPOSITION 5.1. The space $\underline{S}(F,U)$ is a group-space if $|U|\geq 3$ and $U \subseteq K(F)$ or U^* is normal in $(F\setminus[0],\cdot)$. □

We close this section with some further properties.

PROPOSITION 5.2 (Rößler 1985). The following properties hold in any $\underline{S}(F,U)$.

(i) $\underline{S}(F,U)$ is selfadjoint iff $-1 \in U$.

(ii) $\underline{S}(F,U)$ is commutative (i.e. $x \sqcup y = y \sqcup x$ holds for all $x,y\in F$) iff $U \pm U \subseteq U$. □

6. ONE-DIMENSIONAL SPACES OVER NEARRINGS

6.1. Consider a (at first arbitrary) nearring N (with $(a+b)c = ac+bc$) and try to generalize the situation stated in sect.3 after Definition 3.2, at first for $|I|= 1$. (This case is of course trivial if N is a nearfield.) We obtain

$$(6.1) \quad X=N \;, \quad R := \{uN \mid u\in N\setminus\{0\}\} \;, \quad <x,y> := (y-x)N \;.$$

In the space $\underline{S}(N) = (N,R,<,>)$ defined in that way we thus have

$$(6.2') \qquad x \sqcup y = \{x\}\cup\{z\in N \mid (y-x)N = (z-x)N\}$$

due to (1.3').

THEOREM 6.1. In the space $\underline{S}(N)$ defined by (6.1) the line $x \sqcup y$ takes the form

$$(6.3) \qquad x \sqcup y = (y-x)N + x$$

iff N is a geometric nearring (Theobald 1981) i.e. N has the following properties

(N0) N is zero-symmetric, i.e. $0a=a0=0$ for all $a\in N$ (cf. Pilz 1983).

(N1) For any $a\in N$ there exists a unit 1_a with $a1_a = a$.

(N2) $a\in bN\setminus[0]$ implies $aN = bN$.

PROOF. This follows by a simple computation. □

We now state some properties of the space $\underline{S}(N)$ for a geometric nearring N.

THEOREM 6.2. Let N be a geometric nearring. Then the space $\underline{S}(N)$ has the following properties:

(i) The euclidean condition (P1) holds in $\underline{S}(N)$.

(ii) $\underline{S}(N)$ is selfadjoint (cf. Def.1.2) iff N is symmetric, i.e. $aN = -aN$ holds for all $a\in N$.

(iii) Sim_q (cf.sect.2) holds for any $q\in N$. Hence $\underline{S}(N)$ is a group-space if N is finite (cf. Theorem 2.1). In this case the group G consists of all transformations $x\mapsto xa+b$ with $x,a,b\in N$ and $ya\neq 0$ for all $y\in N\setminus\{0\}$.

(iv) $\underline{S}(N)$ is properly determined iff $aN = bN+c$ implies $aN = bN$ for all $a,b,c\in N$ *.

PROOF. (i) and (ii) are obvious.

(iii) Let be $x_0,\dots,x_q,x_0';x_1'\in N$ with $(x_1-x_0)N = (x_1'-x_0')N$, hence $x_1' = (x_1-x_0)a + x_0'$ for a suitable $a\in N$. Define $x_i' := (x_i-x_0)a+x_0'$ for $i\in\{1,\dots,q\}$. Then $x_i'-x_j' = (x_i-x_j)a$ for $i,j\in\{0,1,\dots,q\}$, hence $(x_i'-x_j')N = (x_i-x_j)N$, i.e. the first part of (iii). Thus $\underline{S}(N)$ is a group-space if N is finite (Theorem 2.1). A simple computation yields the second part of (iii), cf. also Theobald 1981.

(iv) $\underline{S}(N)$ p.d. means: $(y-x)N+x = (y'-x')N+x'$ implies $(y-x)N = (y'-x')N$, i.e. (RE). The converse follows similarly. □

6.2. Examples of geometric nearrings are the finite planar nearrings (Pilz 1983,§8, Theobald 1981). In this case the lines of $\underline{S}(N)$ are the blocks of a suitable block-design (Ferrero 1970, Pilz 1983). M.Sasso-Sant (1986) proved that any integral planar nearring is geometric.

Other examples are the derivations of a nearfield discovered by Theobald 1981: Let F be a nearfield and let $\mu:F\to F$ be an idempotent endomorphism of $(F,\cdot)$. Define a new multiplication $\circ$ by

*Theobald 1981b denoted this condition by (RE), the direction uniqueness condition (Richtungseindeutigkeit).

$$(6.4) \qquad a \circ b := \begin{cases} a\mu(b) & b \neq 0\,, \\ 0 & b = 0\,. \end{cases}$$

Then $F_\mu := (F,+,\cdot)$ is a geometric nearring except if μ is the zero-mapping. If F is a skewfield then F_μ is also planar (Theobald 1981).

6.3. We close with some further properties of $\underline{S}(N)$.

PROPOSITION 6.1. *Let* $\underline{S}(N)$ *be a space over a geometric nearring* N. *Then* $U \subseteq N$ *is a subspace* (cf.Def.1.3) *going through* 0 *iff* $U \pm U \subseteq U$ *and* $UN \subseteq U$. *Especially if* $(N,+)$ *is abelian,* U *is a subspace through* 0 *iff it is a right ideal of* N. □

PROPOSITION 6.2. *Let* $\underline{S}(N)$ *be a space over a geometric nearring* N *with abelian addition. Then the parallel-closure-condition holds in* $\underline{S}(N)$, *i.e.* $x,y,z \in N$, *pairwise different, imply* $(y \| x \sqcup z) \cap (z \| x \sqcup y) \neq \emptyset$. □

REMARK. If a geometric nearring N with abelian addition has no nontrivial right ideals then the space $\underline{S}(N)$ is primitive.

7. HIGHER DIMENSIONAL SPACES OVER NEARRINGS

Let N be a geometric nearring, I an index-set. Consider the space $\underline{S}(N^{(I)}) = (X,R,<,>)$ where X, R and $<,>$ are defined analogously to (3.1), (3.2) and (3.3). Then

$$(7.1) \qquad x \sqcup y = (y-x)N + x$$

only holds if all units 1_a occurring in the definition of a geometric nearrings coincide, i.e. N contains a right unit 1 (i.e. $x1 = x$ for all $x \in N$). I call such nearrings *strongly geometric*. (Examples are the derivation rings and the finite planar nearrings.)

We are now able to state a structure theorem, analogous to that for nearfield-spaces $\underline{S}(F^{(I)})$, discovered by H.Ney.

THEOREM 7.1 (Ney 1983). *Let* N *be a strongly geometric nearring with abelian addition and no nontrivial right ideal. Then the space* $\underline{S}(N^{(I)})$ *has the following properties:*

(i) $\underline{S}(N^{(I)})$ *is a properly determined desarguesian skewaffine space with the parallel-closure-condition and at least three proper points on every line.*

(ii) *Any subspace meets any line not contained in it in at most one point.*

(iii) *Any two points can be connected by a finite chain of primitive subspaces.*

(iv) Given two primitive subspaces U and U' with $U \cap U' = \{x\}$, there exist $y \in U \setminus \{x\}$ and $y' \in U' \setminus \{x\}$ and a primitive subspace going through y and y' (regularity condition).

Conversely, any space S with these properties (i) to (iv) is of the form $\underline{S}(N^{(I)})$ for a suitable nearring N with the properties stated obove. $|I|$ is uniquely determined by S. □

REMARKS. (i) If the primitive subspaces are straight lines then N becomes a nearfield and the space $\underline{S}(N^{(I)})$ is nearaffine in the sensse of sect.3.

(ii) The cardinal number $|I|$ is sometimes called the dimension of $\underline{S}(N^{(I)})$. It is possible to generalize the dimension-theory of H.Tecklenburg 1978 (see also Misfeld, Tecklenburg 1979) to such spaces.

REFERENCES

André, J.

1975 Affine Geometrien über Fastkörpern. Mitt.Math.Sem.Gießen 114, 1-99.

1981 Nichtkommutative Geometrie und verallgemeinerte Hughes-Ebenen. Math.Z. 177, 449-462.

1984 Noncommutative geometry and combinatorics. In: Combinatorics and Applications. Proceed.Seminar on Combinatorics and Applications in honour of Prof.S.S.Shrikande, ed. by K.S.Vijayan and N.M.Singhi, Calcutta, pp.21-39.

Arnold, H.J.

1968 Algebraische und geometrische Kennzeichnung der schwachaffinen Vektorräume über Fastkörpern. Abh.Math.Sem.Univ. Hamburg 32, 73-88.

Biliotti, M.

1979 Su una generalizzazione di Dembowski dei piani di Hughes. Boll.Un.Mat.Ital. (5) 16-B, 674-693.

Dembowski, P.

1971 Generalized Hughes-planes. Can.J.Math. 23, 481-494.

Dickson, L.E.

1905 On finite algebras. Nachr.kgl.Ges.Wiss. 1905, 358-393.

Ferrero, G.

1970 Stems planari e BIB-disegni. Riv.Mat.Univ.Parma (2) 11, 79-96.

Kalscheuer, F.

1940 Die Bestimmung aller stetigen Fastkörper. Abh.Math.Sem. Univ.Hamburg 13, 413-435.

Karzel, H.

1968 Zusammenhänge zwischen Fastbereichen, scharf 2-fach transitiven Permutationsgruppen und 2-Strukturen mit Rechtecksaxiom. Abh.Math.Sem.Univ.Hamburg 32, 191-206.

Kerby, W.

1974 On infinite sharply multiply transitive groups. Vandenhoeck & Ruprecht, Göttingen.

Misfeld, J. , Tecklenburg, Helga
1979 Dimension of nearaffine spaces. In: Proceed.Conf.Geometry and Differential Geometry, Haifa. Springer-Verlag, Lecture Notes in Mathematics 792, 97-109.

Ney, H.
1983 Planar nearrings and their relations to some noncommutative spaces. Proceed.Conf. Near-Rings and Near-Fields, Harrisonburg, Va., p.47.

Pfalzgraf, J.
1984 Über ein Modell für nichtkommutative geometrische Räume. Diss. Saarbrücken.

1985a A hierarchy of geometric configurations generalizing the axioms of Desargues and Tamaschke. Preprint: Faculdad de Ciencias, Universidad de Chile, Santiago de Chile.

1985b On a model for non commutative geometric spaces. Journ. Geometry 25, 147-163.

Pilz, G.
1983 Near-Rings. The theory and its applications. Revised ed., North-Holland, Amsterdam, New York, Oxford.

Rößler, K.P.
1985 Gruppenpaarräume zu scharf transitiven Permutationsgruppen. Diplomarbeit, Saarbrücken.

Sasso-Sant, M.
1986 Nichtkommutative Räume und Fastringe. Diplomarbeit, Saarbrücken.

Scheid, E.
1984 Fernstrukturen zu vierdimensionalen Fastkörperräumen über Biliotti-Fastkörpern. Diplomarbeit, Saarbrücken.

Sperner, E.
1960 Affine Räume mit schwacher Inzidenz und zugehörige algebraische Strukturen. Journ.r.u.a.Math. 204, 205-215.

Tecklenburg, Helga
1978 Zur Dimension fastaffiner Räume. Mitt.Inst.f.Math. TU Hannover 92, 1-25.

Theobald, E.
1981a Nichtkommutative Geometrie über Fastringen. Diplomarbeit, Saarbrücken.

1981b Near-rings and noncommutative geometry. In: Proceed.Conf. on near-rings and near-fields, ed.by G.Ferrero and C. Ferrero-Cotti, Sa Benedetto del Trento, pp.211-218.

Veblen, O. , Wedderburn, J.H.M.
1907 Non-Desarguesian and non-Pascalian geometries. Transact. Amer.Math.Soc. 8, 379-388.

Wielandt, H.
1964 Finite permutation groups. Academic Press, New York,London.

Zassenhaus, H.
1935 Über endliche Fastkörper. Abh.Math.Sem.Univ.Hamburg 11, 187-220

1986 Über Fastkörper II: Existenz nicht-Dicksonscher Fastkörper endlichen Ranges über Grundkörpern unendlichen Transzendenzgrades. Manuscript.

Near-rings and Near-fields, G. Betsch (editor)

PSEUDO-FINITE NEAR-FIELDS

Ulrich Felgner

Mathematisches Institut der Universität,
Auf der Morgenstelle 10, D - 74 Tübingen

§ 1. Introduction

The theory of finite near-fields is developed up to a high degree of perfection. Starting from Zassenhaus' fundamental classification of all finite near-fields (1936) a large number of the most important problems in this area have been solved in the last two decades, most prominently by S.Dancs-Groves, H.Karzel and H. Lüneburg.

In the case of infinite near-fields there are only sporadic results known. Most of these results are quite impressive and substantial, yet a general structure theory for infinite near-fields is still missing. Under the assumption of suitable finiteness conditions however some authors had been able to develop a satisfactory structure theory. Let me mention the locally-finite near-fields (S.Dancs Groves) and the locally compact near-fields (Kalscheuer, Grundhöfer). Pseudo-finiteness is another interesting finiteness condition and there is some hope that the class of all pseudo-finite near-fields might have a reasonable structure theory. It is the aim of this paper to do a few initial steps towards such a structure theory.

The notion of a *pseudo-finite commutative field* is due to J.P. Serre [14] , chap.XIII, §2 . A (commutative) field D is called *quasi-finite*, if D is perfect and has precisely one extension of each finite degree. A quasi-finite field D is called *pseudo-finite* if every absolutely irreducible variety defined over D has a D-valued point. J.Ax [1] proved that the pseudo-finite fields are precisely the infinite models of the first-order theory of finite fields. Thus we are led to define analogously

Definition. A near-field F is called *pseudo-finite* if F is an infinite model of the first-order theory of finite near-fields.

Thus, a near-field F is pseudo-finite if F is infinite and elementarily equivalent to an ultraproduct of finite near-fields.

§ 2. Some remarks on finite near-fields

A (left distributive) near-field is a structure $\mathbb{F} = \langle F, +, \circ \rangle$ which satisfies all the axioms for a skew-field except possibly right distributivity (cf. Wähling [16], pp.42-43). $Z(\mathbb{F})$ denotes the center and $K(\mathbb{F})$ the kernel of $\mathbb{F}$, where

$$K(\mathbb{F}) = \{x \in F \;;\; \forall y, z \in F: (y+z) \circ x = y \circ x + z \circ x\}.$$

We need a brief outline of the Dickson-Zassenhaus construction. For a detailed description see Zassenhaus [17] or Ellers-Karzel [7].

Let $q = p^{\ell}$ be a power of a prime p and let n be such that

$$(\star) \quad \begin{cases} \textit{for all primes } \pi: & \textit{if } \; \pi \mid n \; \textit{ then } \; \pi \mid q-1 \\ \textit{and} & \textit{if } \; 4 \mid n \; \textit{ then } \; 4 \mid q-1 . \end{cases}$$

Then $\{(q^i - 1)/(q-1) \;;\; 0 < i \leq n\}$ is a complete set of residues modulo n . Let ω be a generator of the multiplicative group of $GF(q^n)$ and let $\mathfrak{K} = \langle \omega^n \rangle$ be the subgroup generated by ω^n. To each $a \in GF(q^n)$, $a \neq 0$, we associate an automorphism ψ_a of the Galois field $GF(q^n)$ such that

$$\psi_a = \langle x \mapsto x^{q^i} \;;\; x \in GF(q^n) \rangle \iff a \in \mathfrak{K} \cdot \omega^{(q^i - 1)/(q-1)}$$

Finally we introduce on the set $GF(q^n)$ a new multiplication $\circ$ as follows:

$$a \circ b = \begin{cases} 0 & \textit{if} \;\; a = 0 \\ a \cdot \psi_a(b) & \textit{if} \;\; a \neq 0 . \end{cases}$$

It can be shown that the set $F = GF(q^n)$ equipped with the addition of the Galois field $GF(q^n)$ and the new multiplication $\circ$ is a near-field $\mathbb{F} = \langle F, +, \circ \rangle$ such that $Z(\mathbb{F}) = K(\mathbb{F}) \cong GF(q)$. The isomorphism type of $\mathbb{F}$ depends on q, n and ω. The near-fields which are obtained by the Dickson-Zassenhaus construction are called *finite Dickson near-fields*. The fundamental theorem of Zassenhaus [17] states that up to seven exceptions all finite near-fields are either Galois fields or finite Dickson near-fields. The orders of the seven exceptions are 5^2, 7^2, 11^2, 23^2, 29^2 and 59^2.

The subnear-field structure of a finite near-field is analogous to the subfield structure of finite fields. More precisely, the following holds.

THEOREM (S.Dancs Groves [4], [5]): *Let $\mathbb{F}$ be a finite Dickson near-field, $|F| = q^n$, $Z(\mathbb{F}) \cong GF(q)$, where $q = p^\ell$ for some prime p.*
(i) If $\mathbb{H}$ is a subnear-field if $\mathbb{F}$, then $H = p^h$ with $h \mid \ell \cdot n$.
(ii) Conversely, if $h \mid \ell \cdot n$, then $\mathbb{F}$ has a unique subnear-field $\mathbb{H}$ of power p^h. Moreover $\mathbb{H}$ is a Dickson near-field, and if $z = gcd(j\ell, h)$, where $0 < j \leq n$ and $j \equiv (q^n - 1)/(p^h - 1) \pmod{n}$, then $Z(\mathbb{H}) = K(\mathbb{H}) \cong GF(p^z)$.

<u>COROLLARY</u> (S.Dancs Groves): *Let $\mathbb{F}$ be a finite Dickson near-field of power q^n where $Z(\mathbb{F}) \cong GF(q)$ and $q = p^\ell$. Let $\mathbb{H}$ be a subnear-field of $\mathbb{F}$ of power q^h. Then $Z(\mathbb{F}) \subseteq \mathbb{H}$ and $Z(\mathbb{H}) \cong GF(q^z)$ where $z = gcd(n/h, h)$.*

Using these results we prove:

<u>THEOREM 2.1</u>: *Let $\mathbb{F}$ be a finite Dickson near-field of power q^n such that $Z(\mathbb{F}) \cong GF(q)$ where q is a power of a prime. Let λ be maximal such that $\lambda^2 \mid n$. Among all commutative subfields $\mathbb{D}$ of $\mathbb{F}$ with $Z(\mathbb{F}) \subseteq \mathbb{D}$ there is a unique maximal one, called $M(\mathbb{F})$, and we have $M(\mathbb{F}) \cong GF(q^\lambda)$.*

Proof. By the theorem of Dancs-Groves $\mathbb{F}$ has precisely one subnear-field $\mathbb{B}$ of power q^λ. Clearly, $Z(\mathbb{F}) \subseteq \mathbb{B}$. It also follows from $gcd(n/\lambda, \lambda) = \lambda$ by the above corollary that $\mathbb{B}$ is commutative. If, on the other hand, $\mathbb{D}$ is any commutative subfield of $\mathbb{F}$ such that $Z(\mathbb{F}) \subseteq \mathbb{D}$, then $|\mathbb{D}| = q^d$ for some d. Since $\mathbb{D} \cong GF(q^d) = Z(\mathbb{D})$ it follows from the above corollary that $d = gcd(d, n/d)$. Hence $d^2 \mid n$ and since λ is maximal we obtain that $d \mid \lambda$. Since $\mathbb{D}$ is the only subnear-field of power q^d and since $\mathbb{B}$ has a subnear-field of power q^d it follows that $\mathbb{D}$ is contained in $\mathbb{B}$. We define $M(\mathbb{F}) = \mathbb{B}$. □

<u>Notation</u>. $M(\mathbb{F})$ = the unique maximal commutative subfield of the finite Dickson near-field $\mathbb{F}$ which contains the center of $\mathbb{F}$.

In contrast to $Z(\mathbb{F})$ and $K(\mathbb{F})$ no 'internal' first-order description of $M(\mathbb{F})$ is known.

<u>Open problem</u>. Is there a first-order formula $\Phi(x)$ in the language of fields such that for any finite Dickson near-field $\mathbb{F}$ we have

$$M(\mathbb{F}) = \{ a \in F ; \Phi(a) \} ?$$

We shall need the following lemma on near-field extensions.

LEMMA 2.2: *Let $\mathbb{F}$ be a Dickson near-field of power q^n such that $Z(\mathbb{F}) \cong GF(q)$, where q is a power of a prime. For any $t \equiv 1 \pmod n$ there is a Dickson near-field $\mathbb{D}$ which contains $\mathbb{F}$ as a subnear-field such that $Z(\mathbb{F}) \subseteq Z(\mathbb{D}) \cong GF(q^t)$ and $|\mathbb{D}| = q^{tn}$.*

Proof. $n \mid t-1$ implies $q^n - 1 \mid q^{t-1} - 1$. From (☆) it follows that $n \mid q^n - 1$. Hence $q^{tj} \equiv q^j \pmod{q^n - 1}$ for all j and therefore

$$(*) \qquad (q^{tj} - 1)/(q^t - 1) \equiv (q^j - 1)/(q - 1) \pmod{q^n - 1} \text{ for all } j,$$

$$(**) \qquad (q^{tn} - 1)/(q^n - 1) \equiv t \equiv 1 \pmod n$$

Since $\mathbb{F}$ is a Dickson near-field there is a generator ω of the multiplicative group of the field $GF(q^n)$ such that $a \circ b = a \cdot \psi_a(b)$ for all $a, b \in GF(q^n)$ with $a \neq 0$, where $\circ$ denotes the multiplication in $\mathbb{F}$ and $\cdot$ denotes the multiplication in the Galois field $GF(q^n)$ and $\psi_a = \langle x \longmapsto x^{q^i} ; x \in GF(q^n) \rangle \iff a \in \langle \omega^n \rangle \cdot \omega^{(q^i - 1)/(q-1)}$. $\langle \omega \rangle = GF(q^n)^{\#}$ is a cyclic subgroup of $GF(q^{nt})^{\#}$ of index

$$\tau = (q^{nt} - 1)/(q^n - 1).$$

Select a generator σ of $GF(q^{nt})^{\#}$ such that $\sigma^\tau = \omega$ and put $\mathfrak{K} = \langle \omega^n \rangle$ and $\mathfrak{K}^* = \langle \sigma^n \rangle$. Clearly $\mathfrak{K} \subseteq \mathfrak{K}^*$. It follows from $(**)$ that $\tau \equiv 1 \pmod n$, hence $\mathfrak{K}^* \sigma = \mathfrak{K}^* \sigma^\tau = \mathfrak{K}^* \omega$. This implies

$$\mathfrak{K}\omega^{(q^i - 1)/(q-1)} = \mathfrak{K}\omega^{(q^{ti} - 1)/(q^t - 1)} \subseteq \mathfrak{K}^* \sigma^{(q^{ti} - 1)/(q^t - 1)}$$

(using $(*)$). Now we associate to each $a \in GF(q^{nt})$, $a \neq 0$, a field-automorphism δ_a as follows:

$$\delta_a = \langle x \longmapsto x^{q^{ti}} ; x \in GF(q^{nt}) \rangle \iff a \in \mathfrak{K}^* \sigma^{(q^{ti} - 1)/(q^t - 1)}.$$

We introduce a new multiplication $\circ$ on the set $D = GF(q^{nt})$ as follows: $a \circ b = 0$ if $a = 0$, and $a \circ b = a \cdot \delta_a(b)$ if $a \neq 0$. As we have mentioned in the outset $\mathbb{D} = \langle D, +, \circ \rangle$ is a near-field of power q^{nt} such that $Z(\mathbb{D}) \cong GF(q^t)$. By construction ψ_a is the restriction of δ_a, hence $\mathbb{F}$ is a subnear-field of $\mathbb{D}$. Clearly, $Z(\mathbb{D})$ contains a subfield of order q. By the uniqueness result of S. Dancs Groves we have hence $Z(\mathbb{F}) \subseteq Z(\mathbb{D})$. □

LEMMA 2.3: *Let $\mathbb{F}_1 \subset \mathbb{F}_2 \subset \ldots \subset \mathbb{F}_t$ be a tower of finite Dickson near-fields such that $Z(\mathbb{F}_1) = Z(\mathbb{F}_2) = \ldots = Z(\mathbb{F}_t) \cong GF(q)$. If m denotes the number of distinct positive primes which divide $q - 1$, then $t \leq m$.*

Proof. We have $|F_i| = q^{n_i}$ for suitable n_i and $n_i | n_j$ for $i<j$. By the corollary of Dancs Groves we have in addition

$$\gcd(n_j/n_i\ ,\ n_i) = 1 \quad \textit{for all} \quad 1 \leq i < j \leq t.$$

From this our claim readily follows. □

§ 3 . The center of pseudo-finite near-fields

Let $\mathcal{L}_{NF}$ be the first-order language based on $+, -, \circ, ^{-1}$ and $0, 1$, where x^{-1} denotes the inverse of x with respect to the multiplication $\circ$. For any near-field $\mathbb{F} = \langle F,+,-,\circ,^{-1},0,1\rangle$ let $Th(\mathbb{F})$ denote the $\mathcal{L}_{NF}$-theory of $\mathbb{F}$. Thus

$Th(\mathbb{F})$ = the set of all sentences expressible in $\mathcal{L}_{NF}$ which are true in $\mathbb{F}$.

Then

$$Th(fnf) = \bigcap \{ Th(\mathbb{F})\ ;\ \ \mathbb{F} \text{ is a finite near-field}\}$$

is the first-order theory of finite near-fields. A near-field $\mathbb{F}$ is called *pseudo-finite* if $\mathbb{F}$ is an infinite model of $Th(fnf)$.

LEMMA 3.1: *If $\mathbb{F}$ is a pseudo-finite near-field, then*

(i) $\mathbb{F}$ is planar,

(ii) $Z(\mathbb{F}) = K(\mathbb{F})$,

(iii) $Z(\mathbb{F})$ is a pseudo-finite field or a finite field.

Proof. Ad (i): A near-field is *planar* (or *projective*) if it satisfies the following $\mathcal{L}_{NF}$ - sentence: $\forall x\, [\, x \neq 1 \rightarrow \exists y\colon\ x\circ y = y + 1\,]$. This sentence belongs to $Th(fnf)$ since all finite near-fields are planar (cf. Zemmer).

Ad (ii): Since $Z(\mathbb{F})$ and $K(\mathbb{F})$ are first-order definable and since the equality $Z(\mathbb{F}) = K(\mathbb{F})$ holds in all finite near-fields $\mathbb{F}$ of power $> 59^2$, the claim follows.

Ad(iii): If $\mathbb{F}$ is a pseudo-finite near-field, then there is an ultraproduct $\prod \mathbb{D}_i / \mathcal{F}$ of finite near-fields $\mathbb{D}_i$ such that $\mathbb{F} \equiv \prod \mathbb{D}_i / \mathcal{F}$ (here $\equiv$ denotes elementary equivalence, and $\mathcal{F}$ is an ultrafilter on $\mathbb{N}$). Then

$$Z(\mathbb{F}) \equiv Z(\prod_{i\in\mathbb{N}} \mathbb{D}_i / \mathcal{F}) = \prod_{i\in\mathbb{N}} Z(\mathbb{D}_i) / \mathcal{F}$$

and the claim follows from Ax' characterization of pseudo-finite fields. □

The planarity of finite near-fields can be proved quite easily. The function $f\colon y \mapsto f(y) = (y+1)\circ y^{-1}$ is obviously a bijection from $F-\{0\}$ onto $F-\{1\}$ (since F is finite). For each $x \in F$, $x \neq 1$, there is hence a $y \in F$, $y \neq 0$, such that $f(y) = x$, which means that $x \cdot y = y + 1$. □

Our next result is a two-cardinal theorem which concerns the possible sizes of the center of a pseudo-finite near-field.

THEOREM 3.2: *Let p be any prime and $1 \leq m \in \mathbb{N}$ and $\aleph_\alpha$ arbitrary, where $m \geq 2$ is assumed in the case $p \in \{2,3\}$. Then there is a pseudo-finite Dickson near-field $\mathbb{F}$ such that*

$$p^m = |Z(\mathbb{F})| = |K(\mathbb{F})| < |\mathbb{F}| = \aleph_\alpha .$$

Proof. Case 1: $p=2$ and $m \geq 2$. Let π be any prime divisor of $p^m - 1$. Then for each integer i the pair p^m, π^i is a Dickson pair and there is a near-field $\mathbb{F}_i$ of power $p^{m\pi^i}$ such that $Z(\mathbb{F}_i) \cong GF(p^m)$. Let $\mathcal{F}$ be a free ultrafilter on the set $\mathbb{N}$ of all natural numbers and let $\mathbb{D} = \prod \mathbb{F}_i / \mathcal{F}$ be the ultraproduct over all these finite near-fields. Clearly $\mathbb{D}$ is infinite and by Łoś' theorem, $Z(\mathbb{D}) \cong GF(p^m)$. Notice also that $\mathbb{D}$ is a Dickson near-field. Now our claim follows from the Löwenheim-Skolem theorem.

Case 2: $p=3$ and $m \geq 2$. If $p^m - 1 = 3^m - 1 = 2^t$ for some t, then $m=2$ and $t=3$ (see Passman [13], lemma 19.3). Hence, if $m \geq 3$, then there is an odd prime π which divides $p^m - 1$ and for each integer $i \in \mathbb{N}$, p^m, π^i is a Dickson pair. If $m = 2$ then $9, 2^i$ is a Dickson pair for each $i \in \mathbb{N}$. Thus there are Dickson near-fields $\mathbb{F}_i$ of power $p^{m\pi^i}$ such that $Z(\mathbb{F}_i) \cong GF(p^m)$. Now proceed as in case 1.

Case 3: $p \geq 5$: and $1 \leq m$. If p is not a Fermat prime, then there is an odd prime divisor π of $p^m - 1$ and p^m, π^i is a Dickson pair for each $i \geq 1$. If p is a Fermat prime, then $4 | p^m - 1$ (since $5 \leq p$) and hence $p^m, 2^i$ is a Dickson pair for each $i \geq 1$. Now proceed as above. □

It is easy to see (using ultraproducts and the Löwenheim-Skolem theorem), that for each ordinal $\alpha \geq 0$ there are pseudo-finite Dickson near-fields of any prescribed characteristic, $\mathbb{F}$, such that

$$|Z(\mathbb{F})| = |K(\mathbb{F})| = |\mathbb{F}| = \aleph_\alpha .$$

Hence the problem arises whether there are pseudo-finite near-fields $\mathbb{F}$ such that $|Z(\mathbb{F})| = \aleph_0 < |\mathbb{F}| = \aleph_1$. The answer is positive as we shall show in the next theorem. I am grateful to

D.Mundici (Florence) for calling my attention to Keisler's paper [12] .

THEOREM 3.3: *There is an uncountable pseudo-finite near-field* $\mathbb{F}$ *whose center is countably infinite.*

Proof. We consider the language L(Q) which is formed by adding to $\mathcal{L}_{NF}$ a new quantifier (Qx) with the interpretation "there are uncountably many x ". For any prime p let Σ_p be the following countable set of L(Q)-sentences:

$$\Sigma_p = Th(fnf) \cup \{\underbrace{1+1+\ldots+1}_{p} = 0\} \cup \{Qx : x = x\} \cup$$
$$\cup \{\neg Qx\ \forall y: x \circ y = y \circ x\} \cup$$
$$\cup \{\exists x_1,\ldots,x_k \Big[\bigwedge_{i \neq j} x_i \neq x_j \wedge \bigwedge_{i=1}^{k} \big(\forall y: x_i \circ y = y \circ x_i\big)\Big] ; k \in \mathbb{N}\}.$$

It follows from theorem 3.2 that each finite subset of Σ_p has a model. Thus by the Fuhrken-Vaught compactness theorem for L(Q) (cf. Keisler [12] , p.27) Σ_p has a model $\mathbb{F}$. Clearly, $\mathbb{F}$ has characteristic p , is uncountable and has a countably infinite center. □

COROLLARY 3.4: *Assume the generalized continuum hypothesis GCH. For every regular cardinal* $\aleph_\alpha$ *there is a pseudo-finite near-field* $\mathbb{F}$ *of cardinality* $\aleph_{\alpha+1}$ *whose center has cardinality* $\aleph_\alpha$.

Proof. This is a direct consequence of theorem 3.3 and Chang's Two-Cardinal theorem (cf. Chang-Keisler [2] , theorem 7.2.7). □

We do not know whether for any pair of infinite cardinals, say $\aleph_\beta < \aleph_\alpha$, there is always a pseudo-finite near-field $\mathbb{F}$ of power $\aleph_\alpha$ whose center has power $\aleph_\beta$.

§ 4. Locally-finite sub-nearfields

We show that each pseudo-finite near-field $\mathbb{F}$ has a unique maximal locally-finite subnear-field, called the "locally-finite socle" of $\mathbb{F}$. We also prove that each (finite or infinite) locally-finite near-field appears as the locally-finite socle of some pseudo-finite near-field.

THEOREM 4.1: *Let* $\mathbb{F}$ *be a pseudo-finite near-field of characteristic* p , *where* $p \neq 0$. *Then* $\mathbb{F}$ *contains a unique maximal locally-finite subnear-field.*

Proof. Let $S(\mathbb{F})$ be the set of all finite subnear-fields of $\mathbb{F}$. The first-order statement "*if there are more than* 59^2 *elements then there are* p *elements which form a subfield contained in the center*" is true in all finite near-fields and hence in $\mathbb{F}$ too (this follows from Zassenhaus' classification of all finite near-fields). Thus $GF(p) \in S(\mathbb{F})$ and $S(\mathbb{F})$ is non-empty. Let $L(\mathbb{F})$ denote the subnear-field of $\mathbb{F}$ generated by $\bigcup S(\mathbb{F})$. We claim that $S(\mathbb{F})$ is upwards directed.

First step: $\mathbb{F}$ is elementarily equivalent to an ultraproduct of finite near-fields, where $\mathbb{F}$ is infinite. Hence we may assume without loss of generality that $\mathbb{F}$ is elementarily equivalent to an ultraproduct of finite Dickson near-fields. If $A \in S(\mathbb{F})$ then A is a subnear-field of almost all components of the ultraproduct (by the theorem of Łoś). It follows from the theorem of Dancs - Groves that A is a Dickson near-field. Thus we have proved that all elements of $S(\mathbb{F})$ are finite Dickson near-fields.

Second step: By the theorem of S.Dancs Groves the following sentence holds in all finite nearfields:

"*if there are* p^a *elements* $x_1, x_2, \dots$ *forming a subnear-field and if there are* p^b *elements* $y_1, y_2, \dots$ *forming a sub-near-field and if* $c = lcm(a,b)$, *then there are* p^c *elements* $z_1, z_2, \dots$ *forming a subnear-field such that all the* p^a *elements* $x_1, x_2, \dots$ *and all the* p^b *elements* $y_1, y_2, \dots$ *are among these* p^c *elements* $z_1, z_2, \dots$."

($lcm(a,b)$ is the least common multiple of a and b) This sentence, hence, belongs to $Th(fnf)$ and is therefore true in $\mathbb{F}$. Thus, if $A, B \in S(\mathbb{F})$, $|A| = p^a$ and $|B| = p^b$ and $c = lcm(a,b)$, then there is a $C \in S(\mathbb{F})$ such that C contains A and B as subnear-fields and $|C| = p^c$. Thus $S(\mathbb{F})$ is upwards directed and it follows that $\bigcup S(\mathbb{F})$ is a subnear-field, whence $L(\mathbb{F}) = \bigcup S(\mathbb{F})$. If $\mathbb{D}$ is any locally-finite subnear-field of $\mathbb{F}$ then clearly $\mathbb{D} \subseteq L(\mathbb{F})$ and the theorem is proved. □

Let $\mathbb{F}$ be a pseudo-finite near-field. If $\mathbb{F}$ has characteristic 0 then $\mathbb{F}$ has no finite subnear-fields. But if $\mathbb{F}$ has non-zero characteristic p then $GF(p)$ is a (central) subfield of $\mathbb{F}$. In this case the maximal locally-finite subnear-field of $\mathbb{F}$ is non-trivial, finite or countably infinite.

Definition. If $\mathbb{F}$ is a pseudo-finite near-field of non-zero characteristic, then $L(\mathbb{F})$ denotes the maximal locally-finite subnear-field of $\mathbb{F}$.

<u>COROLLARY 4.2</u>: *Let $\mathbb{F}_1$ and $\mathbb{F}_2$ be pseudo-finite near-fields of non-zero characteristic. If $\mathbb{F}_1$ and $\mathbb{F}_2$ are elementarily equivalent, then $L(\mathbb{F}_1) \cong L(\mathbb{F}_2)$.*

Proof. This is a direct consequence of the proof of theorem 4.1.□

The isomorphism type of a locally-finite near-field can be characterized by socalled Steinitz numbers and sequences of polynomials (cf. Dancs Groves [6]). These are first-order characterizations of $L(\mathbb{F})$.

<u>THEOREM 4.3</u>: *If $\mathbb{L}$ is any locally-finite Dickson near-field, then there is a pseudo-finite Dickson near-field $\mathbb{F}$ such that*

$$\mathbb{L} \cong L(\mathbb{F}).$$

Proof. Case 1: $\mathbb{L}$ is infinite. Let $S(\mathbb{L})$ be the set of all finite subnear-fields of $\mathbb{L}$. For $A \in S(\mathbb{L})$ put $t(A) = \{B \in S(\mathbb{L})\ ;\ A \subseteq B\}$. The set $S(\mathbb{L})$ is upwards directed (see the proof of 4.1). Hence, if $A, B \in S(\mathbb{L})$, then there is a $C \in S(\mathbb{L})$ such that $A \subseteq C$ and $B \subseteq C$ and we have $t(C) \subseteq t(A) \cap t(B)$, where $t(C) \neq \emptyset$. Thus $\{t(A)\ ;\ A \in S(\mathbb{L})\}$ has the finite intersection property and can be extended to an ultrafilter $\mathcal{F}$ on the countably infinite set $S(\mathbb{L})$.

Let $\mathbb{F}$ be the ultraproduct of the near-fields in $S(\mathbb{L})$ modulo $\mathcal{F}$, i.e.

$$\mathbb{F} = \prod_{A \in S(\mathbb{L})} A / \mathcal{F} .$$

For any $a \in \mathbb{L}$ let f_a the following function from $S(\mathbb{L})$ into the cartesian product of all $A \in S(\mathbb{L})$: $f_a(A) = a$ if $a \in A$, and $f_a(A) = 0$ otherwise. If $f_a/\mathcal{F}$ denotes the equivalence class of f_a modulo $\mathcal{F}$ then $a \longmapsto f_a/\mathcal{F}$ is an embedding of $\mathbb{L}$ into $\mathbb{F}$. Thus $\mathbb{L} \subseteq L(\mathbb{F})$. It follows easily from Łoś' theorem that a finite subnear-field H of $\mathbb{F}$ is contained in some A for $A \in S(\mathbb{L})$ and hence in $\mathbb{L}$. Thus $\mathbb{L} = L(\mathbb{F})$.

Case 2: $\mathbb{L}$ is finite. We need the following special case of Dirichlet's prime number theorem:

(#) *For every natural number n there are infinitely many primes p such that $p \equiv 1 \pmod{n}$.*

For a direct proof of (#) see Estermann [8] or R.A.Smith [15] .

Since $\mathbb{L}$ is a finite Dickson near-field we have $|\mathbb{L}| = q^n$ and $Z(\mathbb{L}) \cong GF(q)$ where $q = p^\ell$ is a power of a prime p. By (#) there are infinitely many primes π_i $(i \in \mathbb{N})$ such that $\pi_i \equiv 1 \pmod n$ and π_i does not divide ℓ. By lemma 2.2 there are Dickson near-fields $\mathbb{F}_i$ such that

$$|\mathbb{F}_i| = q^{n\pi_i} = p^{\ell n \pi_i} \quad , \qquad Z(\mathbb{F}_i) \cong GF(q^{\pi_i}),$$

such that $\mathbb{L}$ is a subnear-field of $\mathbb{F}_i$ and the center of $\mathbb{F}_i$ contains the center of $\mathbb{L}$. Now let $\mathcal{F}$ be a free ultrafilter on the set $\mathbb{N}$ of all natural numbers and let $\mathbb{F}$ be the ultraproduct of all the $\mathbb{F}_i$ $(i \in \mathbb{N})$ modulo $\mathcal{F}$. The theorem of Łoś implies that $\mathbb{L}$ is a subnear-field of $\mathbb{F}$, whence $\mathbb{L} \subseteq L(\mathbb{F})$.

Now let $\mathbb{H}$ be any finite subnear-field of $\mathbb{F}$. Since the existence of $\mathbb{H}$ can be expressed by a first-order formula and since $\mathbb{H}$ is unique in its order it follows from the theorem of Łoś that there are infinitely many i such that $\mathbb{H}$ is a subnear-field of $\mathbb{F}_i$. Thus, if $\mathbb{H} \subseteq \mathbb{F}_i$ and $\mathbb{H} \subseteq \mathbb{F}_j$, then by the theorem of Dancs Groves we have $h \mid \ell n \pi_i$ and $h \mid \ell n \pi_j$, where $|\mathbb{H}| = p^h$. It follows from the choice of the primes π_i that $h \mid \ell n$ and hence that $\mathbb{H}$ is a subnear-field of $\mathbb{L}$. This proves that $\mathbb{L} = L(\mathbb{F})$. □

Remark. The use of Dirichlet's prime number theorem in the proof of theorem 4.3 can be avoided. Let $\theta_o, \theta_1, \ldots, \theta_n, \ldots$ be the sequence of all primes different from p and coprime to n. Put $\pi_i = \theta_i^{\varphi(n)}$, there φ is Euler's totient function. By Euler's lemma $\pi_i \equiv 1 \pmod n$. Now proceed as above.

Notice that similarly as in theorem 4.3 one can prove that for each finite or locally-finite field $\mathbb{E}$ there is a pseudo-finite field $\mathbb{K}$ such that $\mathbb{E} = L(\mathbb{K})$.

§ 5. Maximal commutative subfields

A well-known theorem of P.Hall, C.R.Kulatilaka and M.I.Kargapolov states that each infinite locally-finite group has an infinite abelian subgroup. We ask whether analogously each pseudo-finite near-field has an infinite commutative subfield. Notice that according to theorem 3.2 the center could be finite!

THEOREM 5.1: *Each pseudo-finite near-field $\mathbb{F}$ has an infinite commutative subfield.*

Proof. If $\mathbb{F}$ has characteristic 0 then $0,1 \in Z(\mathbb{F}) = K(\mathbb{F})$ implies that the field $\mathbb{Q}$ of all rational numbers is contained in $Z(\mathbb{F})$ and hence $Z(\mathbb{F})$ is infinite. Assume from now on that $\mathbb{F}$ has characteristic p where p is a prime, $p \neq 0$.

Case 1: $\mathbb{F}$ is an ultraproduct of finite near-fields,

$$\mathbb{F} = \prod_{i \in \mathbb{N}} \mathbb{K}_i / \mathcal{F} \quad .$$

If $Z(\mathbb{F})$ is infinite then we are done. Assume therefore that $Z(\mathbb{F})$ is finite, $Z(\mathbb{F}) = GF(q)$, where $q = p^{\ell}$ for some $\ell \in \mathbb{N}$. Since $\mathbb{F}$ is infinite we may assume w.l.o.g. that all $\mathbb{K}_i$ are finite Dickson near-fields of characteristic p. Put $|\mathbb{K}_i| = q^{n_i}$. Since $Z(\mathbb{F})$ has q elements we may, by the theorem of Łoś, assume w.l.o.g. that the center of each $\mathbb{K}_i$ has precisely q elements. Hence $\varpi(n_i) \subseteq \varpi(q-1)$, where for any natural number k, $\varpi(k)$ denotes the set of prime divisors of k. Let d_i be the largest natural number such that $d_i^2 | n_i$.

Claim: $\mathbb{F}$ contains an infinite tower of commutative subfields which all contain $Z(\mathbb{F})$.

Proof of the claim: Let $\mathbb{H}$ be any subfield of $\mathbb{F}$ such that $Z(\mathbb{F}) \subseteq \mathbb{H}$. By the theorem of Łoś we may assume that $\mathbb{H}$ is a subfield of each $\mathbb{K}_i$ such that $Z(\mathbb{F}) = GF(q) = Z(\mathbb{K}_i) \subseteq \mathbb{H}$. Put $|\mathbb{H}| = q^h$. Then $h^2 | n_i$ for all $i \in \mathbb{N}$ (cf. theorem 2.1). For each $\pi \in \varpi(q-1)$ define

$$E_\pi = \{ i \in \mathbb{N} ;\ (\pi h)^2 | n_i \}$$

If $E_\pi \notin \mathcal{F}$ for each $\pi \in \varpi(q-1)$, then $E^* = \mathbb{N} - \bigcup \{E_\pi ;\ \pi \in \varpi(q-1)\}$ belongs to $\mathcal{F}$. Let t be the product of all primes in $\varpi(q-1)$. Then $\varpi(n_i) \subseteq \varpi(q-1)$ implies $n_i | (th)^2$ for all $i \in E^*$. Thus the numbers n_i are bounded on a filter set. By the theorem of Łoś it would follow that $\mathbb{F}$ is finite, a contradiction.

There is, hence, a prime $\pi \in \varpi(q-1)$ such that $E_\pi \in \mathcal{F}$. Then by theorem 2.1 $GF(q^{h\pi})$ is a subfield of each $\mathbb{K}_i$ (for $i \in E_\pi$) containing the center of $\mathbb{K}_i$. Now $E_\pi \in \mathcal{F}$ implies that similarly $GF(q^{h\pi})$ is a subfield of $\mathbb{F}$ containing the center of $\mathbb{F}$. Thus $\mathbb{H}$ can be extended and our claim follows.

It follows from the claim that $\mathbb{F}$ contains an infinite locally finite nearfield $\mathbb{D}$ such that $Z(\mathbb{F}) \subseteq \mathbb{D}$, where $\mathbb{D}$ is commutative.

Case 2: $\mathbb{F}$ is an arbitrary pseudo-finite near-field. But then $\mathbb{F}$ is elementarily equivalent to an ultraproduct of finite near-fields and our claim follows easily. □

As a corollary to the proof of theorem 5.1 we obtain the following surprising fact.

COROLLARY 5.2: *Let* $\mathbb{F}$ *be a pseudo-finite near-field of non-zero characteristic.*
(i) if $Z(\mathbb{F})$ *is finite, then* $L(\mathbb{F})$ *is infinite;*
(ii) if $L(\mathbb{F})$ *is finite, then* $Z(\mathbb{F})$ *is infinite.*

Proof. (i) follows from the proof of theorem 5.1 and (ii) is just the contraposition of (i).

We are now able to extend a result from § 4. In theorem 4.3 we proved that for any locally-finite Dickson near-field $\mathbb{L}$ there is at least one pseudo-finite near-field $\mathbb{F}$ of power $2^{\aleph_0}$ such that $\mathbb{L} = L(\mathbb{F})$. A natural question is whether there is only one such $\mathbb{F}$, and if not, how many such $\mathbb{F}$'s are there?

THEOREM 5.3: *Let* $\mathbb{L}$ *be a finite Dickson near-field. Ineach uncountable cardinality* κ *there are precisely* 2^{κ} *pairwise non-isomorphic pseudo-finite near-fields* $\mathbb{F}$ *of power* κ *such that* $\mathbb{L} = L(\mathbb{F})$.

Proof. By theorem 4.3 we know that there is at least one pseudo-finite near-field $\mathbb{F}$ such that $\mathbb{L} = L(\mathbb{F})$. By corollary 5.2 $Z(\mathbb{F})$ is infinite and by lemma 3.1 $Z(\mathbb{F})$ is pseudo-finite. Hence $Z(\mathbb{F})$ is not algebraically closed. It follows from a theorem of Cherlin-Shelah [3] (cf. Felgner [9], Satz 3.11) that $Z(\mathbb{F})$ is not super-stable. It follows readily that also $\mathbb{F}$ is not super-stable (to see this, work with types over subsets of the center). By a theorem of Shelah (cf. Felgner [9], Satz 4.3) there are in each uncountable cardinal κ precisely 2^{κ} pairwise non-isomorphic structures of power κ which are elementarily equivalent to $\mathbb{F}$. □

Pseudo-finite near-fields are not $\aleph_0$-categorical since the multiplicative group is not of bounded exponent (cf. Felgner [9], Satz 3.1). Hence for each finite (or locally-finite) Dickson near-field $\mathbb{L}$ there are at least two (and by a theorem of Vaught - see Felgner [9], p.14 - at least three) pseudo-finite countable near-fields $\mathbb{F}$ such that $L(\mathbb{F}) = \mathbb{L}$. The exact number of such countable $\mathbb{F}$'s is not known to us. We also do not know whether an analogous version of theorem 5.3 holds in the case of infinite locally-finite near-fields $\mathbb{L}$. It seems to be possible to construct for a given locally-finite Dickson near-field $\mathbb{L}$ different pseudo-finite nearfields $\mathbb{F}_1$, $\mathbb{F}_2, \ldots$ with $\mathbb{L} = L(\mathbb{F}_i)$ such that $L(\mathbb{F}_1) \cap Z(\mathbb{F}_1) = Z(\mathbb{L}) \neq L(\mathbb{F}_2) \cap Z(\mathbb{F}_2)$. This would prove that the first-order theory of a pseudo-finite near-field $\mathbb{F}$ is not uniquely determined by $L(\mathbb{F})$. This would be in contrast to a result of Ax [1] which states that the first-order theory of a commutative pseudo-finite field $\mathbb{K}$ is uniquely determined by its subfield $L(\mathbb{K})$ (where $L(\mathbb{K}) = \mathrm{Abs}(\mathbb{K})$ in the notation of Ax).

§ 6. The Dickson Property

According to H.Karzel [11] a near-field $\mathbb{F} = \langle F, +, \circ \rangle$ is called a *Dickson near-field* if a binary operation " $\cdot$ " can be defined on the set F such that $\mathbb{F}^* = \langle F, +, \cdot \rangle$ is a skew-field and for each $a \in F - \{0\}$ the mapping

$$\psi_a = \langle x \longmapsto a^{-1} \cdot (a \circ x) \;\; ; \;\; x \in F \rangle$$

is an automorphism of the skew-field $\mathbb{F}^*$ (here a^{-1} is the inverse of a in the skew-field).

In many situations it is advantageous to know that the near-field one is dealing with is a Dickson near-field. It is unknown whether all infinite near-fields are necessarily Dickson near-fields or not. In the sequel we shall discuss the problem whether pseudo-finite near-fields are Dickson near-fields.

In §3 we have introduced the first-order language $\mathcal{L}_{NF}$ of near-field theory which is based on $+, -, \circ, \overset{-1}{\smile}, 0$ and 1. In contrast to that language we let $\mathcal{L}_{SF}$ be the first-order language of skew-field theory which is based on $+, -, \cdot, {}^{-1}, 0$ and 1. We introduce the following sets of axioms.

Σ_{NF} = the usual set of axioms for near-fields (cf.Wähling [16], p.43) formulated in the language $\mathcal{L}_{NF}$.

Σ_{SF} = the usual set of axioms for skew-fields formulated in the language $\mathcal{L}_{SF}$.

Σ_D = the axiom (formulated in $\mathcal{L}_{NF} \cup \mathcal{L}_{SF}$) saying that for each $a \neq 0$, $x \longmapsto a^{-1} \cdot (a \circ x)$ is an automorphism with respect to "+" and "$\cdot$".

A near-field $\mathbb{F} = \langle F, +, -, \circ, \overset{-1}{\smile}, 0, 1 \rangle$ is hence a Dickson near-field if a multiplication "$\cdot$" can be defined on F such that

$\langle F,+,-,\cdot,^{-1},0,1\rangle \models \Sigma_{SF}$ and $\langle F,+,-,\cdot,^{-1},\circ,\overset{-1}{\circ},0,1\rangle \models \Sigma_D$. Dickson near-fields are hence *reducts* of $\mathcal{L}_{NS} \cup \mathcal{L}_{SF}$ - structures satisfying $\Sigma_{NF} \cup \Sigma_{SF} \cup \Sigma_D$ to the language $\mathcal{L}_{NF}$. The class of all Dickson near-fields is therefore a socalled *projective* or *pseudo-elementary class*. Since all pseudo-elementary classes are closed under the formation of ultraproducts (cf. Chang-Keisler [2] , p.177) we obtain the following

PROPOSITION 6.1: *Ultraproducts of Dickson near-fields are Dickson near-fields.* □

It is a difficult and still open problem whether the class of all Dickson near-fields is an elementary class (i.e. EC_Δ). However the following holds.

LEMMA 6.2: *The class of all near-fields which are embeddable into Dickson near-fields is an elementary class (i.e. EC_Δ).*

Proof. This class is closed under ultraproducts (use proposition 6.1) and closed under elementary equivalence (if $\mathbb{F}_1 \equiv \mathbb{F}_2$ where $\mathbb{F}_1$ is a subnearfield of the Dickson near-field $\mathbb{D}_1$, then $\mathbb{F}_2$ satisfies all universal $\mathcal{L}_{NF}$ -sentences which are provable in $\Sigma_{NF} \cup \Sigma_{SF} \cup \Sigma_D$. A compactness argument shows that $\mathbb{F}_2$ can be embedded into a Dickson near-field). □

THEOREM 6.3: *Every pseudo-finite near-field $\mathbb{F}$ can be embedded into a Dickson near-field.*

Proof. A pseudo-finite near-field $\mathbb{F}$ is elementarily equivalent to an ultraproduct of finite Dickson near-fields. Now apply proposition 6.1 and proceed as in the proof of lemma 6.2 to conclude that $\mathbb{F}$ can be embedded into a Dickson near-field. □

We do not know whether all pseudo-finite near-fields are Dickson near-fields or not. However, if $\mathbb{F}$ is pseudo-finite near-field such that $\mathbb{F}$ **has finite dimension over** $Z(\mathbb{F})$ **then** $\mathbb{F}$ is a Dickson near-field. This follows readily from Grundhöfer [10] since the multiplicative group of $\mathbb{F}$ is solvable of class 2.

REFERENCES

[1] J.AX: *The elementary theory of finite fields.* Annals of Math. 88(1968),pp.239-271.

[2] C.C.CHANG - H.J.KEISLER: *Model Theory.*(North Holland,Amsterdam 1973).

[3] G.CHERLIN - S.SHELAH: *Superstable fields and groups.*Annals of Math.Logic 18(1980),pp.227-270.

[4] S.DANCS: *The sub-near-field structure of finite near-fields.* Bull.Austral.Math.Soc. 5 (1971),pp.275-280.

[5] S.DANCS: *On finite Dickson near-fields.* Abhandl.Math.Seminar Hamburg 37 (1972),pp.254-257.

[6] S.DANCS GROVES: *Locally finite near-fields.* Abhandl.Math.Seminar Hamburg 48 (1979),pp.89-107.

[7] E.ELLERS - H.KARZEL: *Endliche Inzidenzgruppen.* Abhandl.Math.Seminar Hamburg 27 (1964),pp.250-264.

[8] T.ESTERMANN: *Note on a paper of A.Rotkiewicz.* Acta Arithmetica 8 (1963),pp.465-467.

[9] U.FELGNER: *Kategorizität.* Jahresber.d.Dt.Math.Verein. 82(1980) pp.12-32.

[10] T.GRUNDHÖFER: *Transitive linear groups and nearfields with solubility condition.* To appear in the J.of Algebra.

[11] H.KARZEL: *Unendliche Dicksonsche Fastkörper.* Archiv d.Math.16 (1965),pp.247-256.

[12] H.J.KEISLER: *Logic with the quantifier "there exist uncountably many".* Annals of Math.Logic 1(1970),pp.1-93.

[13] D.S.PASSMAN: *Permutation Groups.* (W.A.Benjamin,Inc.,New York-Amsterdam 1968).

[14] J.P.SERRE: *Corps locaux* (Hermann, Paris 1962).

[15] R.A.SMITH: *A note on Dirichlet's theorem.* Canad.Math.Bull. 24 (1981),pp.379-380.

[16] H.WÄHLING: *Bericht über Fastkörper.* Jahresber.d.Dt.Math.Verein. 76(1974),pp.41-103.

[17] H.ZASSENHAUS: *Über endliche Fastkörper.* Abhandl.Math.Seminar Hamburg 11(1936),pp.187-220.

Near-rings and Near-fields, G. Betsch (editor)

ON DERIVATIONS IN NEAR-RINGS

HOWARD E. BELL* and GORDON MASON

The literature on near-rings contains a number of theorems asserting that certain conditions implying commutativity in rings imply multiplicative or additive commutativity in special classes of near-rings. We shall add to this body of results several commutativity theorems for near-rings admitting suitably-constrained derivations.

1. INTRODUCTION

Throughout this paper N will denote a zero-symmetric left near-ring with multiplicative center Z. A *derivation* on N is defined to be an additive endomorphism satisfying the "product rule" $D(xy) = xD(y) + D(x)y$ for all $x,y \in N$; elements x of N for which $D(x) = 0$ are called *constants*. For $x,y \in N$, the symbol $[x,y]$ will denote the commutator $xy - yx$, while the symbol (x,y) will denote the additive-group commutator $x + y - x - y$. The derivation D will be called *commuting* if $[x,D(x)] = 0$ for all $x \in N$. Finally, N will be called *prime* if $a,b \in N$ and $aNb = \{0\}$ imply that $a = 0$ or $b = 0$. (Note that this definition implies the usual definition of prime near-ring. It does not seem to be known whether the two definitions are equivalent.)

2. PRELIMINARY RESULTS

We begin with two quite general and useful lemmas.

<u>LEMMA 1</u> Let D be an arbitrary derivation on the near-ring N. Then N satisfies the following *partial distributive law*:

$$(aD(b) + D(a)b)c = aD(b)c + D(a)bc \text{ for all } a,b,c \in N.$$

<u>Proof</u> Note that $D((ab)c) = abD(c) + (aD(b) + D(a)b)c$, and that $D(a(bc)) = aD(bc) + D(a)bc = a(bD(c) + D(b)c) + D(a)bc = abD(c) + aD(b)c + D(a)bc$. Equating these two expressions for $D(abc)$ now yields the announced partial distributive law.

<u>LEMMA 2</u> Let D be a derivation on N, and suppose $u \in N$ is not a left zero divisor. If $[u,D(u)] = 0$, then (x,u) is a constant for every $x \in N$.

<u>Proof</u> From $u(u + x) = u^2 + ux$, we obtain $uD(u + x) + D(u)(u + x) = uD(u) + D(u)u + uD(x) + D(u)x$, which reduces to $uD(x) + D(u)u = D(u)u + uD(x)$. Since $D(u)u = uD(u)$, this equation is expressible as

$$u(D(x) + D(u) - D(x) - D(u)) = 0 = uD((x,u)).$$

Thus, $D((x,u)) = 0$.

---

*Supported by the Natural Sciences and Engineering Research Council of Canada, Grant No. A3961.

THEOREM 1 Let N have no nonzero divisors of zero. If N admits a nontrivial commuting derivation D, then (N,+) is abelian.

Proof Let c be any additive commutator. Then c is a constant by Lemma 2. Moreover, for any $w \in N$, wc is also an additive commutator, hence also a constant. Thus, $0 = D(wc) = wD(c) + D(w)c$ and $D(w)c = 0$. Since $D(w) \not\equiv 0$ for some $w \in N$, we conclude that $c = 0$.

3. PRIME NEAR-RINGS

Since prime rings are the setting in which derivations in rings have been most fruitfully studied, we proceed to consider prime near-rings.

LEMMA 3 Let N be a prime near-ring.

(i) If $z \in Z \setminus \{0\}$, then z is not a zero divisor.

(ii) If Z contains a nonzero element z for which $z + z \in Z$, then (N,+) is abelian.

(iii) Let D be a nonzero derivation on N. Then $xD(N) = \{0\}$ implies $x = 0$, and $D(N)x = \{0\}$ implies $x = 0$.

(iv) If N is 2-torsion-free and D is a derivation on N such that $D^2 = 0$, then $D = 0$.

Proof

(i) If $z \in Z \setminus \{0\}$ and $zx = 0$, then $zNx = \{0\}$, hence $x = 0$.

(ii) Let $z \in Z \setminus \{0\}$ be an element such that $z + z \in Z$, and let $x,y \in N$. Since $z + z$ is distributive we get $(x+y)(z+z) = x(z+z) + y(z+z) = xz + xz + yz + yz = z(x+x+y+y)$. On the other hand, $(x+y)(z+z) = (x+y)z + (x+y)z = z(x+y+x+y)$. Thus, $x + x + y + y = x + y + x + y$ and therefore $x + y = y + x$.

(iii) Let $xD(N) = \{0\}$, and let r, s be arbitrary elements of N. Then $0 = xD(rs) = xrD(s) + xD(r)s = xrD(s)$. Thus, $xND(N) = \{0\}$; and since $D(N) \not\equiv \{0\}$, $x = 0$. A similar argument works if $D(N)x = \{0\}$ since Lemma 1 provides enough distributivity to carry it through.

(iv) For arbitrary $x,y \in N$, we have $0 = D^2(xy) = D(xD(y) + D(x)y) = xD^2(y) + D(x)D(y) + D(x)D(y) + D^2(x)y = 2D(x)D(y)$. Since N is 2-torsion-free, $D(x)D(N) = \{0\}$ for each $x \in N$, and (iii) yields $D = 0$.

THEOREM 2 If a prime near-ring N admits a nontrivial derivation D for which $D(N) \subseteq Z$, then (N,+) is abelian. Moreover, if N is 2-torsion-free, then N is a commutative ring.

Proof Let c be an arbitrary constant, and let x be a non-constant. Then $D(xc) = xD(c) + D(x)c = D(x)c \in Z$. Since $D(x) \in Z \setminus \{0\}$, it follows easily that $c \in Z$. Since $c + c$ is a constant for all constants c, it follows from Lemma 3(ii) that (N,+) is abelian, provided that there exists a nonzero constant.

Assume, then, that 0 is the only constant. Since D is obviously commuting, it follows from Lemma 2 that all u which are not zero divisors belong to the center of (N,+), denoted by $\xi(N)$. In particular, if $x \neq 0$, $D(x) \in \xi(N)$. But then for all $y \in N$, we get $D(y) + D(x) - D(y) - D(x) = D((y,x)) = 0$, hence $(y,x) = 0$.

We complete the proof by assuming that N is 2-torsion-free and showing that N is commutative. By Lemma 1, $(aD(b) + D(a)b)c = aD(b)c + D(a)bc$ for all $a,b,c \in N$; and using the fact that $D(ab) \in Z$, we get $caD(b) + cD(a)b = aD(b)c + D(a)bc$. Since (N,+) is abelian and $D(N) \subseteq Z$, this equation can be rearranged to yield

$$D(b)[c,a] = D(a)[b,c] \text{ for all } a,b,c \in N.$$

Suppose now that N is not commutative. Choosing $b,c \in N$ with $[b,c] \neq 0$ and letting $a = D(x)$, we get $D^2(x)[b,c] = 0$ for all $x \in N$; and since the central element $D^2(x)$ cannot be a nonzero divisor of zero, we conclude that $D^2(x) = 0$ for all $x \in N$. But by Lemma 3(iv), this cannot happen for nontrivial D.

The next theorem extends to near-rings a theorem of Herstein [2].

<u>THEOREM 3</u> Let N be a prime near-ring admitting a nontrivial derivation D such that $[D(x),D(y)] = 0$ for all $x,y \in N$. Then (N,+) is abelian; and if N is 2-torsion-free as well, then N is a commutative ring.

<u>Proof</u> The argument used in the proof of Lemma 3(ii) shows that if both z and z + z commute elementwise with D(N), then $zD(c) = 0$ for all additive commutators c. Thus, taking $z = D(x)$, we get $D(x)D(c) = 0$ for all $x \in N$, so $D(c) = 0$ by Lemma 3(iii). Since wc is also an additive commutator for any $w \in N$, we have $D(wc) = 0 = D(w)c$; and another application of Lemma 3(iii) gives $c = 0$.

Assume now that N is 2-torsion-free. By the partial distributive law, $D(D(x)y)D(z) = D(x)D(y)D(z) + D^2(x)yD(z)$ for all $x,y,z \in N$; hence $D^2(x)yD(z) = D(D(x)y)D(z) - D(x)D(y)D(z) = D(z)(D(D(x)y) - D(x)D(y)) = D(z)D^2(x)y = D^2(x)D(z)y$. Thus

$$D^2(x)(yD(z) - D(z)y) = 0 \text{ for all } x,y,z \in N.$$

Replacing y by yt, we obtain

$$D^2(x)ytD(z) = D^2(x)D(z)yt = D^2(x)yD(z)t$$

for all $x,y,z,t \in N$, so that $D^2(x)N[t,D(z)] = \{0\}$ for all $x,t,z \in N$. The primeness of N now shows that either $D^2 = 0$ or $D(N) \subseteq Z$; and since the first of these conditions is impossible by Lemma 3(iv), the second must hold and N is a commutative ring by Theorem 2.

In the proof of the next result we use a construction from [1], called the localization of N at $Z \setminus \{0\}$, which allows an embedding of N in a near-ring N* with 1. Specifically, if Z is nonzero, consider the set of ordered pairs (x,z) with $x \in N$ and $z \in Z \setminus \{0\}$. Obtain an equivalence relation ~ by defining $(x_1,z_1) \sim (x_2,z_2)$ to mean that $x_1z_2 = x_2z_1$. Let N* be the set of equivalence classes $\langle x,z\rangle$ with addition and multiplication defined by $\langle x_1,z_1\rangle + \langle x_2,z_2\rangle = \langle x_1z_2 + x_2z_1, z_1z_2\rangle$ and $\langle x_1z_1\rangle\langle x_2,z_2\rangle = \langle x_1x_2,z_1z_2\rangle$; and embed N in N* by mapping x to $\langle xz,z\rangle$.

<u>THEOREM 4</u> Let N be any prime near-ring admitting a derivation D such that $x - D(x) \in Z$ for all $x \in N$. Then (N,+) is abelian. If in addition D is commuting and N is 2-torsion-free, then N can be embedded in a near-field.

<u>Proof</u> It is clear that constants are in Z; hence, if there exists a nonzero constant, (N,+) is abelian by the argument of Theorem 2. Thus, we assume that 0 is the only constant. We can then apply Lemma 2 to any nonzero u of the form $x - D(x)$, and conclude that $x - D(x) \in \xi(N)$. It follows that for each $x \in N$, $x - D(x) + x - D(x) = x + x - D(x + x) \in Z$; hence we get (N,+) abelian by Lemma 3(ii) once we demonstrate that there exists a nonzero element of the form $x - D(x)$.

Assume, therefore, that $x - D(x) = 0$ for all $x \in N$. Then for all $x,y \in N$, $xy = D(xy) = xD(y) + D(x)y = xy + xy$, so $xy = 0$. But this is impossible in a prime near-ring, hence (N,+) is abelian.

We now introduce the additional hypotheses that D is commuting and N is 2-torsion-free. Since Z has been shown to be nonzero, we can localize at $Z \setminus \{0\}$, embedding N in a near-ring N* with 1. Note that in N*, the element $\langle x,z\rangle$ has a right inverse if there exists $y \in N$ such that $xy \in Z \setminus \{0\}$. Call all other elements $x \in N$ *exceptional*. Note that if D is the trivial derivation, there are no nonzero exceptional elements, so N* is a near-field; hence, we assume that $D \neq 0$.

If x is any exceptional element, then $x^2 - D(x^2) = x(x - D(x) - D(x)) = 0$. Thus, for every $y \in N$, we have $x^2y - D(x^2y) = x^2y - (x^2D(y) + D(x^2)y) = -x^2D(y) = x(-xD(y)) \in Z$; hence $x^2D(N) = \{0\}$, and by Lemma 3(iii), $x^2 = 0$. We now get $D(x^2) = 0 = xD(x) + xD(x)$, and the absence of 2-torsion yields $xD(x) = 0$. Thus $x(x - D(x)) = 0$; and if $x \neq 0$, the fact that Z contains no non-zero divisors of zero forces $x - D(x) = 0$. But for every $y \in N$, xy is exceptional also, so $xy - D(xy) = xy - (xD(y) + D(x)y) = -xD(y) = 0$; consequently, $xD(N) = \{0\}$ and therefore $x = 0$. Thus, N has no nonzero exceptional elements and N* is a near-field.

<u>COROLLARY 1</u> Let N be a prime distributively-generated near-ring admitting a derivation D such that $x - D(x) \in Z$ for all $x \in N$. Then N is a commutative ring.

<u>Proof</u> By the theorem, (N,+) is abelian, which in the setting of distributively generated near-rings forces N to be a ring. In the presence of distributivity our hypothesis on D implies that D is commuting; hence if $D \neq 0$ we can invoke a result of Posner [4] to the effect that a prime ring admitting a nontrivial commuting derivation must be commutative. In the event that $D = 0$, Corollary 1 is obvious.

4. NEAR-RINGS WITHOUT NILPOTENT ELEMENTS

In our final section we capitalize on the good behavior of annihilator ideals with respect to derivations.

LEMMA 4 Let N have no nonzero nilpotent elements. Then, for any derivation D, annihilators are D-invariant. Thus, there exists a family of completely prime ideals $\{P_\alpha | \alpha \in \Lambda\}$ such that N is a subdirect product of the near-rings N/P_α, and such that, for each $\alpha \in \Lambda$, the definition $\tilde{D}(x + P_\alpha) = D(x) + P_\alpha$ yields a derivation $\tilde{D}$ on N/P_α.

Proof Recall that in near-rings without nilpotent elements, there is no distinction between left and right annihilators. Suppose that $xy = 0$. Then $0 = D(xy) = xD(y) + D(x)y$. Left-multiplying by x and noting that N is an IFP-near-ring, we get $x^2D(y) = 0$, so that $(xD(y))^2 = 0$ and hence $xD(y) = 0$.

The remaining conclusions follow from the fact that N is a subdirect product of N/P_α for a family $\{P_\alpha\}$ of completely prime ideals which are annihilators [3, 9.36].

THEOREM 5 Let N have no nonzero nilpotent elements, and suppose that N admits a derivation D such that $x - D(x) \in Z$ for all $x \in N$. Then $(N,+)$ is abelian.

Proof Use the subdirect-product representation from Lemma 4, and let $\tilde{N}$ be a typical factor near-ring N/P_α. Then $\tilde{N}$ has no nonzero divisors of zero, hence is certainly prime and thus $(\tilde{N},+)$ is abelian by Theorem 4.

REFERENCES

[1] H. E. Bell, Certain near-rings are rings, J. London Math. Soc. (2) 4 (1971), 264-270.

[2] I. N. Herstein, A note on derivations, Canad. Math. Bull. 21 (1978), 369-370.

[3] G. Pilz, Near-rings, 2nd Ed., North-Holland, Amsterdam, 1983.

[4] E. C. Posner, Derivations in prime rings, Proc. Amer. Math. Soc. 8 (1957), 1093-1100.

Howard E. Bell
Mathematics Dept.
Brock University
St. Catherines, Ontario
Canada L2S 3A1

Gordon Mason
Dept. of Math.
University of New Brunswick
Fredericton, New Brunswick
Canada

Near-rings and Near-fields, G. Betsch (editor)

EMBEDDING OF A NEAR-RING INTO A NEAR-RING WITH IDENTITY

Gerhard BETSCH

Universität Tübingen, Mathematisches Institut
D-7400 Tübingen, Fed. Rep. Germany

It is well known, that an arbitrary near-ring N may be embedded into a near-ring $\overline{N}$ with identity. Details and references are to be found, e.g., in [3; § 1, section c]. Also, it is very well known, that any ring A is an ideal of a ring A* with identity [2; p. 11]. In this note we present a class of near-rings, which cannot be one-sided ideals in a near-ring with identity. And we characterize those near-rings, which do have an embedding as a one-sided ideal into a near-ring with identity.

We follow the terminology and notation of [3], except that we consider left near-rings, and consequently our near-rings act from the right on N-groups. Furthermore, we restrict our consideration to zero-symmetric near-rings. Hence, "near-ring" in the sequel always means "zero-symmetric left near-ring". N will always denote a near-ring. We are looking for "nice" embeddings of N into a near-ring $\overline{N}$ with identity. If $(\Gamma,+)$ is a (not necessarily commutative) group, then $M_0(\Gamma)$ denotes the near-ring (with identity) of all mappings of Γ into itself which leave the neutral element of Γ fixed. The additive groups of near-rings N, $\overline{N}$,... are denoted by N_+, $\overline{N}_+$, ...

1. For the sake of convenience we recall a few known facts.

a) If Γ is a faithful N-group, then N is embeddable into $M_0(\Gamma)$, and will be considered as a subnear-ring of $M_0(\Gamma)$. Conversely, if N is embedded into a near-ring $\overline{N}$ with identity, then $\overline{N}_+$ is a faithful N-group. Consequently, embedding a near-ring into a near-ring with identity and finding a faithful N-group are equivalent (easy) problems.

b) Let $(\Gamma,+)$ be a group properly containing N_+. Then the definition

$$\gamma n := \begin{cases} n, & \text{if } \gamma \in \Gamma \setminus N \\ \gamma n, & \text{if } \gamma \in N \end{cases}$$

turns Γ into a faithful N-group.

(This remarkably simple construction was discovered independently by several authors. See [3] for references. I learned it long ago from unpublished manuscripts of H. Wielandt, which date back to 1937-1952.)

Now any subnear-ring $\overline{N}$ of $M_0(\Gamma)$, which contains N and id_Γ, is a near-ring with identity containing N.

2. Definition. An ideal embedding (IE) of N is a near-ring monomorphism $\varphi: N \longrightarrow \overline{N}$ of N into a near-ring $\overline{N}$ with identity such that $\varphi(N)$ is a two-sided ideal of $\overline{N}$. Usually, we identify N and $\varphi(N)$. Right and left ideal embeddings (RIE and LIE) are defined in an analogous way.

3. Remarks. 1) For every group $(\Gamma,+)$ the near-ring $M_0(\Gamma)$ has only trivial ideals [1; Theorem 1] and [3; Theorem 7.30]. Hence the embedding of N into $M_0(\Gamma)$ according to section 1 is never an ideal embedding.
2) The motivation for our interest in IE's is the following: We can expect a closer relationship between the properties of N and $\overline{N}$ in the case, where N is an ideal of $\overline{N}$ than in the case, where N is merely a subnear-ring of $\overline{N}$.
One example: Let Γ be a faithful N-group of type 1. If N is an ideal of a subnear-ring $\overline{N}$ of $M_0(\Gamma)$ with identity, then Γ is a faithful $\overline{N}$-group of type 2, and the "best" density theorem is available for $\overline{N}$. Cf. [3; § 4, sections c) and d)].

4. Proposition. Let N be a near-ring with non-commutative additive group N_+ and multiplication given by

$$ab = \begin{cases} b, & \text{if } a \neq 0 \\ 0, & \text{if } a = 0. \end{cases}$$

If $\overline{N}$ is a near-ring with identity 1 containing N, then N_+ cannot be a normal subgroup of $\overline{N}_+$. Consequently, N is not a one-sided ideal in $\overline{N}$.

Proof. We choose $n, n' \in N$ such that $n + n' \neq n' + n$.
Now we consider $t := n + 1 + n - 1$.
We have $nt = n + n$,
$n't = n + n' + n - n'$.
If $t \in N$, then $t = nt = n't$, hence $n = n' + n - n'$, which is a contradiction to the choice of n and n'. Hence $t \notin N$, and N_+ cannot be a normal subgroup of $\overline{N}_+$.

Remark. The author does not know an example of a near-ring with commutative additive group, which has no LIE and no RIE.

5. Theorem. Let N be a near-ring. The following statements are equivalent:
a) N has a LIE.
b) There exists a faithful N-group Γ (hence $N \subseteq M_0(\Gamma)$) such that

i) the mapping $x \to -1 + x + 1$ of $M_0(\Gamma)$ into itself induces an automorphism of N_+ (1 = identity of $M_0(\Gamma)$).

ii) $(n + 1z)m \in N$ for all $n,m \in N$ and $z \in \mathbb{Z}$.
(We consider the cyclic subgroup <1> of $M_0(\Gamma)_+$ generated by 1 as $\mathbb{Z}$-module, which explains $1z$).

c) N has an IE.

Proof. a) $\Longrightarrow$ b). If $\overline{N}$ is a near-ring with identity containing N, then $\Gamma := \overline{N}_+$ is a faithful N-group as well as a faithful $\overline{N}$-group. By suitable identification we obtain $N \subseteq \overline{N} \subseteq M_0(\Gamma)$, and the identities of $\overline{N}$ and $M_0(\Gamma)$ coincide. Now if N is a left ideal of $\overline{N}$, then i) and ii) are true.
b) $\Longrightarrow$ c). Let Γ be a faithful N-group with i) and ii). We have $N \subseteq M := M_0(\Gamma)$. Let $\overline{N}_+$ be the subgroup of M_+ generated by N and {1}. From i) we infer, that $\overline{N}_+ = \{n + 1z \mid n \in N,\ z \in \mathbb{Z}\}$ and that N_+ is a normal subgroup of $\overline{N}_+$. Now ii) implies, that $\overline{N}_+$ is multiplicatively closed with respect to the multiplication in M. Hence $\overline{N}_+$ is the additive group of a subnear-ring $\overline{N}$ of M containing N, and $1 \in \overline{N}$.
Furthermore, N is a left ideal of $\overline{N}$. This follows from ii) and the fact, that N_+ is a normal subgroup of $\overline{N}_+$.
Now let $n,m,x \in N$ and $z_1,z_2 \in \mathbb{Z}$. Repeatedly using ii) and the fact, that N_+ is a normal subgroup of $\overline{N}_+$ we obtain, for suitable elements $n',n'' \in N$, the equations

$$\begin{aligned}&(n+m+1z_1)(x+1z_2) - (m+1z_1)(x+1z_2)\\ &= (n+m+1z_1)x + (n+m+1z_1)(1z_2) - (m+1z_1)(1z_2) - (m+1z_1)x\\ &= n' + (n+m+1z_1)(1z_2) - (m+1z_1)(1z_2)\\ &= n' + n'' + (m+1z_1)(1z_2) - (m+1z_1)(1z_2) = n' + n'' \in N.\end{aligned}$$

Hence N is also a right ideal in $\overline{N}$.
c) $\Longrightarrow$ a) is trivial.
The proof is completed.

Corollary. If N has a LIE, then N also has a RIE.

Remark. The author does not know, whether the converse of this corollary is true or false.

6. Theorem. Let N be a near-ring. Then the following statements are equivalent:

a) N has a RIE.

b) There exists a faithful N-group Γ (hence $N \subseteq M_0(\Gamma)$) such that

i) the mapping $x \to -1 + x + 1$ of $M_0(\Gamma)$ into itself induces an automorphism of N_+ (1 = identity of $M_0(\Gamma)$).

ii) $(n + 1z)m - (1z)m \in N$ for all $n, m \in N$ and $z \in \mathbb{Z}$.

The proof of this theorem follows the lines of section 5 and will be omitted.

REFERENCES

[1] Berman, G. and Silverman, R.J., Simplicity of near-rings of transformations. Proc. Amer. Math. Soc. 10 (1959), 456-459.

[2] Jacobson, N., Structure of Rings. AMS Coll. Publ. vol.XXXVII. 2nd Ed. Providence R.I. 1964.

[3] Pilz, G., Near-rings.2nd Ed. Amsterdam: North-Holland 1983.

Near-rings and Near-fields, G. Betsch (editor)

THE NEAR-RING OF SOME ONE DIMENSIONAL NONCOMMUTATIVE FORMAL GROUP LAWS

James R. CLAY
Department of Mathematics
University of Arizona
Tucson, AZ 85721

1. INTRODUCTION

Hazewinkel shows in [1] that, for a commutative ring A with identity, if there is to be a one dimensional noncommutative group law over A, then A must have an element $b \neq 0$ that is nilpotent and of finite additive order. He goes on to show that this is sufficient. In fact, if A is a commutative ring with identity, and b is a nonzero element of A that is nilpotent and of finite order, then there is an element c in A such that $c \neq 0$, $c^2 = 0$, and $pc = 0$ for some prime p, and

$$F(X, Y) = X + Y + cXY^p$$

defines a one dimensional noncommutative formal group law. Examples of such rings are provided by $A = K[t]/(t^2)$, where K is a field of characteristic p and t is an indeterminant, and by $A = Z_{p^2}$, the integers modulo p^2.

If $F(X, Y)$ defines a one dimensional commutative formal group law, then the endomorphisms of F form a ring. However, if F is noncommutative, the sum of two endomorphisms need not be an endomorphism, so there is no hope that the endomorphisms of F will be a ring. But for $+_F$ defined by

$$\gamma_1(X) +_F \gamma_2(X) = F(\gamma_1(X), \gamma_2(X)),$$

the power series over A, with zero constant terms, become a group $(A_0[[X]], +_F)$, and when one adds composition $\circ$, one gets a near-ring $(A_0[[X]], +_F, \circ)$ with identity $\iota(X) = X$. It is in this setting that the endomorphisms of a noncommutative group law become interesting. They form a monoid with respect to composition, and each endomorphism is a distributive element in the near-ring $(A_0[[X]], +_F, \circ)$. Hence, the endomorphisms generate, with respect to $+_F$, a distributively generated (d.g.) near-ring, with identity $\iota(X) = X$ [2,3].

For the examples of A provided above, these d.g. near-rings turn out to be commutative rings with identity. It is the purpose of this paper to demonstrate this, to determine the structure of these rings, and the structure of their groups of units.

2. PRELIMINARIES

For a commutative ring A with identity 1, the <u>one dimensional formal group laws over A</u> are the formal power series in two indeterminates $F(X, Y) \in A[[X, Y]]$ of the form

$$F(X, Y) = X + Y + \sum_{i,j\geq 1} c_{ij} X^i Y^j$$

such that the "associative" condition

$$F(X, F(Y, Z)) = F(F(X, Y), Z)$$

is valid. If, in addition, $F(X, Y) = F(Y, X)$, we say F is <u>commutative</u>. Otherwise F is <u>noncommutative</u>.

Let $A_0[[X]]$ denote all power series over A of the form

$$\alpha(X) = a_1 X + a_2 X^2 + \dots .$$

So $A_0[[X]]$ is just the power series over A with zero for the constant term.

If one defines $+_F$ on $A_0[[X]]$ by

$$\alpha(X) +_F \beta(X) = F(\alpha(X), \beta(X)),$$

then $(A_0[[X]], +_F)$ is a group. For $f, g \in A_0[[X]]$, if

$$f \circ g(X) = f(g(X)),$$

then $(A_0[[X]], +_F, \circ)$ is a right distributive near-ring with identity $\iota(X) = X$.

The <u>endomorphisms</u> of F are power series $\alpha \in A_0[[X]]$ satisfying

$$\alpha(F(X, Y)) = F(\alpha(X), \alpha(Y)).$$

If F is commutative, and if End F denotes the endomorphisms of F, then (End F, $+_F$, $\circ$) is a ring with identity. However, if F is noncommutative, and if hom(F, F) denotes the endomorphisms of F, then for $\alpha, \beta \in$ hom(F, F), the power series $\alpha +_F \beta$ need not be in hom(F, F), though it is in $A_0[[X]]$, of course. So hom(F, F) generates, under $+_F$, a subgroup (Hom(F, F), $+_F$) of $(A_0[[X]], +_F)$.

Since the semigroup (hom(F, F), $\circ$) consists of elements α that satisfy $\alpha \circ (f +_F g) = (\alpha \circ f) +_F (\alpha \circ g)$ for all $f, g \in A_0[[X]]$, the semigroup (hom(F, F), $\circ$), under $+_F$, generates a distributively generated (d.g.) near-ring (Hom(F, F), $+_F$, $\circ$), a sub-near-ring of $(A_0[[X]], +_F, \circ)$.

These observations are easily proven.

3. THE ELEMENTS OF hom(F, F)

Throughout the rest of this paper, A will denote either the ring of integers modulo p^2, p a prime, or $K[t]/(t^2)$, where K is a field of characteristic p, and t is an indeterminant.

If $A = Z_{p^2}$, then F will denote the noncommutative formal group law

$$F(X, Y) = X + Y + pXY^p, \tag{I}$$

and if $A = K[t]/(t^2)$, then F will denote the noncommutative formal group law

$$F(X, Y) = X + Y + tXY^p. \tag{II}$$

In this latter case, there is no harm in thinking of the elements of A to be of the form $B = b + \beta t$, where $b, \beta \in K$, and where $t^2 = 0$. We have done this already in (II).

We want to determine the elements of hom(F, F) for F as defined in (I) and in (II). Since the results are very close for each of the two laws (I) and (II), and since the proofs are very nearly parallel, we shall develop our results concurrently.

We shall see that the elments of hom(F, F) for (I) are exactly the elements of the form

$$\alpha(X) = aX + p \sum_{k=0}^{\infty} a_k X^{p^k}, \tag{I-a}$$

where $a \in \{0, 1\}$ and each $a_k \in Z_p = \{0,1, \ldots, p-1\}$.

For (II), the elements of hom(F, F) will be shown to be exactly the elements of the form

$$\alpha(X) = aX + t \sum_{k=0}^{\infty} a_k X^{p^k}, \tag{II-a}$$

where $a \in \{0, 1\}$ and each $a_k \in K$.

Towards this goal, we shall see first that in each case, if $\alpha \in \text{hom}(F, F)$, then

$$\alpha(X) = a_1 X + \sum_{k=1}^{\infty} a_{kp} X^{kp}, \tag{1}$$

where $a_i \in A$.

When we consider $\alpha \in \text{hom}(F, F)$ and F as in (I), we shall write

$$\alpha(X) = a_1 X + a_2 X^2 + \ldots .$$

However, for F as in (II), we shall write

$$\alpha(X) = B_1 X + B_2 X^2 + \dots .$$

In each case, we study the implications of the condition

$$(2) \qquad \alpha(F(X, Y)) = F(\alpha(X), \alpha(Y)).$$

Now $\alpha(F(X, Y)) = \sum_{k=1}^{\infty} a_k F(X, Y)^k$ or $\alpha(F(X, Y)) = \sum_{k=1}^{\infty} B_k F(X, Y)^k$. For $k \geq 2$, we have

$$F(X, Y)^k = \begin{cases} (X + Y)^k + k(X + Y)^{k-1}(pXY^p), & \text{for (I)}, \\ (X + Y)^k + k(X + Y)^{k-1}(tXY^p), & \text{for (II)}. \end{cases}$$

So we get

$$(\text{I-b}) \quad \alpha(F(X, Y)) = \sum_{k=1}^{\infty} a_k (X + Y)^k + (pXY^p) \sum_{k=1}^{\infty} k a_k (X + Y)^{k-1}$$

and

$$(\text{II-b}) \quad \alpha(F(X, Y)) = \sum_{k=1}^{\infty} B_k (X + Y)^k + (tXY^p) \sum_{k=1}^{\infty} k B_k (X + Y)^{k-1}.$$

The right hand side of (2) yields

$$(\text{I-c}) \quad F(\alpha(X), \alpha(Y)) = \sum_{k=1}^{\infty} a_k (X^k + Y^k) + p \sum_{\substack{i \geq 1 \\ j \geq 1}} a_i a_j^p X^i Y^{jp}$$

and

$$(\text{II-c}) \quad F(\alpha(X), \alpha(Y)) = \sum_{k=1}^{\infty} B_k (X^k + Y^k) + t \sum_{\substack{j \geq 1 \\ j \geq 1}} B_i B_j^p X^i Y^{jp}.$$

Suppose $1 < i < p$. Collecting terms of degree i, we conclude $a_i = 0$ and $B_i = 0$. Collecting terms of degree $p + i$, we conclude $a_{p+i} = 0$ and $B_{p+i} = 0$. Also $pa_p = 0$ and $tB_p = t(b_p + t\beta_p) = tb_p$.

By a rather messy induction argument, one shows that $a_{kp+i} = 0$ and $B_{kp+i} = 0$ if $1 < i < p$, and that $pa_{kp} = 0$ and $tB_{kp} = tb_{kp}$.

This gives α the form described in (1). However, to get α in the form described by (I-a) and (II-a) takes considerable calculation. We shall sketch the steps for (I-a). The proceedure for (II-a) is parallel but more involved. For that case, we shall point out a few adjustments that must be made.

With $\alpha(X) = a_1 X + \sum_{k=1}^{\infty} a_{kp} X^{kp}$, where each $pa_{kp} = 0$, we get, from insisting that (2) be valid, the equation

$$(3)\quad pa_1\, XY^p + \sum_{k=1}^{\infty} a_{kp}\, (X + Y)^{kp} = pa_1^{p+1}\, XY^p + \sum_{k=1}^{\infty} a_{kp}\, (X^{kp} + Y^{kp}).$$

Using $pa_{kp} = 0$, we get

$$a_{kp}(X^{kp} + Y^{kp}) = a_{kp} \sum_{j=0}^{k} \binom{k}{j} (X^p)^{k-j} (Y^p)^j,$$

which forces $ka_{kp} = 0$. So, if $a_{kp} \neq 0$, then p divides k, $p|k$. Setting $k = k_0\, p^{s-1}$ with $(k_0, p) = 1$, one shows that $k_0 = 1$ and so $k = p^{s-1}$. This in turn makes

$$\alpha(X) = a_1\, X + \sum_{k=1}^{\infty} a_{p^k} X^{p^k}.$$

Returning our attention to (3), we now consider the consequence of $pa_1 = pa_1^{p+1}$. Certainly this can be if $pa_1 = 0$ and so $\alpha(X) = \sum_{k=0}^{\infty} a_{p^k} X^{p^k}$ with $pa_{p^k} = 0$. Alternatively, if $pa_1 \neq 0$, we express $a_1 = mp + i$ with $(i, p) = 1$, $1 \leqslant i \leqslant p - 1$. So $pa_1 = pi$ and $a_1^p \equiv (mp + i)^p \equiv i^p \bmod p^2$. Hence $pa_1^{p+1} = (pi)i^p = pi^{p+1}$. This means $pi = pi^{p+1}$ and consequently $p = pi^p$, which makes $i^p = 1$ and so $i = np + 1$ for some n. Thus $a_1 = (m + n)p + 1$, which in turn makes $\alpha(X) = X + \sum_{k=0}^{\infty} a_{p^k} X^{p^k}$, where $pa_{p^k} = 0$ for each k.

This completes the proof of half of the following theorem, and the proof of the other half is direct.

THEOREM I-1. For F defined by (I), we have

$$\hom(F, F) = \{aX + p \sum_{k=0}^{\infty} a_k\, X^{p^k} \,|\, a \in \{0, 1\},\ a_k \in Z_p\}.$$

We will get a similar theorem for (II). Above, where we saw that $a_1 = (m + n)p + 1$ if $pa_1 \neq 0$, we also get from the analogous condition $B_1\, t = B_1^{p+1}\, t$, that $B_1 = 1 + \beta_1 t$ or $B_1 = \beta_1 t$. Consider these two cases separately, with $B_1 = \beta_1 t$ considered first.

In this first case, assume that (2) is valid. We are forced to have

$$\sum_{k=1}^{\infty} B_{kp}\, (X + Y)^{kp} = \sum_{k=1}^{\infty} B_{kp}\, (X^{kp} + Y^{kp}) + t(\sum_{k=1}^{\infty} B_{kp}\, X^{kp})(\sum_{k=1}^{\infty} B_{kp}^p\, Y^{kp^2}).$$

For $1 < i < p$, conclude that $B_{ip} = 0$, and $B_p = \beta_p t$. Set up the following conditions for an induction:

first segment- $\quad p^0 < i < p^1$ implies $B_i = 0$,

$\qquad B_{p^0} = \beta_{p^0}\, t$;

second segment- $p^1 < i < p^2$ implies $B_i = 0$,

Assume $B_{p^1} = \beta_{p^1} t$;

k th segment- $p^{k-1} < i < p^k$ implies $B_i = 0$,

$B_{p^{k-1}} = \beta_{p^{k-1}} t$.

Then prove the

(k+1)th segment- $p^k < i < p^{k+1}$ implies $B_i = 0$,

$B_{p^k} = \beta_{p^k} t$.

So, in the case $B_1 = \beta_1 t$, we get

$$\alpha(X) = t \sum_{k=0}^{\infty} \beta_{p^k} X^{p^k}.$$

In the case $B_1 = 1 + \beta_1 t$, one uses calculations parallel to those used for (I) to get the desired results. So we also have

THEOREM II-1. For F defined by (II), we have

$$\text{hom}(F, F) = \{aX + t \sum_{k=0}^{\infty} a_k X^{p^k} \mid a \in \{0, 1\}, a_k \in K\}.$$

4. THE ELEMENTS OF Hom(F, F).

The elements of hom(F, F) form a monoid with respect to $\circ$, and if $\alpha \in \text{hom}(F, F)$ and $f, g \in A_0[[X]]$, then $\alpha \circ (f +_F g) = (\alpha \circ f) +_F (\alpha \circ g)$. So hom(F, F) is a semigroup of distributive elements in the near-ring $(A_0[[X]], +_F, \circ)$. With respect to $+_F$, the semigroup hom(F, F) will generate a group $(\text{Hom}(F, F), +_F)$ which is closed with respect to $\circ$, thereby yielding a distributively generated (d.g.) near-ring $(\text{Hom}(F, F), +_F, \circ)$. We now proceed to identify the elements of Hom(F, F).

For $\alpha, \beta \in \text{hom}(F, F)$, we set

$$\alpha = aX + \lambda \sum_{k=0}^{\infty} a_k X^{p^k},$$

and

$$\beta = bX + \lambda \sum_{k=0}^{\infty} b_k X^{p^k},$$

where λ is either p or t, depending upon whether F is of form (I), or or (II), respectively. In each case, $\lambda^2 = 0$ and we get

(4) $$\alpha +_F \beta = \alpha + \beta + \lambda ab X^{p+1}.$$

In deriving this, we use the fact that $b \in \{0, 1\}$ and so $b^p = b$. So $\alpha +_F \beta = \beta +_F \alpha$ making $(\mathrm{Hom}(F, F), +_F, \circ)$ a ring with identity $\iota(X) = X$.

Similarly,

$$\alpha \circ \beta = abX + \lambda \sum_{k=0}^{\infty} (ab_k + ba_k)X^{p^k}, \tag{5}$$

so $(\mathrm{hom}(F, F), \circ, \iota(X) = X)$ is a commutative monoid, thus making $(\mathrm{Hom}(F, F), +_F, \circ, \iota(X) = X)$ a commutative ring with identity. This gives us

THEOREM 2. For F defined by (I) or (II), we have that

$$(\mathrm{Hom}(F, F), +_F, \circ, \iota(X) = X)$$

is a commutative ring with identity $\iota(X) = X$.

Equation (4) shows that $\mathrm{hom}(F, F)$ is properly contained in $\mathrm{Hom}(F, F)$.

For $\alpha \in \mathrm{Hom}(F, F)$, let $m\alpha$ be the sum of m α's with respect to the usual $+$, and let

$$m \cdot \alpha = [(m - 1) \cdot \alpha] +_F \alpha.$$

One easily sees that

$$m \cdot \alpha = m\alpha + \frac{m(m-1)}{2} \lambda a X^{p+1} \tag{6}$$

for any $\alpha \in \mathrm{hom}(F, F)$. So, if p is an odd prime, each $\alpha \in \mathrm{hom}(F, F)$ has order p with respect to $+_F$, but for $p = 2$, each $\alpha \in \mathrm{hom}(F, F)$ has order 2, if $a = 0$, and has order 4 if $a = 1$.

Consider the case with $p \neq 2$. One easily gets

$$(m \cdot \alpha) +_F (n \cdot \beta) = (ma + nb)X + \lambda \sum_{k=0}^{\infty} (ma_k + nb_k)X^{p^k} + [\frac{m(m-1)}{2} a + \frac{n(n-1)}{2} b + mnab]\lambda X^{p+1}. \tag{7}$$

Recall that $a, b \in \{0, 1\}$, so $ma + nb = s \in Z_p$, and

$$\frac{m(m-1)}{2} a + \frac{n(n-1)}{2} b + mnab = \frac{s(s-1)}{2}.$$

That is, if $s + \lambda u$ is the coefficient of X for $\gamma \in \mathrm{Hom}(F, F)$, then $[s(s - 1)/2]\lambda$ is the coefficient of X^{p+1} for γ. So we have

THEOREM 3. For an odd prime p, the ring $\mathrm{Hom}(F, F)$ consists of all power series of the form

$$mX + \lambda \sum_{i=0}^{\infty} a_i X^{p^i} + \frac{m(m-1)}{2} \lambda X^{p+1}$$

where $m \in Z_p$, and $a_i \in K$ or $a_i \in Z_p$, depending upon whether $\lambda = t$ or $\lambda = p$, respectively.

Take $\gamma = maX + \lambda \sum_{i=0}^{\infty} a_i X^{p^i} + [m(m - 1)/2]a\lambda X^{p+1}$ and

$\delta = nbX + \lambda \sum_{i=0}^{\infty} b_i X^{p^i} + [n(n-1)/2]b\lambda X^{p+1}$ arbitrarily from Hom(F, F), where $a, b \in \{0, 1\}$. Then

$$8)\quad \gamma \circ \delta = mnabX + \lambda \sum_{i=0}^{\infty} (nba_i + mab_i)X^{p^i} + [mn(mn-1)/2]ab\lambda X^{p+1}.$$

This again shows that $\circ$ is commutative in Hom(F, F).

Now consider the case where $p = 2$. For $\alpha = aX + \lambda \sum_{k=0}^{\infty} a_k X^{p^k}$, we get $2 \cdot \alpha = \lambda a X^{p+1}$ and $3 \cdot \alpha = aX + \lambda \sum_{k=0}^{\infty} a_k X^{p^k} + \lambda a X^{p+1}$. One easily gets that Hom(F, F) consists of all power series of the form

$$aX + \lambda \sum_{k=0}^{\infty} a_k X^{2^k} + \lambda b X^{p+1}$$

where $a, b \in Z_2$, and $a_k \in K$ or $a_k \in Z_2$, depending upon whether $\lambda = t$ or $\lambda = 2$, respectively. Again (Hom(F, F), $+_F$, $\circ$) is a commutative ring with identity $\iota(X) = X$.

5. THE STRUCTURE OF THE RING Hom(F, F)

If R is a commutative ring with identity and M is a unitary (left) R-module, then the idealization of M, denoted by RXM, is a ring on the set $R \times M$, where

$$(9)\quad \begin{aligned} (r, m) + (r', m') &= (r + r', m + m') \\ (r, m) \cdot (r', m') &= (rr', rm' + r'm). \end{aligned}$$

Now $(R \times M, +, \cdot)$ is a commutative ring with identity (1, 0). The group of units U of RXM consists of the elements (r, m) where r is a unit of R, and $(r, m)^{-1} = (r^{-1}, -r^{-2}m)$.

Let $A = \{(r, 0) | (r, 0) \in U\}$ and $B = \{(1, m) | m \in M\}$. Then $A \cap B = \{(1, 0)\}$, A and B are normal subgroups of U, and $U = AB$, since $(r, m) = (r, 0)(1, r^{-1}m)$. So U is a direct product of A and B. Notice that B, as a multiplicative group, is isomorphic to M, as an additive group.

For $(r, m) \in R \times M$, $(r, m)^k = (r^k, kr^{k-1}m)$ so the nilpotent elements are the elements of $J(R) \times M$, where $J(R)$ is the radical of R. So $J(R) \times M$ is the radical of RXM.

It is easy to see that the ideals of RXM are the $R' \times M'$, where R' is an ideal of R, and M' is a submodule of M with the additional property that $R'M \subseteq M'$.

Our rings (Hom(F, F), $+_F$, $\circ$) have this structive if $p \neq 2$. Take $\alpha = mX + \lambda \sum_{i=0}^{\infty} a_i X^{p^i} + [m(m-1)/2]\lambda X^{p+1}$ from Hom(F, F). It is formed from

$(m, (a_i)) \in Z_p \times M$, where $M = Z_p^N$, if $\lambda = p$, and $M = K^N$, if $\lambda = t$. Here $N = \{0,1,2, \ldots\}$ and U^V denotes all mappings from V to U. Note that M is a unitary Z_p-module. Define the map $\psi: Z_p X M \to \mathrm{Hom}(F, F)$ by

$$\psi[(m, (a_i))] = mX + \lambda \sum_{i=0}^{\infty} a_i X^{p^i} + [m(m - 1)/2]\lambda X^{p+1}.$$

Then

$$\psi[(m, (a_i)) + (n, (b_i))] = \psi[(m + n, (a_i + b_i))]$$

and

$$= \psi[(m, (a_i))] +_F \psi[(n, (b_i))]$$

$$\psi[(m, (a_i)) \cdot (n, (b_i))] = \psi[(mn, m(b_i) + n(a_i))]$$

$$= \psi[(m, (a_i))] \circ \psi[(n, (b_i))].$$

It is easy to see that ψ is an isomorphism. We now easily have

THEOREM 4. For an odd prime p, the ring $(\mathrm{Hom}(F, F), +_F, \circ)$ is isomorphic, via ψ, to the idealiation of the Z_p-module M, where $M = Z_p^N$ if $\lambda = p$, and $M = K^N$ if $\lambda = t$. The group of units of $\mathrm{Hom}(F, F)$ is isomorphic to the direct product $Z_p^* \times M$. The radical J of $\mathrm{Hom}(F, F)$ is all endomorphisms of the form $\lambda \sum_{i=0}^{\infty} a_i X^{p^i}$, is a maximal ideal, and $\mathrm{Hom}(F, F)/J \cong Z_p$. The ideals of $\mathrm{Hom}(F, F)$ are the isomorphic images, via ψ, of the submodules of M.

At the end of Section 3, the case where $p = 2$ was considered. An element $\alpha \in \mathrm{Hom}(F, F)$ is of the form $\alpha = aX + \lambda \sum_{k=0}^{\infty} a_k X^{2^k} + \lambda b X^3$, where $a, b \in Z_2$, $a_k \in Z_2$ if $\lambda = 2$, and $a_k \in K$ if $\lambda = t$. Take $\beta = cX + \lambda \sum_{k=1}^{\infty} c_k X^{2^k} + \lambda d X^3$. Then we can identify

$$\alpha \longleftrightarrow [(a, b), (a_k)], \qquad \beta \longleftrightarrow [(c, d), (c_k)],$$

$$\alpha +_F \beta \longleftrightarrow [(a + c, b + d + ac), (a_k \pm c_k)]$$

$$\alpha \circ \beta \longleftrightarrow [(ac, ad + bc), (ca_k + ac_k)].$$

So, for $p = 2$, $(\mathrm{Hom}(F, F), +_F, \circ)$ is isomorphic to the idealization of the Z_4-module M, where $M = Z_2^N$ if $\lambda = 2$, and $M = K^N$ if $\lambda = t$.

Since $J(Z_4) = \{0, 2\}$ is a proper ideal of Z_4, it follows that the radical J of $\mathrm{Hom}(F, F)$ is all power series of the form $\lambda \sum_{k=0}^{\infty} a_k X^{2^k} + \lambda b X^3$, and $\mathrm{Hom}(F, F)/J \cong Z_2$. Since $J(Z_4)M = \{0\}$, we again have the submodules M' of M identified with ideals of $\mathrm{Hom}(F, F)$, but in this case, one also gets ideals from the ideals of the form $J(Z_4) \times M'$ in $Z_4 X M$. The group of units is isomorphic to $Z_2 \times M$.

6. AN INTERESTING EXTENSION.

We have seen in theorem 4 that $(\mathrm{Hom}(F, F), +_F, \circ)$ motivates the idea of idealization of an R-module M. We also noticed, if $p \neq 2$, that the elements of Hom(F, F) are of the form

$$\alpha = kX + \lambda \sum_{i=0}^{\infty} a_i X^{p^i} + [k(k-1)/2]\lambda X^{p+1}$$

where $k \in Z_p$, and each $a_i \in Z_p$ if $\lambda = p$, and each $a_i \in K$ if $\lambda = t$. All the powers of X are of the form p^i except for X^{p+1}, and its coefficient is derived from the coefficient $k + \lambda a_0$ of X.

If $p = 2$, we have seen the elements of Hom(F, F) to be of the form

$$\alpha = kX + \lambda \sum_{i=0}^{\infty} a_i X^{p^i} + m\lambda X^{p+1}$$

where $k, m \in Z_p$, and m does not depend upon the coefficient of X. This motivates looking at H, the set of all power series of the form

$$kX + \lambda \sum_{i=0}^{\infty} a_i X^{p^i} + m\lambda X^{p+1}$$

where $k, m \in Z_p$. For $\alpha = kX + \lambda \sum_{i=0}^{\infty} a_i X^{p^i} + \lambda m X^{p+1}$ and $\beta = uX + \lambda \sum_{i=0}^{\infty} b_i X^{p^i} + \lambda v X^{p+1}$, we get

$$\alpha +_F \beta = \alpha + \beta + \lambda k u X^{p+1}$$

and

$$\alpha \circ \beta = kuX + \lambda \sum_{i=0}^{\infty} (ua_i + kb_i) X^{p^i} + (kv + mu^2)\lambda X^{p+1}.$$

Taking direction from Section 5, we have the bijection $\alpha \longleftrightarrow [m, (k, a)]$, $\beta \longleftrightarrow [v, (u, b)]$ where $a = (a_i)$, $b = (b_i) \in M$. Then

$$\alpha +_F \beta \longleftrightarrow [m + v + ku, (k + u, a + b)]$$

and

$$\alpha \circ \beta \longleftrightarrow [kv + mu^2, (ku, kb + ua)].$$

This, as in Theorem 4, motivates the following construction.

Let R be a commutative ring with identity, and let M be a unitary R-module. As before, RXM is the idealization of M. Consider the set $R \times (RXM)$. Taking guidance from the last paragraph, define

$$[a, (b, m)] + [c, (d, n)] = [a + c + bd, (b + d, m + n)]$$

and

$$[a, (b, m)] \cdot [c, (d, n)] = [bc + ad^2, (bd, bn + dm)].$$

It is direct to prove

THEOREM 5. $(R \times (RXM), +, \cdot)$ is a near-ring with identity $[0, (1, 0)]$ in which the left distributive law is not valid.

COROLLARY 6. $(H, +_F, \circ)$ is a right distributive near-ring with identity $\iota(X) = X$.

THEOREM 7. The units U of $R \times (RXM)$ is the set

$$U = \{[a, (u, m)] | a, u \in R, u \text{ is a unit of } R, m \in M\},$$

and $[a, (u, m)]^{-1} = [-u^3 a, (u^{-1}, -u^{-2}m)]$.

In discussing group structure, we shall use the notation and terminology found in [4].

It is easy to see that the additive group $(R \times (RXM), +)$ is an extension of R^+ by the additive group of RXM realizing $\theta:(RXM)^+ \to \text{Aut } R^+$, where $\theta(b, m) = 1_R$ for each $(b, m) \in RXM$, and with factor set $f:(RXM)^+ \times (RXM)^+ \to R^+$ defined by $f[(b, m), (d, n)] = bd$.

What is not so obvious is that the group of units, U, of a near-ring $R \times (RXM)$, is a semidirect product of the normal subgroup $S = \{[a, (1, 0)] | a \in R\}$, - which is isomorphic to the additive group R^+ of the ring R, - by the subgroup $T = \{[0, (u, m)] | (u, m) \in U(RXM) = U(R) \times M\}$, where U(RXM), U(R) are the units of the rings RXM and R, respectively. Here, T is isomorphic to U(RXM). It is direct to see that S is a normal subgroup of U, that T is a subgroup of U, and that $S \cap T = \{[0, (1, 0)]\}$. Observing that $[a, (u, m)] = [au^{-2}, (1, 0)][0, (u, m)]$, we get $U = ST$, so U is a semidirect product of S by T. We can write the extension by

$$0 \to R^+ \to U \to U(RXM) \to 1,$$

so $U \cong R^+ \times_\theta U(RXM)$, where

$$(a, (u, m)) * (b, (v, n)) = (a + (u, m)b, (u, m)(v , n))$$

$$= (a + u^{-1}b, (uv, un + vm)).$$

The isomorphism is defined by $F[a, (u, m)] = (au^{-2}, (u, m))$.

ACKNOWLEDGEMENT

The author wishes to thank Professor A. Fröhlich, who initially suggested this line of research, and with whom the author has had many stimulating discussions.

REFERENCES

[1] Hazewinkel, M., Formal Groups and Applications (Academic press, New York, 1978).

[2] Pilz, G., Near-rings (North-Holland, Amsterdam, 1983).

[3] Fröhlich, A., Distributively Generated Near-rings I. Ideal Theory, Proc. London Math. Soc. 8 (1958), 76-94.

[4] Rotman, J. J., Theory of Groups (Allyn and Bacon, Boston, 1984).

Near-rings and Near-fields, G. Betsch (editor)

ON THE EXISTENCE OF NIL IDEALS IN DISTRIBUTIVE NEAR-RINGS

S.De Stefano and S.Di Sieno*

ABSTRACT

In this paper we show first that the sum of the nil right ideals of a distributive near-ring N and the sum of the nil left ideals of N have the same two-sided annihilator. This ideal - here denoted by $\alpha(N)$ - proves useful in studying the existence of nil non-nilpotent ideals when the near-ring is weakly semiprime. In particular, we show that N has a nil non-nilpotent left ideal if and only if $\alpha(N)$ is properly contained in N and that every one-sided non-nilpotent ideal contains a nil non-nilpotent one-sided ideal if and only if $\alpha(N)$ coincides with the (two-sided) annihilator of N. Finally we prove that N has a two-sided nil non-nilpotent ideal if and only if the ring $N/\alpha(N)$ has one.

1. PRELIMINARIES

Throughout this paper N will be a distributive near-ring; standard terminology and relative notations can be found in [4]. If it is not stated otherwise, it will be understood that definitions and results relative to left structures hold also for the corresponding right ones.

A left non-zero ideal L of N is called *nilpotent* if there is a positive integer k such that $L^k=(0)$; more in general, L is called *nil* if for every x in L there exists a positive integer h such that $x^h=0$; finally, L is called *maximal nil left ideal* if it is maximal in the family of all nil left ideals of N. By Zorn's lemma, every nil left ideal is contained in a maximal nil left ideal: in particular, this is true of the (two-sided) annihilator of N

$$A(N) = \{n\in N \mid nx = xn = 0 \ \forall x\in N\}.$$

Moreover every maximal nil left ideal of N is the inverse image of a maximal nil left ideal of $N/A(N)$ under the canonical epimorphism $\pi: N \twoheadrightarrow N/A(N)$ and hence contains $\ker\pi$. This proves

REMARK 1 : *Every maximal nil left ideal of* N *contains the annihilator of* N.

Now, if the near-ring N is a non-ring, A(N) is different from zero (since A(N) contains the commutator subgroup of the additive group of N): thus every non-ring N has a (trivial) nil non-zero one-sided ideal. Our interest is in those nil (one-sided) ideals which are not contained in A(N).

---

* Both authors are members of the G.N.S.A.G.A. of the Italian C.N.R.

THEOREM 2 : *The near-ring* N *has a nil left ideal not contained in* A(N) *if and only if it has a nil right ideal not contained in* A(N). *Precisely, if* L *is a nil left ideal of* N *not contained in* A(N), *every right ideal generated by an element of* L *which is not in* A(N) *will work; and similarly on the other side.*

PROOF : Let L be a nil left ideal not contained in A(N) and consider the right ideal $I_r(x)$ generated by an element x of L which does not belong to A(N).

By the commutativity of the sum in N^2, the square y^2 of any element y of $I_r(x)$

$$y = \sum_{i=1}^{s} (-m_i+z_ix+xn_i+m_i) \qquad \text{where } s\in \mathbb{N},\ z_i\in \mathbb{Z},\ m_i,n_i\in N$$

has the form xn, with n in N. Since nx belongs to the nil left ideal L, $(nx)^h$ is zero for a suitable positive integer h; this implies $y^{2(h+1)}=0$, i.e. $I_r(x)$ is nil. On the other side, $I_r(x)$ is not contained in A(N) and is therefore an ideal of the required kind.

In the same way one proves the "if" part of the statement.

From now on it will be supposed that N has a left ideal not contained in A(N).

2. THE NIL-ANNIHILATOR

Let $\mathcal{L}$ be the family of the maximal nil left ideals of N, and $\mathcal{R}$ be the family of the maximal nil right ideals of N. Evidently the sum of the ideals of $\mathcal{L}$ (respectively of $\mathcal{R}$) coincides with the sum of all the nil left (respectively right) ideals of N.

The following result holds:

THEOREM 3 : *The left annihilator of the sum of the ideals of* $\mathcal{L}$ *coincides with the left annihilator of the sum of the ideals of* $\mathcal{R}$.

PROOF : Denote the left annihilator of a set X by $A_\ell(X)$ and - for sake of brevity - by S_ℓ the left annihilator of the sum of the ideals of $\mathcal{L}$ and by S_r the left annihilator of the sum of the ideals of $\mathcal{R}$. Clearly

$$S_\ell = \cap A_\ell(L) \qquad \text{and} \qquad S_r = \cap A_\ell(R)$$

where L ranges in $\mathcal{L}$ and R in $\mathcal{R}$.

In order to prove that S_r is contained in S_ℓ, take L in $\mathcal{L}$ and x in L. By theorem 2 the right ideal $I_r(x)$ is nil even if x does not belong to A(N): thus there is a maximal right ideal R_x of $\mathcal{R}$ containing $I_r(x)$. Since R_x contains x, one gets $A_\ell(x) \supseteq A_\ell(R_x)$; therefore, if x ranges all over L,

$$A_\ell(L) \supseteq \cap A_\ell(R_x),$$

being $A_\ell(L) = \cap A_\ell(x)$. Now let L vary in $\mathcal{L}$ and x in L: then R_x ranges in (a subset of) $\mathcal{R}$, so that one can deduce

$$S_\ell = \cap A_\ell(L) \supseteq \cap A_\ell(R_x) \supseteq S_r.$$

The same kind of arguments applied to an ideal R of $\mathcal{R}$ shows that the opposite inclusion holds too: that is, S_r coincides with S_ℓ.

In a completely analogous way one proves that the right annihilator D_r of the sum of the ideals of $\mathcal{R}$ coincides with the right annihilator D_ℓ of the sum of the ideals of $\mathcal{L}$.

This implies that the sum of the ideals belonging to the family $\mathcal{L}$ and the sum of the ideals belonging to the family $\mathcal{R}$ have the same (two-sided) annihilator, which will be called *nil-annihilator of* N and denoted by $\alpha(N)$. It is a two-sided ideal of N and can be described as the intersection of S_ℓ and D_r or, equivalently, as $S_\ell \cap D_\ell$, $S_r \cap D_r$, $S_r \cap D_\ell$.

It is useful to point out the following

REMARK 4 : *Let* B *be a nil left ideal of* N. *If there exists a positive integer* k *such that* B^k *is contained in* $\alpha(N)$, *then* B *is nilpotent of class* (k+1).

Indeed, the (two-sided) annihilator of B contains $\alpha(N)$ and hence B^k.

3. RELATIONS BETWEEN THE NIL-ANNIHILATOR AND THE ANNIHILATOR OF N

First of all note that, under the assumption (made at the end of section 1) that N has a nil left ideal not contained in A(N), the ideal $\alpha(N)$ is different from N. In fact, if $\alpha(N) = N$, for every nil left ideal L it happens that $LN = NL = (0)$, that is, L is contained in A(N).

On the other side, if N is a non-ring, then $\alpha(N)$ is different from (0) too, since it contains A(N).

In general the inclusion $\alpha(N) \supseteq A(N)$ is proper. Restricting our attention to the class of weakly semiprime near-rings, it is possible to characterize the near-rings N for which $\alpha(N)$ and A(N) coincide.

Recall that, if N is weakly semiprime, A(N) contains - besides the nilpotent two-sided ideals of N - all the nilpotent one-sided ideals of N (see [1]): so the assumption that N has a nil left ideal which is not contained in A(N) is equivalent to the assumption that N has a nil left ideal which is not nilpotent.

Let now prove

THEOREM 5 : *Let* N *be a weakly semiprime near-ring. The following are equivalent:*
i) every non-nilpotent left ideal contains a nil non-nilpotent left ideal;
ii) $A(N) = \alpha(N)$.

PROOF : *i)* ⟹ *ii)*. If $\alpha(N) \neq A(N)$, the ideal $\alpha(N)$ is non-nilpotent and, by remark 4, $\alpha(N)$ cannot contain nil non-nilpotent left ideals, so that *i)* is false.

ii) => i). Suppose $A(N) = \alpha(N)$ and let B be a left ideal of N which is not nilpotent. Then B^2 is not contained in A(N) (and more in general no power B^k of B is contained in A(N)): indeed it will be shown that there exists a maximal nil right ideal R of N such that $RB^2 \neq (0)$.

In order to prove this result, first of all observe that in the semiprime ring $N' = N/A(N)$ the right annihilator $D_r(N')$ coincides with $\alpha(N')$ (see [2]) and that the image of $D_r(N)$ under the natural epimorphism $\pi: N \longrightarrow N'$ is contained in $D_r(N')$.

Consequently, if it were true that $RB^2=(0)$ for every maximal nil right ideal R, then the set $\pi(B^2)$ would be contained in $\alpha(N')$; thus, for every $x' \in \pi(B^2)$ and for every maximal nil right ideal R' of N', it would result $x'R'=(0)$. But this would imply that for every maximal nil right ideal R of N the product B^2R is contained in A(N), because every maximal nil right ideal R of N is the inverse image under π of a maximal nil right ideal R' of N'. Hence, for every maximal nil right ideal R of N, it would be $B^3R=(0)=RB^3$, and therefore B^3 would be contained in $\alpha(N) = A(N)$, contrary to the initial remark.

Now let $\bar{R}$ be a maximal nil right ideal such that $\bar{R}B^2 \neq (0)$: then $\bar{R}B$ has an element x such that $xN \neq (0)$. Since $\bar{R}B$ is contained in $\bar{R} \cap B$, the left ideal generated by x is contained in B, is nil (by theorem 2 applied to $\bar{R}$) and is non-nilpotent (as x does not belong to A(N)), that is $I_\ell(x)$ is the required ideal.

4. TWO-SIDED NIL IDEALS OF N

If N is a weakly semiprime near-ring with $\alpha(N) = N$, everyone of its nil ideals is nilpotent and two-sided, since it is contained in A(N).

On the contrary, if $\alpha(N)$ is a proper ideal of N, then N has certainly a nil non-nilpotent one-sided ideal, but it is not known if this fact implies that N has nil non-nilpotent two-sided ideals. An answer to this question will come down from the following

THEOREM 6 : *Let* N *be a weakly semiprime near-ring.* N *has a nil non-nilpotent two-sided ideal if and only if* $N/\alpha(N)$ *has one.*

PROOF : Call ψ the natural epimorphism of N onto $N/\alpha(N)$.

Suppose that N has a nil non-nilpotent two-sided ideal I: since I is not contained in $\alpha(N)$ (see remark 4), $\psi(I)$ is a nil two-sided ideal of $N/\alpha(N)$ different from zero. Actually it is the required ideal as it is not nilpotent, by remark 4.

Suppose now that $N/\alpha(N)$ has a nil non-nilpotent two-sided ideal I' and take $I = \psi^{-1}(I')$. Since I^2 is not contained in $\alpha(N)$, there is either a maximal nil right ideal R of N such that $I^2R \neq (0)$ or a maximal left ideal L of N such that $LI^2 \neq (0)$.

In the first case IR is different from zero and generates a two-sided ideal B which is contained in I. Since I' is nil, for every x of B there is a positive integer k such that x^k lies in $\alpha(N)$: therefore $IRx^k=(0)$ and, by the commutativity of the sum in N^2, $Bx^k=(0)$. It follows that every element x of B is nilpotent. On the other side, B is not nilpotent itself since, being $I^2R\neq(0)$, the product IB is different from zero, which implies that B is not contained in A(N). Therefore B is the required ideal.

In the second case an analogous argument proves that the two-sided ideal generated by LI is nil and non-nilpotent.

Now, if N has a nil non-nilpotent left ideal L, then also the ring $N/\alpha(N)$ has a left ideal of the same kind (for instance the image $\psi(L)$ of L under the natural epimorphism) and viceversa (see the initial remark of this section). Hence, the problem stated at the beginning of this section admits a positive answer if and only if every ring having a nil non-nilpotent one-sided ideal has a nil non-nilpotent two-sided one (i.e. the Köthe's conjecture for rings (see [3]) is true).

REFERENCES

[1] De Stefano S. and Di Sieno S., Anelli e quasi-anelli debolmente semiprimi, *Atti Acc. Sc. Torino* 117 (1983).

[2] Di Franco F., On the annihilator of one-sided maximal nil ideals in semiprime rings, *Boll. Un. Mat. Ital.* (6), 4 - A (1985), 65-70.

[3] Herstein I.N., *Non commutative rings* (The Mathematical Association of America, 1968).

[4] Pilz G., *Near-rings* (North-Holland, Amsterdam, 1983).

Dipartimento di Matematica
Università degli Studi
Via Saldini, 50
20133 MILANO - ITALY

Near-rings and Near-fields, G. Betsch (editor)

DISTRIBUTIVE NEAR-RINGS WITH MINIMAL SQUARE

S.De Stefano and S.Di Sieno *

ABSTRACT

There exist many examples of distributive near-rings such that their square generates a minimal left N-subgroup of N. Here it is shown that a near-ring belongs to this class if and only if it is the semi-direct sum of a zero near-ring and a skew-field.

1. PRELIMINARIES

Let N be a distributive near-ring: terminology and notations are those of [2]. In particular, denote with E(N) the right near-ring which is distributively generated by the set End(N) of the endomorphisms of the additive group of N and with $C_N(N)$ the centralizer of N when it is regarded as a (left) N-group:

$$C_N(N) = \{\psi \in \mathrm{End}(N) \mid \psi(xy) = x\psi(y) \text{ for all } x,y \in N\}.$$

The centralizer of N is a monoid, with respect to composition, which generates (distributively) a subnear-ring of E(N), which will be denoted with $C(N)$.

Every right product by an element n of N (which will be denoted with h_n) is an element of $C_N(N)$ and therefore of $C(N)$; nevertheless, if N is a non-ring, the set $\mathrm{Ass}(N) = \{h_n, n \in N\}$ does not exhaust $C(N)$ nor $C_N(N)$, since it does not contain any automorphism of N. The set Ass(N) is a subring of $C(N)$ which will be called the *associated ring of* N. An easy computation shows that Ass(N) is anti-isomorphic to the quotient ring of N by its right annihilator

$$(\mathrm{Ass}(N))^{op} \simeq \frac{N}{A_r(N)}.$$

2. THE RING "SQUARE OF N"

Let S be the subgroup of the additive group of N generated by N^2: it is easily seen that S is a ring with respect to the product defined in N, which will be called ring *square of* N. Therefore E(S) coincides with the set End(S) of its generators.

Moreover, S is both a left and a right N-subgroup of N, surely different from

---

* Both authors are members of the G.N.S.A.G.A. of the Italian C.N.R.

zero if N is not a nilpotent near-ring.

From now on it will be supposed that N is non-nilpotent and S is a minimal (left) N-subgroup.

Under these assumptions, the right annihilator $A_r(S)$ of the ring S is zero. In fact $A_r(S)$ is a (left) N-subgroup of S - since it is an ideal of the N-group S - and therefore is zero, because S is minimal and non-nilpotent.

Hence the ring Ass(S) is anti-isomorphic to S, which implies that any two distinct elements s and s' of S give rise to distinct products h_s and $h_{s'}$. Moreover, since the minimal N-subgroup S is idempotent, there exists in S an idempotent element e which acts in S as a right identity (see [1] proposizione 1.1). It follows that every endomorphism of the (left) N-group S is a right product by an element of S, that is

$$C(S) = C_N(S) = Ass(S).$$

This proves

PROPOSITION 1 : *Let* S *be a minimal (left) N-subgroup of* N. *The centralizer of* S *coincides with the ring associated with* S *and is anti-isomorphic to* S.

3. RELATIONS BETWEEN THE CENTRALIZERS OF N AND S.

Let now ψ be an element of C(N). For every element s of S it holds $\psi(s) = \psi(se) = s\psi(e)$, which means that $\psi(S)$ is contained in S.

THEOREM 2 : *Let the ring* S *be a minimal (left) N-subgroup of* N. *The correspondence* α *which associates with each element* ψ *of* C(N) *its restriction to* S *is an epimorphism of* C(N) *over* $C_N(S)$, *which induces a ring isomorphism between* Ass(N) *and* $C_N(S)$. *The kernel of* α *is formed by those* ψ *of* C(N) *whose images lie in the right annihilator of* N.

PROOF : The restriction $\alpha(\psi)$ of ψ to S is an endomorphism of the additive group of S since, for any two elements s and s' of S,

$$\psi(s+s') = (s+s')\psi(e) = \psi(s)+\psi(s')$$

and obviously $\alpha(\psi)$ is also an element of $C_N(S)$.

The correspondence α is evidently a near-ring homomorphism; actually it is an epimorphism, since $C_N(S)$ coincides with Ass(S).

Moreover, for every h_n in Ass(N) and every x in N it results

$$h_n(x) = xn = (xn)e = x(ne) = h_{ne}(x).$$

Hence the correspondence between the elements h_n of Ass(N) and the elements h_{ne} of $C_N(S)$ is one-to-one, that is, α is an isomorphism between these two rings.

Finally, if ψ belongs to kerα, then $n\psi(N) = \psi(nN) = (0)$, for every n in N, so that the inclusion $\psi(N) \subseteq A_r(N)$ holds. Conversely, this inclusion implies that

$\psi(s)$ is zero for every s in S, q.e.d.

COROLLARY 3 : *Let the ring* S *be a minimal (left) N-subgroup of* N. *The near-ring* C(N) *generated by the centralizer of* N *is the semidirect sum of* $\ker\alpha$ *and a subring isomorphic to* $C_N(S)$. *It coincides with the subnear-ring* C *of those elements* ψ *of* E(N) *such that* $\psi(xy) = x\psi(y)$ *for any two* x,y *belonging to* N.

PROOF : The near-ring C(N) contains the subring H of right products by elements of S, which is isomorphic to $C_N(S)$. Obviously, $C(N) = \ker\alpha + H$. Furthermore, the intersection of $\ker\alpha$ with H is zero, because, if the N-subgroup $h_s(N) = Ns$ is contained in $A_r(N)$, then it is zero (since S is minimal and N is non-nilpotent) and this implies s=0.

The second part of the statement follows from a proposition similar to theorem 2, concerning the near-ring C instead of C(N).

4. A CHARACTERIZATION OF NEAR-RINGS WITH MINIMAL SQUARE

The image by α of the identical endomorphism 1 of N is the right product by the right identity e of S, being

$$\alpha(1)(s) = s = se = h_e(s)$$

for every s in S. This remark and preceding theorem 2 imply that e *is also a left identity in* S. In fact, if ψ is an element of $C_N(N)$ such that $\alpha(\psi) = h_t$ (with t in S), for every s of S

$$h_t h_e(s) = \alpha(\psi 1)(s) = \alpha(\psi)(s) = h_t(s).$$

Hence

$$S(et-t) = (0),$$

and consequently $et = t$. Since α is surjective, such an equality holds for every t in S, which means that e is a left identity in S.

Moreover, every element s of S different from zero has an inverse because the N-subgroup Ss coincides with S. This proves

PROPOSITION 4 : *If* S *is a minimal (left) N-subgroup, then* S *is a (possibly skew) field.*

As a consequence, S is minimal as a right N-subgroup too.

Moreover, under the initial assumptions, the following statement holds:

PROPOSITION 5 : *The near-ring* N *is the semidirect sum of* $A_r(N)$ *and* S.

PROOF : The ring $N/A_r(N)$ is anti-isomorphic to Ass(N), which in turn is anti-isomorphic to S (see proposition 1 and theorem 2). Since $A_r(N) \cap S$ is zero, this proves the statement.

Propositions 4 and 5 show that every near-ring N which is non-nilpotent and has minimal square is a semidirect sum of a zero near-ring and a (skew) field which is both a left and a right N-subgroup of N. This is a characterization of near-rings with minimal square, since the following result holds:

PROPOSITION 6 : *If the distributive near-ring* N *is a semidirect sum of a normal subgroup* B *which is a zero near-ring and of a two-sided N-subgroup* T *which is a skew-field, then* N *is non-nilpotent and has a square which is minimal and coincident with* T.

PROOF : Since T is a skew-field, it has an identity e which is an idempotent element of N: therefore N is non-nilpotent. Furthermore, for any two elements $n_1 = b_1+t_1$ and $n_2 = b_2+t_2$ of N, the product $n_1n_2 = t_1b_2+b_1t_2+t_1t_2$ is an element of the two-sided N-subgroup T; thus S is a two-sided non-zero ideal of the skew-field T and hence S coincides with T. This proves also that S is a minimal (left) N-subgroup.

FINAL REMARKS

In spite of corollary 3 and proposition 5, one cannot prove that C(N) is the semidirect sum of the centralizer of S and $C(A_r(N)) = E(A_r(N))$. See, for instance, the distributive near-ring over the symmetric group S_3.

This same example shows also that the near-ring generated by the centralizer is not left distributive even if the distributive near-ring has minimal square.

REFERENCES

[1] De Stefano S. and Di Sieno S., Anelli e quasi-anelli debolmente semiprimi, *Atti Acc. Sc. Torino* 117 (1983).

[2] Pilz G., *Near-rings* (North-Holland, Amsterdam, 1983).

Dipartimento di Matematica
Università degli Studi
Via Saldini, 50
20133 MILANO - ITALY

Near-rings and Near-fields, G. Betsch (editor)

NEAR-RINGS WITH E-PERMUTABLE TRANSLATIONS

Celestina COTTI FERRERO

Dipartimento di Matematica, Università degli Studi, 43100
Parma, Italy *

Abstract. Our purpose is to study rings and near-rings N such that their endomorphisms are permutable with the left (right) translations. This situations is related to rigidity questions, and we prove that in a lot of interesting cases such an N is $\mathcal{A}$-rigid. We obtain characterizations of such near-rings with further conditions too.

1. INTRODUCTION

It is well known that the study on rings with given group of automorphisms has been very intensive.Lately, the rigid rings (see f.e. [8],[9], [13],[14],[15]), that are the ones whose only endomorphisms are the trivial ones, have been studied. In [4] we find a work on P_h-rings, i.e. on rings R that have a function h (of R in R) such that xh(y)=h(x)y, and Z_h-rings in wich h(x) belongs to the centre of R.

Here we study,with the name ELT(ERT)-near-rings,the near-rings whose endomorphisms are permutable with left(right) translations.In particular,we state that a reduced ELT-ring is $\mathcal{A}$-rigid,and it is either rigid or a direct sum of reduced ELT-rings.For reduced artinian ELT-rings we obtain a characterization as direct sum of complete local rings.Moreover,if a reduced ELT-ring has a zero-square endomorphism,then it is a P_h-ring and a Z_h-ring,in the sense of[4],and it turns out to be a ring with involution if and only if it is anti-commutative and the characteristic of its annihilator is two.

At last we completely characterize the ELT-rings such that the set of their nilpotent elements is a minimal ideal.

2. GENERAL REMARKS

Let N be a left near-ring;concerning notations and elementary results we often refer to [11]without any explicit remark.

Also,we denote by $A_s(M)$, $A_d(M)$,$A(M)$ respectively the left,right,two-sided

*This research was partially supported by italian MPI.

annihilator of the subset M of N; by s_a the left translation ($x \longrightarrow ax$) defined by the element a belonging to N, and by d_a the right translation defined by a.

Moreover N^+ denotes the additive group of N, and N° its multiplicative semigroup; as in ring theory we will call a near-ring without non-zero nilpotent elements *reduced* .

Let us remember that a near-ring N is said to be *rigid* (see [8]) if the only endomorphisms are the trivial ones; it is called $\mathcal{A}$-*rigid* (see [12]) if the only automorphism of N is the identity ;we will call an endomorphism f of N *proper* if Kerf≠0 and Imf≠N.

A proper endomorphism f of N is said to be a *near-projection* of N if there exists a subnear-ring H of N such that Im f $\subseteq$ H and f is the identity on H. Let I be an ideal of N ;the *radical* of I is the set $\sqrt{I}$ of the elements of N a power of which belongs to I.

As usual, N^2 will denote the subnear-ring of N generated by the products of two elements of N.

Definition A. A near-ring is an ELT-near-ring (ERT-near-ring) if its left (right) translations commute with its endomorphisms. It will be called an ELT^+-near-ring (ERT^+-near-ring) if its left(right) translations commute with the endomorphisms of its additive group.

Near-rings with "E-permutable left(right) translations " are ELT(ERT)-near-rings. *The near-ring N is ERT^+ if every endomorphism of N^+ is one of N_N.*

Definition B. Let N be a left near-ring, and let f be an endomorphism of N; f is called t_s (t_d) -permutable if f commute with all the left (right) translations of N.

The following propositions are obvious consequences of definitions A,B.

Proposition 1. If f is a t_s (t_d) -permutable endomorphism of a near-ring N, then $xf(y)=f(x)f(y)$ ($f(x)y=f(x)f(y)$), $\forall x,y \in N$.

Proposition 2. Let N be a distributive near-ring and let f be a t_s-permutable endomorphism of N. Then $x-f(x) \in A_s(\mathcal{I}m\ f)$ and $f(f(y)x)=yf(x)$, $\forall x,y \in N$.

Proposition 3. *Let N be a distributive near-ring such that xy=0 implies yx=0; then for each t_s-permutable endomorphism f of N, $x-f(x) \in A(\mathrm{Im}\, f)$ and f is the identity on Im fN and on NIm f.*

From Prop.2 follows $x-f(x) \in A_s(\mathrm{Im}\ f)$, and so $x-f(x) \in A(\mathrm{Im}\ f)$ since now xy=0 implies yx=0. We have the result observing that $f(f(x)y)-f(x)y=(f^2(x)-f(x))f(y)+$ $+f(x)(f(y)-y)$ and also $f(yf(x))-yf(x)=f(y)(f^2(x)-f(x))+(f(y)-y)f(x)$.

From now, the zero symmetric part (the constant part) of a near-ring N will be denoted by $N_o(N_c)$.

Proposition 4. *Let f be a t_d-permutable endomorphism of a near-ring N; then*

1. $x-f(x) \in A_d(\mathrm{Im}\ f)$ *and* $f(xf(y))=f(x)y$, $\forall x,y \in N$;
2. *Ker* $f \in A_d(\mathrm{Im}\ f) \in N_o$;
3. $N_c \subseteq \mathrm{Im}\ f$ *and f is the identity on* N_c;
4. *if* N_o *is a reduced near-ring, then for all* $x,y,n \in N_o$, *xy=0 implies yx=0 and xny=0.*

1. Immediate from definitions A,B.
2. Since $y-f(y) \in A_d(\mathrm{Im}\ f)$, $\forall y \in N$, if $y \in \mathrm{Ker}\ f$ then $y \in A_d(\mathrm{Im}\ f)$.
3. Let $y \in N_c$, then $0y=y$ and $f(0)f(y)=f(y)$: it follow that $f(y)=f(0)f(y)=f(0)y=$ $=0y=y$ and our assertion is shown.
4. We note that N_o (in consequence of our hypothesis) is a zero-symmetric reduced near-ring; the assertion follow from [2] lemma 1.

3. NEAR-RINGS WITH E^+-PERMUTABLE LEFT TRANSLATIOS

A near-ring is called *weakly commutative* (see [11]) if, $\forall x,y,z \in N$, xyz=yxz. Such a near-ring is obviously *medial* (see [10]) , and it is also an *IFP-near-ring* .

Proposition 5. Let N be an ELT^+*-near-ring; then*

1. *N is weakly commutative;*
2. N^2 *is additively permutable with* $A_d(N)$;
3. *if N is distributive then* N^2 *is an ideal of N contained in the centre of* N^+.

For 1. observe that left translations commute within themselves. For 2. observe that left translations commute with the inner automorphisms defined by the elements of $A_d(N)$. To prove the third assertion it is enough to remember that the distributivity of N implies that N^2 is a ring.

Theorem 1. Let N be an ELT^+-near-ring and let e be an idempotent element of N. Then $N=A_d(e)+eN$, where $A_d(e)$ is an ideal of N, with $eNA_d(e)=0$, eN is a left ideal and N-subgroup of N and $A_d(e) \cap eN=0$. Also, $A_d(e)\neq 0$ if e is not a left identity.

If e is not a left identity of N, then there exists an $x \in N$ such that ex $\neq x$; it follow that $A_d(e) \neq 0$, in fact $\forall x \in N$ $ex-x \in A_d(e)$. Furthemore $A_d(e)$ is an ideal of N because (Prop.5) N is an IFP-near-ring. We see immediately that $eNA_d(e)=0$ and that $eN \cap A_d(e)=0$. From Prop.5 it follows that $NeN \subseteq eN$, and $eNN \subseteq eN$. Now obviously, $x=x-ex+ex$, and $N=A_d(e)+eN$. Let us prove, now, that eN is a normal subgroup of N^+: indeed, since now s_e commutes with the inner automorphisms induced by the elements of $A_d(e)$, we have that $\forall a \in A_d(e)$ $a+en=en+a$. It follows that $\forall x \in N$, since we can write x as $a+en'$, we have $-x+en+x \in eN$. Hence eN is a left ideal of N; it is obviously an N-subgroup of N, too, and N^+ turns out to be directly reducible.

Corollary 1. Let N be a ELT^+-near-ring without left identity, and let e be an idempotent element of N; then N is sum of ELT-near-rings and there exists an endomorphism f of N such that $A_d(e)=Ker\ f$ and $eN=Im\ f$.

4. NEAR-RINGS WITH E-PERMUTABLE LEFT TRANSLATIONS

We start our study considering (as a particular case) the ELT-ring.

Theorem 2. A reduced ELT-ring R is $\mathcal{A}$-rigid and it is either rigid or a direct sum of reduced ELT-ring.

Via lemma 1 of [2] we already know that R is an IFP-ring. Let f be a non-zero endomorphism of R. Then $A(Im\ f)=Ker\ f$: in fact (Prop.3) $Ker\ f \subseteq A(Im\ f)$ and $A(Im\ f) \subseteq Ker\ f$ because if $z \in A(Im\ f)$ then, from Prop.2, $f(z) \in A(Im\ f)$ and $f^2(z) \in A(Im\ f)$. It follow that $f^2(z)f(y)=0=f(z)y$, $\forall y \in R$ (Prop.3), and so $f(z)=A(R)=0$: i.e. $z \in Ker\ f$. From this it derives that R is $\mathcal{A}$-rigid; moreover R does not have non-identical monomorphisms or epimorphisms: indeed if $Ker\ f=0$ then $A(Im\ f)=0$ and $\forall x \in R$ $x=f(x)$ (Prop.2). If $Im\ f=R$ then $A(Im\ f)=A(R)=0$, and f is the identity even in this case.

Now, if R is not rigid, let f be a proper endomorphism of R, then (Prop.3), Im f is an ideal of R and Ker f ∩ Im f=0, being R a reduced ring: therefore R=Ker f ⊕ Im f. Now Ker f and Im f are reduced rings and ELT-rings because any endomorphism of theirs can be extended in a trivial way to an endomorphism of R; the same for their left translations.

There exists obvious examples of reduced non-rigid ELT-rings.

Corollary 2. A reduced ELT-ring is rigid if, and only if, it is directly irreducible.

Corollary 3. The direct summands of a reduced ELT-ring R are exatly the kernels of the proper endomorphisms of R. Besides, the endomorphisms of R are projections of R upon suitable ideals of R.

Corollary 4. A reduced ring R has a proper t_s-permutable endomorphism f if and only if it is directly reducible and f is a projection of R on one of its direct summands.

Theorem 3. A non-rigid artinian reduced ELT-ring is a direct sum of finitely many complete local rings.

If R is non-rigid, then (Th.2), it possess at least one proper endomorphism, and therefore it is a direct sum of two reduced ELT-rings. Finally, R being artinian it is a direct sum of finitely many reduced rigid rings which are artinian, too. Thus, from th.3.1 of [8], R is direct sum of complete local rings.

If the ring R has non-zero nilpotent elements then we let Q(R) denote the set of the nilpotent elements of R.

Theorem 4. Let R be a ring such that: $0 \neq Q(R) \neq R$, xy=0 impies yx=0 and R is without non-zero nilpotent endomorphisms. The ring R is a ELT-ring with at least one proper non-zero endomorphism, Q(R) minimal and R'=R/Q(R) subdirectly irreducible, if and only if it is a direct sum of a nil simple ELT-ring and a subdirectly irreducible rigid integral domain.

Since xy=0 implies yx=0, Q(R) is a proper ideal of R and R/Q(R) is a reduced ring. Let f be a non-zero and non nilpotent proper endomorphism of R, and let R'=R/Q(R) be subdirectly irreducible, with Q(R) minimal. It is easy to see that the endomorphism f' induced by f on R' is t_s-permutable.If f'was proper then R' would be directly reducible;therefore it must be either Ker f'=0 or Im f'=0. Recall now that (Prop.3) Im f is an ideal of R.If Ker f'=0 then (Th.2) f'is the identity on R' and therefore $\forall x \in R$ $x-f(x) \in Q(R)$.From here we get Kerf $\subseteq Q(R)$:indeed if f(y)=0 then $y \in Q(R)$;hence it must be Ker f=Q(R),as now it cannot be Ker f=0:in fact,if it is so,then it would be either $Imf \cap Q(R)=0$ or Im $f \supseteq Q(R)$.But if Im $f \supseteq Q(R)$,then from $x-f(x) \in Q(R)$ we would get $N \subseteq$ Im f,and f would turn out to be an automorphism,but this is excluded by our hypothesis. Then,let Ker f=Q(R);from $x-f(x) \in Q(R)=$Ker f we get $f^2(x)=f(x)$, $\forall x \in R$,and so R=Ker f $\oplus$ Im f.It follows that Im f is a reduced ELT-ring,subdirectly irreducible,and thus it is rigid (Th.2) and it is an integral domain by lemma 3 of [2],while Ker f is a nil simple ELT-ring.

At last,let Im f'=0,and therefore Im $f \subseteq Q(R)$.At once we have $R=\sqrt{\text{Ker } f}$ because $\forall x \in R$ there exists a natural number m=m(x) such that $f(x)^m=f(x^m)=0$: thus $x^m \in$ Ker f. Since Im $f \neq 0$, we get Im f=Q(R) and Ker $f \neq 0$, because otherwise R would be nil,in contradiction to what is supposed.Moreover, being $A(R) \subseteq Q(R)$,we have A(R)=0 or A(R)=Q(R). If A(R)=0 then Ker f= A(Im f) (similary to what we did in Th.2), and R=Im f $\oplus$ Ker f again, because Ker f $\cap$ Im f=0, since otherwise it should be Im $f \subseteq$ Ker f, and f would be nilpotent,in contradiction to what it is supposed to.Finally if Im f=Q(R)=A(R), always with A(R) $\cap$ Ker f=0 ,we get f(xy)=xf(y)=0, therefore $R^2 \subseteq$ Ker f, and Ker f is maximal, since Im f is now simple. Again R=A(R) $\oplus$ Ker f. In any case, N is a direct sum of a simple nil ELT-ring and of a subdirectly irreducible reduced ELT-ring: thus it is a subdirectly irreducible rigid (Th.2) integral domain (see [2]).

Vice versa, let R=A $\oplus$ B, where A is a simple nil ELT-ring and B a subdirectly irriducible rigid integral domain. Let f be a non-zero endomorphism of R: it induces an endomorphism f' in $R/A \simeq B$. As now B is rigid, then f' is either the identity or the zero endomorphism. In the first case $x-f(x) \in A$, and Ker $f \subseteq A$.Therefore Ker f =0 or Ker f =A: if Ker f=A, then Im $f \simeq B$ and R=Ker f $\oplus$ Im f, with $f^2(x)=f(x)$, and f is t_s-permutable; if Ker f=0, we have Im f=R, and f is an automorphism. With easy calculation we show that f is t_s-permutable.

Now, let f' be the zero-endomorphism: then Im f=A. Since here f cannot be nilpotent, we get A $\cap$ Ker f=0, and f induces the identity on A. Then we have $\forall x \in R$, $f(x) \in A$, and so $f^2(x)=f(x)$. It follows that $x-f(x) \in \mathrm{Ker}\, f$, so $R=\mathrm{Im}\, f \oplus \mathrm{Ker}\, f$, hence f is t_s-permutable. Moreover A is minimal, and contains all the nilpotent elements of R, and R has at least one proper endomorphism: for instance the second projection.

We recall that if f is a function of a ring to itself, usually S denotes the set of symmetric elements, i.e. the $x \in R$ such that $f(x)=x$, and K denotes the set of skew-symmetric elements, i.e. the $y \in R$ such that $f(y)=-y$.

Theorem 5. Let R be a ring such that $xy=0$ implies $yx=0$, and let $0\neq Q(R)\neq R$ be minimal in R; then R has a zero-square t_s-permutable endomorphism f if and only if there exists an automorphism g of R^+ that is t_s and t_d-permutable, and whose set of skew-symmetric elements is an ideal which contains $Q(R)+R^2$. In particular, g is an involution if and only if R is anticommutative and char $Q(R)=2$.

Indeed, now, Q(R) is an ideal of R, and it is minimal because of our hypothesis. Let f be a zero-square non-zero t_s-permutable endomorphism of R. Then $\mathrm{Im}\, f \subseteq \mathrm{Ker}\, f$, and furthermore $A(R) \subseteq Q(R)$. From Prop.3 we get $x-f(x) \in A(\mathrm{Im}\, f)$, and so $f(x) \in A(\mathrm{Im}\, f)$. It follows that Im f is a zero-ring, and we get at once $\mathrm{Im}\, f=Q(R)$: since $\forall x,y \in R$ it is $xf(y)=f(xy)=f(x)f(y)=0$, we have $\mathrm{Im}\, f \subseteq A(R) \subseteq Q(R)$ and so $\mathrm{Im}\, f=A(R)=Q(R)$: hence Ker f is maximal in R. Let us consider the function $g=f-i$: at once we get $g(xy)=-xy$, $g(xy)=xg(y)$ $=g(x)y$, and $g(x)=-x$, $\forall x \in A(R)$. Now g is an additive homomorphism (Prop. 3); Im g is an ideal of R, and obviously $\mathrm{Ker}\, f \subseteq \mathrm{Im}\, g$ .Then we have Img=R, because Ker f=Im g implies $f(x)=0$, $\forall x \in R$, that is excluded by our hypotesis. On the other hand, Ker g=0 since $x \in \mathrm{Ker}\, g$ if and only if $f(x)=x$, i.e. iff $x \in A(R) \subseteq \mathrm{Ker}\, f$. At last the set of skew-symmetrical elements of g is Ker f. Vice versa, let g be an additive automorphism of R as it was said in the statement, and let $f=g+i$: f is the required endomorphism.
Now it is sufficient to recall th.3 of [5] in order to complete the proof.

Let N be a near-ring; by C(N) we will denote the ideal of N generated by the elements xy-yx.
In the near-ring case we have

Theorem 6. Let N be a near-ring such that Q(N) is an ideal which contains the commutator C(N). Let f be a proper t_s-permutable endomorphism. Then either $N=\sqrt{\mathrm{Ker}\, f}$ or $N=\mathrm{Im}\, f+\sqrt{\mathrm{Ker}\, f}$, with $\mathrm{Im}\, f\cap\sqrt{\mathrm{Ker}\, f}\subseteq Q(N)$.

Indeed now $N'=N/Q(N)$ is reduced and commutative, hence it is a ring; the function f' induced by f on N' is a t_s-permutable endomorphism of its, since f is it. By Th.2 we have f'=0, or f'=i, or f' is proper. In the first case we obtain $N=\sqrt{\mathrm{Ker}\, f}$; in the second one f is an epimorphism, that is excluded. At last we note that obiously $\sqrt{\mathrm{Ker}\, f}$ is an ideal of N. Finally we obtain $\forall x\in R$, $x-f(x)\in \mathrm{Ker}\, f$, and so $N=\sqrt{\mathrm{Ker}\, f}+\mathrm{Im}\, f$.

By J_2 we will denote the (Betsch) radical (see [11]) of the near-ring N (concerning the distributive case, see also [6]), and we will call N a radical near-ring if it concides with its radical J_2. Moreover (see [12]) an ideal I of a near-ring N is called prime of type 0 if for all K,J ideals of N, $KJ\subseteq I$ implies either $K\subseteq I$ or $J\subseteq I$; it is called prime of type 1 if $xNy\subseteq I$ implies $x\in I$ or $y\in I$. Finally, by $P_0(N)$ and $P_1(N)$ (see [12]) we will denote the prime radical or type 0 and 1, respectively.

Proposition 6. Let N be a non-radical distributive ELT-near-ring with $N\neq Q(N)\neq 0$; then we have

1. $P_0=P_1=Q(N)$;
2. $N/\mathcal{J}_2(N)$ *ia a subdirect sum of fields.*

1. We start proving that every prime ideal of type 0 is of type 1. Let I be a proper ideal of N of type 0, and let $xNy\subseteq I$; N being zero-symmetric, we have $xNyN\subseteq I$. Also xN and yN are ideals of N: indeed $xN\subseteq N^2$ is contained in the centre of N^+ (Prop. 5), and N being weakly commutative it is $NxN\subseteq xN^2\subseteq$ $\subseteq xN$, too. It follows that either $xN\subseteq I$ or $yN\subseteq I$. If $xN\subseteq I$, let $<x>$ be the ideal of N generated by x, since N is distributive and N^2 is a ring, then $<x>N\subseteq xN\subseteq I$, hence $<x>\subseteq I$: therefore we obtain $x\in I$. In the same way we can proceed in the case $yN\subseteq I$. So every prime ideal of type 0 is of type 1: at once we get $P_0(N)=P_1(N)$. Since now (Prop. 5) N is an IFP-near-ring, th. 3.9 of [12] confirms our statement.
2. From the previous point, Q(N) turns out to be an ideal of N, and N/Q(N)

is a reduced commutative ring; it follows that $N/J_2(N)$ is semisimple and commutative, and so it is subdirect sum of fields.

We note that in the case of rings or distributive near-rings, the conditions of permutability with endomorphisms of right or left translations, give results absolutely similar.

Theorem 7. Let N be an ERT-near-ring, and let N_o be reduced; then N is $\mathcal{A}$-rigid and either it is rigid or any proper endomorphism of N is a near-projection of N to one of its N-subgroup.

Let f be a non-zero endomorphism of N; we see immediately that Ker $f=A_d(\text{Im } f)$, and this implies (Prop. 4) that f is idempotent, and that f cannot be neither an epimorphism nor a monomorphism different from the identity. Consequently N is $\mathcal{A}$-rigid, and if f is proper, then $f^2=f$ implies that f is a near-projection of N on Im f, which now is an N-subgroup of the near-ring N.

REFERENCES

[1] Abu-Khuzman, H. and Yaqub, A., Structure and commutativity of rings with constraints on nilpotent elements, II, in: Math. J. Oklayoma Univ. 21 (1979), pp. 165-166.

[2] Bell, H.E., Near-rings in which each element is a power of itself, in: Bull. Austral. Math. Soc. 2 (1970), pp. 363-368.

[3] Bergen, J., Automorphisms with unipotent values, in: Rend. Circ. Mat. Palermo, (2) 30 (1982), pp. 225-232.

[4] Blass, A.R and Stanojevic C.V., On certain classes of associative rings, in: Math. Balcanica 1 (1971), pp 19-21.

[5] Ferrero Cotti, C. and Suppa Modena, A., Sugli stems con involuzione, in: Riv. Mat. Univ. Parma (4) 7 (1981).

[6] De Stefano, S. and Di Sieno, S., Sul radicale di Jacobson di un quasi-anello distributivo, in: Rend. Ist. Lombardo (A) 112 (1978), pp 192-204.

[7] Malone, J.J., More on groups in which each element commutes with its endomorphic images, in: Proc. Amer. Math. Soc. 65 (1977), pp.209-214.

[8] Maxon, C.J., Rigid rings, in: Proc. Edinburgh Math. Soc. (1978), 21 pp. 95-101.

[9] McLean, K.R., Rigid artinian rings, in: Proc. Edinburgh Math. Soc. (1982) 25, pp. 97-99.

[10] Pellegrini Manara, S., On Medial near-rings, this volume.

[11] Pilz, G., Near-rings (North-Holland, N.Y. 1977).

[12] Ramakotaiah, D. and Koteswara Rao, G., IFP near-rings, in: J. Austral. Math. Soc. (series A) 27 (1979), pp. 365-370.

[13] Suppa Modena, A., Sui quasi-anelli distributivi $\mathcal{A}$-rigidi, in print.

[14] Suppa Modena, A., Sugli anelli q-rigidi, in print.

[15] Suppa Modena, A., Sugli anelli i-rigidi, in print.

SUMMARY

Si studiano anelli e quasi-anelli N tutti i cui endomorfismi sono permutabili con le traslazioni sinistre (destre). Questa situazione è collegata con questioni di rigidità, e si prova che in molti casi notevoli un tale N è $\mathcal{A}$-rigido. Si ottengono caratterizzazioni di tali quasi-anelli sotto ulteriori condizioni.

Near-rings and Near-fields, G. Betsch (editor)

ENDOMORPHISM NEAR-RINGS OF A DIRECT SUM OF ISOMORPHIC FINITE SIMPLE NON-ABELIAN GROUPS

Y. FONG and J. D. P. MELDRUM

Department of Mathematics,
University of Edinburgh,
Mayfield Road,
EDINBURGH EH9 3JZ,
SCOTLAND.

Department of Mathematics,
National Cheng Kung University,
TAINAN,
TAIWAN.

Let G be an arbitrary finite simple non-abelian group and $H = \oplus^n G$ the external direct sum of n copies of the given group G. In this paper we study the structure of E(H), the distributively generated near-ring which is generated additively by all the endomorphisms of H. Here we show that $E(H) \supseteq \oplus^n M_0(G)$ where $M_0(G) = \{f: G \to G;\ 0f=0\}$. We also prove the key result that E(H) is a 2-primitive near-ring.

1. INTRODUCTION.

A near-ring is a set R with two binary operations + and • such that (R,+) is a not necessarily abelian group with identity 0, (R,•) is a semigroup and $x(y+z) = xy + xz$ for all $x,y,z \in R$. In general the extra axiom $0x = 0$ for all $x \in R$ is imposed to give a zero-symmetric near-ring. An element $s \in R$ is called distributive if $(x+y)s = xs + ys$ for all $x,y \in R$. If there exists a multiplicative semigroup of distributive elements S such that $R = Gp\langle S\rangle$, i. e. R is additively generated by S as a group, we say that R is a distributively generated near-ring, in short a d. g. near-ring.

Let R be a near-ring and G an R-module. We call G monogenic if there exists $g \in G$ such that $gR = G$, that is $G = \{gr;\ r \in R\}$. Let G be a monogenic R-module. Following **[1]**, we say that G is an R-module of type 0 if G is simple, that is it has no non-trivial proper R-ideals; it is an R-module of type 1 if G is simple and for all $g \in G$, either $gR = G$ or $gR = \{0\}$; it is an R-module of type 2 if G has no non-trivial proper R-submodules. If R has an identity it is clear that type 1 and type 2 modules coincide. For $\nu = 0,1,2$, a near-ring R is called ν-primitive on G if G is a faithful R-module of type ν. The near-ring R is called ν-primitive if it is ν-primitive on some R-module G.

Examples of near-rings:

(1) If we let $M(G) = \{f: G \to G\}$ where G is an arbitrary group (not necessarily abelian) and define the product $f \cdot g$ of the two mappings $f,g \in M(G)$ by the rule $x(f \cdot g) = (xf)g$ for all $x \in G$ and the sum $f + g$ by the rule $x(f+g) = xf + xg$ for all $x \in G$, then $(M(G),+,\cdot)$ forms a near-ring.

(2)Let $M_0(G) = \{f \in M(G);\ 0f = 0\}$. Then $(M_0(G),+,\cdot)$ is a zero symmetric near-ring.

The distributive elements of $M_0(G)$ are End G, the semigroup of all the endomorphisms of G. Here in this paper, E(G) the d. g. near-ring generated by End G, is of special interest. More precisely, E(G) is called the endomorphism near-ring of the group G. For general results and definitions in near-rings, we refer to Pilz **[10]**. We use left near-rings where he uses right near-rings, but otherwise the notations and definitions are similar.

2. THE STRUCTURE OF End H.

Here we quote the following theorem which can easily be found in any standard text on group theory.

Theorem 2.1. *Let* $\{G_i;\ i \in I\}$ *be a family of groups. The external direct sum* $\oplus_{i\in I} G_i$ *is equal to the internal direct sum* $\Sigma_{i\in I} G_i^*$ *where* G_i^* *is the image of the canonical injection* ι_i: $G_i \to \oplus_{i\in I} G_i$. *Conversely the internal direct sum* $\Sigma_{i\in I} H_i$ *of a family of normal subgroups of a group is isomorphic to the external direct sum* $\oplus_{i\in I} H_i$.

In view of theorem 2.1, we shall agree to identify x in G_i with $x\iota_i$ in G_i^*, so that $G_i = G_i^*$ and the internal direct sum and external direct sum coincide. So readers should keep in mind that we are going to change notations between external and internal direct sums from time to time for the sake of convenience. Now we are going to determine all the endomorphisms of $\oplus^n G$.

Let G be a finite simple non-abelian group and $F = \{G_i;\ G_i \cong G,\ i \in \{1,2,\ldots,n\}\}$. Write $H = G_1 \oplus G_2 \oplus \ldots \oplus G_n$ or $\oplus_{i=1}^n G_i$. Thus we tacitly assume $G_i \subseteq H$ for all $i \in I$.

Lemma 2.2. *If* G *is a finite simple non-abelian group then* End G = $\{0\} \cup$ Aut G *where* 0 *is the zero endomorphism of* G *and* Aut G *is the automorphism group of* G.

Proof. Elementary.

Take any arbitrary element $\alpha \in$ End H and a typical element $h = (g_{i(1),1}, g_{i(2),2}, \ldots, g_{i(n),n}) \in H$ where the second suffix indicates which copy of G_t and the first suffix indicates which element of G_t. Thus $g_{i(t),t} \in G_t$ for all $t \in \{1,2,\ldots,n\}$. Then

$$\begin{aligned}(g_{i(1),1}, g_{i(2),2}, \ldots, g_{i(n),n})\alpha &= (g_{i(1),1} + g_{i(2),2} + \cdots + g_{i(n),n})\alpha \\ &= g_{i(1),1}\alpha + g_{i(2),2}\alpha + \ldots + g_{i(n),n}\alpha.\end{aligned}$$

Here we note that for all $t \in I$, $g_{i(t),t}\alpha \in H$. Therefore

$$g_{i(t),t}\alpha = (g'_{i,1}, g'_{i,2}, \ldots, g'_{i,n}) \tag{1}$$

where $g_{i(t),t} \to g'_{t,j}$ is an endomorphism of G (since $G_t \cong G$ for all $t \in I$). For if $g_{i(t),t} \in G_t$, then $g_{i(t),t}\iota_t\alpha\pi_j$ is in G_j where ι_t: $G_t \to H$ is the canonical injection and π_j: $H \to G_j$ is the canonical projection. Thus $\iota_t\alpha\pi_j$ is a homomorphism from G_t to G_j for all $t,j \in I$. Since $G_t \cong G_j \cong G$, so $\iota_t\alpha\pi_j$ is just equivalent to an element of End G. Here we write $\alpha_{t,j} = \iota_t\alpha\pi_j$. Thus we have

Lemma 2.3. *With the assumptions and notations as above, we have* $g_{i(t),t}\alpha = g_{i(t),t}(\alpha_{t,1} + \alpha_{t,2} + \ldots + \alpha_{t,n})$ *for all* $g_{i(t),t} \in G_t$, $t \in I$.

Proof. The proof is immediate from the fact $g_{i(t),t}\alpha_{t,j} = g_{i(t),t}\iota_t\alpha\pi_j = g'_{t,j}$ and equation (1).

Thus we have

Theorem 2.4. *Let* G *be a finite simple non-abelian group and* H *the external direct sum of* n *copies of* G. *Then* End H = $\{(\alpha_{i,j});\ \alpha_{i,j} \in \text{End } G,\ 1 \leq i,j \leq n\}$ *is the set of all* $n \times n$ *matrices which have at most one non-zero entry in each column.*

Proof. For all $h = (g_{i(1),1}, g_{i(2),2}, \ldots, g_{i(n),n}) \in H$ and $\alpha \in$ End H, by repeatedly applying lemma 2.3 and theorem 2.1, we have

$$\begin{aligned}
h\alpha &= g_{i(1),1}\alpha + g_{i(2),2}\alpha + \ldots + g_{i(n),n}\alpha \\
&= g_{i(1),1}(\alpha_{1,1}+\alpha_{1,2}+\ldots+\alpha_{1,n}) + \ldots + g_{i(n),n}(\alpha_{n,1}+\alpha_{n,2}+\ldots+\alpha_{n,n}) \\
&= (g'_{1,1}+\ldots+g'_{1,n}) + (g'_{2,1}+\ldots+g'_{2,n}) + \ldots + (g'_{n,1}+\ldots+g'_{n,n}) \\
&= (g'_{1,1}+\ldots+g'_{n,1}, g'_{1,2}+\ldots+g'_{n,2}, \ldots, g'_{1,n}+\ldots+g'_{n,n}) \\
&= (g_{i(1),1}\alpha_{1,1}+\ldots+g_{i(n),n}\alpha_{n,1}, \ldots, g_{i(1),1}\alpha_{1,n}+\ldots+g_{i(n),n}\alpha_{n,n}) \\
&= (g_{i(1),1}, g_{i(2),2}, \ldots, g_{i(n),n}) \begin{bmatrix} \alpha_{1,1} & \alpha_{1,2} & \cdots & \alpha_{1,n} \\ \alpha_{2,1} & \alpha_{2,2} & \cdots & \alpha_{2,n} \\ \cdots & \cdots & \cdots & \cdots \\ \alpha_{n,1} & \cdots & \cdots & \alpha_{n,n} \end{bmatrix}.
\end{aligned}$$

Suppose that for some $i \in I$, there exists $j \neq k$ such that $\alpha_{j,i} \neq 0 \neq \alpha_{k,i}$. Then for every $g_j, g_k \in G$

$$\begin{aligned}
&(0,\ldots,0,g_j,0,\ldots,0,g_k,0,\ldots,0)(\alpha_{i,t}) \\
&= (g_j\alpha_{j,1}+g_k\alpha_{k,1}, g_j\alpha_{j,2}+g_k\alpha_{k,2}, \ldots, g_j\alpha_{j,n}+g_k\alpha_{k,n}) \\
&= ((0,\ldots,0,g_k,0,\ldots,0) + (0,\ldots,0,g_j,0,\ldots,0))(\alpha_{i,t}) \\
&= (g_k\alpha_{k,1}, \ldots, g_k\alpha_{k,n}) + (g_j\alpha_{j,1}, \ldots, g_j\alpha_{j,n}) \\
&= (g_k\alpha_{k,1}+g_j\alpha_{j,1}, g_k\alpha_{k,2}+g_j\alpha_{j,2}, \ldots, g_k\alpha_{k,n}+g_j\alpha_{j,n}).
\end{aligned}$$

Hence $g_j\alpha_{j,i} + g_k\alpha_{k,i} = g_k\alpha_{k,i} + g_j\alpha_{j,i}$ for all $g_j, g_k \in G$. Since G is finite simple non-abelian and $\alpha_{j,i} \neq 0$, $\alpha_{k,i} \neq 0$, so $\alpha_{j,i}, \alpha_{k,i} \in$ Aut G. Fix $g_k \neq 0$, then for every $g_j \in G$, $g_j\alpha_{j,i}$ commutes with $g_k\alpha_{k,i}$. But $g_j\alpha_{j,i}$ runs through the whole of G since $\alpha_{j,i} \in$ Aut G. Therefore $g_k\alpha_{k,i}$ is in the centre of G which is a contradiction because G has no centre. Hence result.

Since End H contains all the endomorphisms of H, it forms a multiplicative semigroup under composition of mappings. As those enodomorphisms in End H have just been given in the form of matrices as in theorem 2.4, so the product of any two endomorphisms can then be carried out in the form of matrix multiplication. The next result gives a complete multiplication table of End H.

Theorem 2.5. *The correspondence given in theorem 2.4 between* End H *and a set of* $n \times n$ *matrices is an isomorphism of multiplicative semigroups.*

Proof. The proof is just a matter of simple routine checking.

Here one immediately realises that addition for End H is not closed in general because G is non-abelian. As in general near-ring theory, one knows that End H can generate an endomorphism near-ring E(H) additively. With this, we attempt to investigate the structure of E(H) in the following section.

3. THE STRUCTURE OF E(H).

Here I(G), (A(G),E(G)) denotes the near-ring generated additively by all inner automorphisms (automorphisms, endomorphisms) of a given group G. We now quote the following result which is due to A. Fröhlich **[6]**.

Theorem 3.1. *If* G *is a finite simple non-abelian group, then* I(G) = A(G) = E(G) = $M_0(G)$.

Here for all i,j ∈ {1,2, . . . ,n}, we define $E^{\alpha}_{i,j}$ to be the matrix with α ∈ End G at the intersection of the **ith** row and **jth** column and the zero mappings elsewhere. Then these n × n matrices $E^{\alpha}_{i,j}$ (i,j ∈ {1,2, . . . ,n}) still sit inside End H. Thus the following corollary is obviously a special case of theorem 2.5.

Corollary 3.2. *For all* $E^{\alpha}_{i,j}, E^{\beta}_{s,t}$ ∈ End H ($1 \leq i,j,s,t \leq n$),

$$E^{\alpha}_{i,j}E^{\beta}_{s,t} = \begin{cases} E^{\alpha\beta}_{i,t} & \text{if } j = s \\ 0 & \text{if } j \neq s. \end{cases}$$

Thus it is easy to verify that $<E^{\alpha}_{i,i};\ \alpha \in \text{End } G>$ for all i ∈ {1,2, . . . ,n}, the near-rings which are additively generated by $E^{\alpha}_{i,i}$ where α ∈ End G and $1 \leq i \leq n$, are subnear-rings of the endomorphism near-ring E(H). In fact all $<E^{\alpha}_{i,i};\ \alpha \in \text{End } G>$ are isomorphic to E(G). Hence we have

Theorem 3.3. *Let* G *be a finite simple non-abelian group. Then* $<E^{\alpha}_{i,i};\ \alpha \in \text{End } G> \cong M_0(G)$ *for all* i ∈ I. *Here* ≅ *denotes "is near-ring isomorphic to".*

Proof. Immediate from the above remark, corollary 3.2, the simplicity of G and theorem 3.1.

We have been tacitly identifying $M_0(G)$ with its isomorphic copies in E(H). For notational convenience, we denote by diag($\alpha_1, \ldots, \alpha_n$) the n × n matrix with $\alpha_1, \ldots, \alpha_n$ down the main diagonal, and zeros elsewhere. Here we remark that for all diag($\alpha_1, \ldots, \alpha_n$) ∈ End H, $\text{diag}(\alpha_1, \ldots, \alpha_n) = E^{\alpha(1)}_{1,1} + E^{\alpha(2)}_{2,2} + \ldots + E^{\alpha(n)}_{n,n}$ and so if $i \neq j$ $\{E^{\alpha}_{i,i};\ \alpha \in \text{End } G\} \cap \{E^{\alpha}_{j,j};\ \alpha \in \text{End } G\} = \{0\}$. Now we claim that $Gp<\text{diag}(\alpha_1, \ldots, \alpha_n);\ \alpha_i \in \text{End } G>$ is just the direct sum of its normal subgroups $Gp<E^{\alpha}_{i,i};\ \alpha \in \text{End } G>$ where $1 \leq i \leq n$. For if $x \in Gp<\text{diag}(\alpha_1, \ldots, \alpha_n);\ \alpha_i \in \text{End } G>$, then there exists a finite number of elements $\text{diag}(a^1_{1,1}, \ldots, a^1_{n,n}), \ldots, \text{diag}(a^n_{1,1}, \ldots, a^n_{n,n}) \in \{\text{diag}(\alpha_1, \ldots, \alpha_n);\ \alpha_i \in \text{End } G\}$ such that $x = \text{diag}(a^1_{1,1}, \ldots, a^1_{n,n}) + \ldots + \text{diag}(a^n_{n,n}, \ldots, a^n_{n,n}) = \text{diag}(\Sigma^n_{j=1} a^j_{1,1}, \ldots, \Sigma^n_{j=1} a^j_{n,n}) = \text{diag}(\Sigma^n_{j=1} a^j_{1,1}, 0, \ldots, 0) + \ldots + \text{diag}(0, \ldots, 0, \Sigma^n_{j=1} a^j_{n,n})$. Thus $Gp<\text{diag}(\alpha_1, \ldots, \alpha_n);\ \alpha_i \in \text{End } G> = Gp<E^{\alpha}_{1,1};\ \alpha \in \text{End } G> + \ldots + Gp<E^{\alpha}_{n,n};\ \alpha \in \text{End } G>$. Again by the above remark, one can easily see that $Gp<E^{\alpha}_{i,i};\ \alpha \in \text{End } G> \cap Gp<E^{\alpha}_{j,j};\ \alpha \in \text{End } G> = \{0\}$ if $i \neq j$ and all the subgroups $Gp<E^{\alpha}_{i,i};\ \alpha \in \text{End } G>$ $1 \leq i \leq n$ of $Gp<\text{diag}(\alpha_1, \ldots, \alpha_n);\ \alpha_i \in \text{End } G>$ are in fact normal. Hence $Gp<\text{diag}(\alpha_1, \ldots, \alpha_n);\ \alpha_i \in \text{End } G> = \oplus^n_{i=1} Gp<E^{\alpha}_{i,i};\ \alpha \in \text{End } G>$. Thus we have

Theorem 3.4. *Let* G *be a finite simple non-abelian group. Then* $<\text{diag}(\alpha_1, \ldots, \alpha_n);$

$\alpha_i \in \mathrm{End}\ G>$, *the near-ring generated additively by* $\{\mathrm{diag}(\alpha_1, \ldots, \alpha_n);\ \alpha_i \in \mathrm{End}\ G\}$ *is isomorphic to a direct sum of* n *copies of the transformation near-ring* $M_0(G)$, *i. e.*

$$<\mathrm{diag}(\alpha_1, \ldots, \alpha_n);\ \alpha_i \in \mathrm{End}\ G> \cong \oplus^n M_0(G).$$

Proof. The only thing which really needs proof is the multiplicative closure. Let $\mathrm{diag}(\alpha_{1,1}, \ldots, \alpha_{n,n})$, $\mathrm{diag}(\beta_{1,1}, \ldots, \beta_{n,n}) \in <\mathrm{diag}(\alpha_1, \ldots, \alpha_n);\ \alpha_i \in \mathrm{End}\ G>$. Applying theorem 2.5, we have $\mathrm{diag}(\alpha_{1,1}, \ldots, \alpha_{n,n})\mathrm{diag}(\beta_{1,1}, \ldots, \beta_{n,n}) = \mathrm{diag}(\alpha_{1,1}\beta_{1,1}, \ldots, \alpha_{n,n}\beta_{n,n}) \in <\mathrm{diag}(\alpha_1, \ldots, \alpha_n);\ \alpha_i \in \mathrm{End}\ G>$. The rest follows from theorems 3.1, 3.3 and the above remarks.

Here we come to an important result.

Theorem 3.5. *Let* G *be a finite simple non-abelian group and* $H = \oplus^n G$. *Then we have* $E(H) \supseteq \oplus^n M_0(G)$.

Proof. Immediate from theorem 3.4 and the fact that $<\mathrm{diag}(\alpha_1, \ldots, \alpha_n;\ \alpha_i \in \mathrm{End}\ G>$ is a subnear-ring of E(H).

Denote by $\mathrm{antidi}(\alpha_1, \ldots, \alpha_n)$ the n x n matrix with α_i in the (n+1−i,i)th position for $1 \leq i \leq n$, and 0 elsewhere. Here we wish to point out that in general $Gp<\mathrm{antidi}(\alpha_1, \ldots, \alpha_n);\ \alpha_i \in \mathrm{End}\ G>$ is only group isomorphic to $M^+{}_0(G) + \ldots + M^+{}_0(G)$ (n copies). $Gp<\mathrm{antidi}(\alpha_1, \ldots, \alpha_n);\ \alpha_i \in \mathrm{End}\ G>$ itself does not form a near-ring. Take the special case when n = 2: if $\mathrm{antidi}(\alpha,\beta), \mathrm{antidi}(\gamma,\delta) \in Gp<\mathrm{antidi}(\alpha,\beta);\ \alpha,\beta \in \mathrm{End}\ G>$, we have $\mathrm{antidi}(\alpha,\beta)\mathrm{antidi}(\gamma,\delta) = \mathrm{diag}(\beta\gamma,\alpha\delta)$ which does not lie inside $Gp<\mathrm{antidi}(\alpha,\beta);\ \alpha,\beta \in \mathrm{End}\ G>$.

Now we come to our key result.

Theorem 3.6. *Let* G *be a finite simple non-abelian group and* $H = \oplus^n G$. *Then* H *is an* E(H)*-module of type* 2 *and* E(H) *is* 2*-primitive.*

Proof. It suffices to show that for every $g \in H - \{0\}$, $gE(H) = H$. Take any $g = (g_{i(1),1}, g_{i(2),2}, \ldots, g_{i(n),n}) \in H - \{0\}$. Now we only need to show that for every $g_j^* \in G_j^*$, $j \in I$, there exists $\beta \in E(H)$ such that $g\beta = g_j^*$. Then the theorem follows immediately from the result on the external direct sum of groups. Take $\alpha = (\alpha_{i,j}) \in \mathrm{End}\ H$, $0 \neq g = (g_{i(1),1}, g_{i(2),2}, \ldots, g_{i(n),n}) \in H$. Assume $g_{i(s),s} \in G_s - \{0\}$. Here we have π_s (projection): $H \to G_s^*$ that sends g into $(0, \ldots, 0, g_{i(s),s}, 0, \ldots, 0)$. Thus we have

$$(0, \ldots, 0, g_{i(s),s}, 0, \ldots, 0)\alpha = g_{i(s),s}\alpha_{s,1} + g_{i(s),s}\alpha_{s,2} + \ldots + g_{i(s),s}\alpha_{s,n}.$$

Choose $\alpha_{s,1} = \alpha_{s,2} = \ldots = \alpha_{s,j-1} = \alpha_{s,j+1} = \ldots = \alpha_{s,n} = 0$. Thus

$$(0, \ldots, 0, g_{i(s),s}, 0, \ldots, 0)\alpha = (0, \ldots, 0, g_{i(s),s}\alpha_{s,j}, 0, \ldots, 0)$$

where $g_{i(s),s}\alpha_{s,j}$ is in the **jth** component. Since $\alpha_{s,j}$ is an endomorphism from G_s into G_j, $G_s \cong G_j$ for all $s,j \in I$ and $E(G) = M_0(G)$, and hence there exists a map $\lambda: G_s \to G_j$ that sends $g_{i(s),s}$ into g_j. Here, take $\lambda = \varepsilon_1\alpha_1 + \varepsilon_2\alpha_2 + \ldots + \varepsilon_r\alpha_r$ where $\alpha_i \in \mathrm{End}\ G$, $\varepsilon_i = \pm 1$ and write $\alpha_i^* = (\alpha_{k,j}^{(i)})$ with

$$\alpha_{k,j}^{(i)} = \begin{cases} \alpha_i & \text{if } k = i_s,\ j = s, \\ 0 & \text{otherwise.} \end{cases}$$

We immediately obtain the map $\beta_j = \varepsilon_1\alpha_1^* + \varepsilon_2\alpha_2^* + \ldots + \varepsilon_r\alpha_r^*$ that sends $(g_{i(1),1}, g_{i(2),2}, \ldots, g_{i(n),n})$ into $(0, \ldots, 0, g_j, 0, \ldots, 0)$ where g_j is in the **jth** component. Thus we have the mapping $\beta_1 + \beta_2 + \ldots + \beta_n$: $(g_{i(1),1}, g_{i(2),2}, \ldots, g_{i(n),n}) \to (g_1, g_2, \ldots, g_n)$ and hence the result.

ACKNOWLEDGEMENT.

The first author would like to acknowledge financial support from the National Science Council, Republic of China for visiting the Mathematics Department of Edinburgh University for the year 1984-85.

REFERENCES.

[1] Betsch, G. *Primitive near-rings.* Math. Z. 130 (1973), 351-361.
[2] Fong, Y. *The endomorphism near-rings of symmetric groups.* Ph. D. thesis, Edinburgh University, 1979.
[3] Fong, Y. *Endomorphism near-rings of a direct sum of isomorphic finite simple non-abelian groups.* Technical Report for National Science Council, Republic of China, 1984-1985.
[4] Fong, Y. and Meldrum, J. D. P. *The endomorphism near-rings of symmetric groups of degree at least five.* J. Austral. Math. Soc. 30A (1980), 37-49.
[5] Fong, Y. and Meldrum, J. D. P. *The endomorphism near-ring of the symmetric group of degree four.* Tamkang J. Math. 12 (1981), 193-203.
[6] Fröhlich, A. *The near-ring generated by the inner automorphisms of a finite simple group.* J. London Math. Soc. 33 (1958), 95-107.
[7] McCoy, N. H. *The theory of rings.* (Macmillan, New York, 1964).
[8] Meldrum, J. D. P. *On the structure of morphism near-rings.* Proc. Royal Soc. Edinburgh 81A (1978), 287-298.
[9] Meldrum, J. D. P. *Near-rings and their links with groups.* (Pitman Research Notes No. 134, London, 1985).
[10] Pilz, G. *Near-rings.* (North-Holland, Amsterdam, 1977 (Revised Edition 1983)).
[11] Scott, W. R. *Group Theory.* (Prentice-Hall, Englewood Cliffs, New Jersey 1964).

Near-rings and Near-fields, G. Betsch (editor)

ON THE IDEAL STRUCTURE IN ULTRAPRODUCTS OF AFFINE NEAR-RINGS

Peter Fuchs

Section 1 contains some general results about ideals in ultraproducts of Ω-groups. We shall obtain necessary and sufficient conditions such that an ultraproduct of Ω-groups has the acc or dcc on ideals and give an example to show that this situation does actually arise. In section II we then investigate the ideal structure in ultraproducts of endomorphism rings and affine near-rings.

1. ULTRAPRODUCTS WHICH HAVE ASCENDING CHAIN CONDITION (acc) OR DESCENDING CHAIN CONDITION (dcc) ON IDEALS.

It is well known that direct products over an infinite index set do not fulfill any chain conditions. The following example however shows that there exist ultraproducts which do have chain conditions.

<u>1.1 Example</u>: Let S be an infinite set, $\mathcal{U}$ an ultrafilter on S. For each $\sigma \in S$ let G_σ be a group and A_σ be a fixed point free group of automorphisms on G_σ. Following the notation in [6] we denote by $M^\circ_{A_\sigma}(G_\sigma)$ the near-ring of transformations on G_σ and by $\dim_{A_\sigma}(G_\sigma)$ the number of orbits under the action of A_σ. Define $G := \prod_{\mathcal{U}} G_\sigma$ and $H := \{h: G \to G \mid \exists (h_\sigma) \in \prod A_\sigma \ \forall (\gamma_\sigma)/\mathcal{U} \in G: \ h((\gamma_\sigma)/\mathcal{U}) = (h_\sigma(\gamma_\sigma))/\mathcal{U}\}$. It is easy to see that H is a f.p.f. group of automorphisms on G. By Theorem 2 in [4] $N := \prod_{\mathcal{U}} M^\circ_{A_\sigma}(G_\sigma)$ is 2-primitive on (G,μ) where $\mu: N \times G \to G$, $\mu((n_\sigma)/\mathcal{U},(\gamma_\sigma)/\mathcal{U}) := (n_\sigma(\gamma_\sigma))/\mathcal{U}$. Let $h: N \to M^\circ_H(G)$, $h((n_\sigma)/\mathcal{U})((\gamma_\sigma)/\mathcal{U}) := (n_\sigma(\gamma_\sigma))/\mathcal{U}$. By 4.52 in [6] Im(h) is dense in $M^\circ_{\mathrm{Aut}_N(G)}(G)$. Clearly $H \subseteq \mathrm{Aut}_N(G)$. Conversely if $g \in \mathrm{Aut}_N(G)$ then by Theorem 2 in [4] $g((\gamma_\sigma)/\mathcal{U}) = (h_\sigma(\gamma_\sigma))/\mathcal{U}$ for some $(h_\sigma)/\mathcal{U} \in \prod_{\mathcal{U}} \mathrm{Aut}_{M^\circ_{A_\sigma}(G_\sigma)}(G_\sigma)$. But $\mathrm{Aut}_{M^\circ_{A_\sigma}(G_\sigma)}(G_\sigma) = A_\sigma$ by 7.13 in [6]. Thus $g \in H$ and Im(h) is dense in $M^\circ_H(G)$. Suppose $\exists n \in \mathbb{N}: \{\sigma \mid \dim_{A_\sigma}(G_\sigma) \leq n\} \in \mathcal{U}$. Then $\{\sigma \mid \dim_{A_\sigma}(G_\sigma) = m\} \in \mathcal{U}$ for some $m \leq n$. Now we can show that $\dim_H(G) = m$. By 4.60 in [6] $N \cong M^\circ_H(G)$ and N is simple.

In our main result of this section we shall characterize all those ultraproducts of Ω-groups with countably many operations which have acc or dcc on ideals. We start by recalling some definitions and results about ideals in Ω-groups. If V is a variety of Ω-groups, $A \in V$, then a polynomial p is an equivalence class of terms (words). For a detailed survey of polynomials and terms see [5]. Following the terminology in [6] we let $A^V[x]$ denote the set of all polynomials in the variable x of A. If + denotes the group operation in A then $(A^V[x],+,o)$ turns out to be a near-ring, where o is the composition of polynomials. To each polynomial $p = [w(a_1 \dots a_n,x)] \in A^V[x]$ we can associate a polynomial function $\overline{p}: A \to A$ $\overline{p}(a) = w(a_1 \dots a_n,a)$ $\forall a \in A$. If $P(A) := \{\overline{p} \mid p \in A^V[x]\}$ then $(P(A),+,o)$ is also a near-ring where o is the usual composition of functions. For a subset $B \subseteq A$ let (B) denote the ideal generated by B.

The following theorem shows that ideals of Ω-groups can be described by polynomial functions.

1.2 Theorem: ([6]) Let A be an Ω-group, $a \in A$.

1) $(a) = \{\overline{p}(a) \mid \overline{p} \in (P(A))_o\}$
2) $\forall B \subseteq A$: $(B) = \sum_{b \in B} (b)$

In all of the following let V be a variety of Ω-groups with countably many operations, S be a set, U an ultrafilter on S and $A_\sigma \in V$, $\forall \sigma \in S$.

Suppose $p \in (\prod_U A_\sigma)^V[x]$, $p = [w((a_1^\sigma)/U \dots (a_k^\sigma)/U,x)]$. Let $\sigma \in S$. If we replace in $w((a_1^\sigma)/U \dots (a_k^\sigma)/U,x)$ each ω_n by ω_n^σ and each $(a_j^\sigma)/U$ by a_j^σ then we get a term $w^\sigma(a_1^\sigma \dots a_k^\sigma,x)$ in A_σ and a polynomial $p^\sigma = [w^\sigma(a_1^\sigma \dots a_k^\sigma,x)] \in A_\sigma^V[x]$. The following proposition shows how $(\prod_U A_\sigma)^V[x]$ is related to $\prod_U A_\sigma^V[x]$.

1.3 Proposition:

a) $h: (\prod_U A_\sigma)^V[x] \to \prod_U A_\sigma^V[x]$, $h(p) := (p^\sigma)/U$ is a n.r homomorphism.

b) $\overline{h}: P(\prod_U A_\sigma) \to \prod_U P(A_\sigma)$ $\overline{h}(\overline{p}) := (\overline{p^\sigma})/U$ is a n.r homomorphism.

1.4 Notation: Let $A \in V$ with operations $\Omega = \{\omega_m \mid m \in \mathbb{N}\}$. For $\Omega' \subseteq \Omega$ let

1) $P^{\Omega'} := \{[w] \in (A^V[x])_0 \mid$ all $\omega \in \Omega$ which occur in w are elements of $\Omega'\}$

2) If w is a term in the language of V then $\ell(w)$ shall denote the number of symbols in w.

3) $\forall n \in \mathbb{N}$: $P_n^{\Omega'} := \{[w] \in P^{\Omega'} \mid \ell(w) \leq n\}$

If $\Omega' = \Omega$ we write P_n for P_n^{Ω}.

For $A' \subseteq A$, $n \in \mathbb{N}$, we consider (if existent) a sequence $x_1 \ldots x_k \ldots$ in A' such that

$$x_2 \neq \overline{p}(x_1) \qquad \text{for all } p \in P_n^{\{\omega_1\ldots\omega_n\}}$$

$$x_3 \neq \sum_{j=1}^{2} \overline{p_j}(x_{i_j}) \qquad \text{for all } i_1, i_2 \in \{1,2\},\ p_1, p_2 \in P_n^{\{\omega_1\ldots\omega_n\}}$$

$$\vdots$$

$$x_k \neq \sum_{j=1}^{k-1} \overline{p_j}(x_{i_j}) \qquad \text{for all } \{i_1\ldots i_{k-1}\} \in \{1\ldots k-1\},\ p_1\ldots p_{k-1} \in P_n^{\{\omega_1\ldots\omega_n\}}$$

$$\vdots$$

We say that $A' \subseteq A$ fulfills $C_{n,k}$, $n,k \in \mathbb{N}$ (denoted by $A' \models C_{n,k}$) if every sequence $x_1 \ldots x_j \ldots$ in A' as defined above stops after at most k steps. If A' does not fulfill $C_{n,k}$ we write $A' \not\models C_{n,k}$. The interest in the conditions $C_{n,k}$ stems from the following result.

<u>1.5 Proposition</u>: If $F := \{\sigma \mid A_\sigma \models C_{n,k}\} \in \mathcal{U}$ for some $n,k \in \mathbb{N}$, then every chain of ideals in $\prod_{\mathcal{U}} A_\sigma$ has length $\leq k$.

We are now ready to state our characterization result. Using the fact that $\prod_{\mathcal{U}} A_\sigma$ is $\aleph_1$-saturated if $\mathcal{U}$ is an ω-incomplete ultrafilter (see [2], pg. 305) one can prove:

<u>1.6 Theorem</u>:

1) If $\mathcal{U}$ is ω-incomplete the following statements are equivalent:

   a) $\prod_{\mathcal{U}} A_\sigma$ has the acc for ideals

   b) $\prod_{\mathcal{U}} A_\sigma$ has the dcc for ideals

   c) $\prod_{\mathcal{U}} A_\sigma$ has a composition series

   d) $\exists k \in \mathbb{N}$: Every chain of ideals in $\prod_{\mathcal{U}} A_\sigma$ stops after at most k steps

e) $\exists n,k \in \mathbb{N}$: $\{\sigma | A_\sigma$ fulfills $C_{n,k}\} \in \mathcal{U}$

2) If $\mathcal{U}$ is ω-complete then

a) $\prod_{\mathcal{U}} A_\sigma$ fulfills the acc for ideals $\Longleftrightarrow$ $\{\sigma | A_\sigma$ fulfills the acc for ideals$\} \in \mathcal{U}$

b) $\prod_{\mathcal{U}} A_\sigma$ fulfills the dcc for ideals $\Longleftrightarrow$ $\{\sigma | A_\sigma$ fulfills the dcc for ideals$\} \in \mathcal{U}$

c) $\prod_{\mathcal{U}} A_\sigma$ has a composition series $\Longleftrightarrow$ $\{\sigma | A_\sigma$ has a composition series$\} \in \mathcal{U}$

It should be pointed out that a similar result can also be obtained for universal algebras with countably many operations.

Let $\mathcal{U}$ be an ω-incomplete ultrafilter on S and suppose that $\prod_{\mathcal{U}} A_\sigma$ does not fulfill acc or dcc. In this case it is possible to show the existence of a chain of ideals which is an η_1 set (see [2], pg 264) and therefore has cardinality at least $2^{\aleph_0}$.

1.7 Theorem: Let $\mathcal{U}$ be ω-incomplete. Then there exists either $k \in \mathbb{N}$ such that each chain of ideals in $\prod_{\mathcal{U}} A_\sigma$ stops after $\leq k$ steps or there exists a chain of ideals in $\prod_{\mathcal{U}} A_\sigma$ which is an η_1 set.

2. THE IDEAL STRUCTURE OF $\prod_{\mathcal{U}} M_{aff}(V_\sigma)$ AND $\prod_{\mathcal{U}} Hom_{F_\sigma}(V_\sigma)$ FOR AN ULTRAFILTER $\mathcal{U}$ ON S.

Let V be a vector space over a field F. If o denotes the composition of functions then $(Hom_F(V),+,o)$ is a ring. (The endomorphism ring of the vector space V). Moreover, if $\delta \in V$ and m_δ: $V \to V$, $m_\delta(v) := \delta$ then $M_{aff}(V) := \{f\colon V \to V | f = h+m_\delta$ for some $h \in Hom_F(V)$, $\delta \in V\}$ is a near-ring w.r.t. +,o. (The affine near-ring of the vector space V) In this section we present the ideals of ultraproducts $\prod_{\mathcal{U}} Hom_{F_\sigma}(V_\sigma)$, $\prod_{\mathcal{U}} M_{aff}(V_\sigma)$ in some detail and apply results from the previous section. Detailed proofs can be found in [3]. For the remainder of the paper let S be a set, $\mathcal{U}$ an ultrafilter on S and ${}_{F_\sigma}V_\sigma$ be a vector space over the field F_σ for all $\sigma \in S$. The next result shows that it suffices to look at ideals of $\prod_{\mathcal{U}} Hom_{F_\sigma}(V_\sigma)$ in order to find all ideals of $\prod_{\mathcal{U}} M_{aff}(V_\sigma)$. The proof is similar to a result in [7] and is omitted.

<u>2.1 Proposition:</u> [3] Let $\{0\} \neq I \trianglelefteq \prod M_{aff}(V_\sigma)$. Then

$$I = I_0 + \prod_{\mathcal{U}} M_c(V_\sigma) \quad \text{for some} \quad I_0 \trianglelefteq \prod_{\mathcal{U}} \mathrm{Hom}_{F_\sigma}(V_\sigma)$$

To study ideals in $\prod_{\mathcal{U}} \mathrm{Hom}_{F_\sigma}(V_\sigma)$ we introduce the following concept. Let $(\tau_\sigma)_{\sigma\in S}$ be a sequence of cardinals and $\mathcal{U}$ an ultrafilter on S. Let K be the class of all cardinals τ such that $\{\sigma \mid \tau_\sigma \leq \tau\} \in \mathcal{U}$. K is nonempty and has therefore a least element τ'.

<u>2.2 Definition:</u> τ' is called the <u>$\mathcal{U}$-limit</u> of the sequence $(\tau_\sigma)_{\sigma\in S}$, denoted by $\tau' = \lim_{\mathcal{U}}(\tau_\sigma)$.

We note that $\lim_{\mathcal{U}}(\tau_\sigma)$ is always a limit cardinal if $\{\sigma \mid \tau_\sigma = \tau\} \notin \mathcal{U}$ for all cardinals τ. If ${}_F V$ is a vector-space then the ideals of $\mathrm{Hom}_F(V)$ are well known.

<u>2.3 Theorem:</u> [1] Let ${}_F V$ be a vector space over a field F and for each ordinal λ let $T_\lambda := \{h \in \mathrm{Hom}_F(V) \mid \dim \mathrm{Im}(h) < \aleph_\lambda\}$

a) I is a nonzero ideal of $\mathrm{Hom}_F(V)$ if and only if $I = T_\lambda$ for some ordinal λ.

b) Let $h, h_1 \in \mathrm{Hom}_F(V)$ with $\dim \mathrm{Im}(h_1) \leq \dim \mathrm{Im}(h)$. Then there exists $h_2, h_3 \in \mathrm{Hom}_F(V)$ such that $h_2 \circ h \circ h_3 \circ h_1 = h_1$.

Let $(\lambda_\sigma)_{\sigma\in S}$ be a sequence of ordinals. Then clearly $\prod_{\mathcal{U}} T_{\lambda_\sigma}$ is an ideal of $\prod_{\mathcal{U}} \mathrm{Hom}_{F_\sigma}(V_\sigma)$. If λ is an ordinal and $T_\lambda^* := \{(h_\sigma)/\mathcal{U} \mid \lim_{\mathcal{U}} (\dim \mathrm{Im}(h_\sigma)) < \aleph_\lambda\}$ then T_λ^* is an ideal of $\prod_{\mathcal{U}} \mathrm{Hom}_{F_\sigma}(V_\sigma)$ as well. It is easy to show that $T_\lambda^* \subseteq \prod_{\mathcal{U}} T_{\lambda_\sigma}$ where $\lambda_\sigma := \lambda \quad \sigma \in S$. We remark that this inclusion can be strict ([3]). It can be shown that for each ideal I of $\prod_{\mathcal{U}} \mathrm{Hom}_{F_\sigma}(V_\sigma)$ there is a certain ordinal λ such that $T_\lambda^* \subseteq I \subseteq \prod_{\mathcal{U}} T_\lambda$. More precisely:

<u>2.4 Theorem:</u> [3] Let $\{0\} \neq I \trianglelefteq \prod_{\mathcal{U}} \mathrm{Hom}_{F_\sigma}(V_\sigma)$. Then there exists an ordinal λ such that $I = \prod_{\mathcal{U}} T_\lambda$ or $I = T_\lambda^*$ or $T_\lambda^* \subsetneqq I \subsetneqq \prod_{\mathcal{U}} T_\lambda$.

If λ is a successor ordinal then $T^*_\lambda = \prod_U T_\lambda$. In general, however, there are many ideals strictly betweeen T^*_λ and $\prod_U T_\lambda$. The next result shows how many principal ideals are between $\{0\}$ and $\prod_U T_0$. Let $\aleph_0^{|S|}/U$ denote the cardinal of $\mathbb{N}^S/U$.

2.5 Theorem: [3] Let S be an infinite set and $\{\sigma | \dim (V_\sigma) \geq \aleph_0\} \in U$

1) If U is ω-complete, then there is no ideal strictly between $\{0\}$ and $\prod_U T_0$

2) If U is ω-incomplete, then there are exactly $\aleph_0^{|S|}/U$ many principal ideals between $\{0\}$ and $\prod_U T_0$.

The proof, which is lengthy, will be omitted. Applying our results from section 1 we now determine when $\prod_U \mathrm{Hom}_{F_\sigma}(V_\sigma)$ or $\prod_U M_{aff}(V_\sigma)$ has acc (dcc) provided that U is ω-incomplete.

2.6 Theorem: Let U be ω-incomplete.

1) $\prod_U \mathrm{Hom}_{F_\sigma}(V_\sigma)$ has the dcc (acc) for ideals $\Longleftrightarrow$ $\prod_U \mathrm{Hom}_{F_\sigma}(V_\sigma)$ is a simple ring.

2) If $\prod_U M_{aff}(V_\sigma)$ has dcc (acc) on ideals then $\prod_U M_{aff}(V_\sigma)$ has exactly one proper ideal I and $I = \prod_U M_c(V_\sigma)$.

Proof: 1) By Theorem 1.6 we can choose $n,k \in \mathbb{N}$ such that $\{\sigma | \mathrm{Hom}_{F_\sigma}(V_\sigma)$ fulfills $C_{n,k}\} \in U$. Suppose that $F := \{\sigma | \dim(V_\sigma) > (n+1)^k\} \in U$. If $\sigma \in F$ there are elements $x_1^\sigma \dots x_{k+1}^\sigma \in \mathrm{Hom}(V_\sigma)$ with

$$\begin{cases} \dim \mathrm{Im}\,(x_1^\sigma) = 1 \\ \dim \mathrm{Im}\,(x_2^\sigma) = n+1 \\ \vdots \\ \dim \mathrm{Im}\,(x_{k+1}^\sigma) = (n+1)^k \end{cases}$$

If $x_1, x_2 \in \mathrm{Hom}_{F_\sigma}(V_\sigma)$, $x_2 = \overline{p}(x_1)$ for some $p \in P_n$ then $\dim \mathrm{Im}\,(x_2) \leq n \dim \mathrm{Im}\,(x_1)$. Now one can easily show by induction that $x_1^\sigma \dots x_{k+1}^\sigma$ stops after k+1 steps for all $\sigma \in F$. This is a contradiction since $\{\sigma | \mathrm{Hom}_{F_\sigma}(V_\sigma)$ fulfills $C_{n,k}\} \in U$. Consequently $\{\sigma | \dim(V_\sigma) \leq (n+1)^k\} \in U$.

Similar arguments as in Example 1.1 show that $\prod_{\mathcal{U}} \mathrm{Hom}_{F_\sigma}(V_\sigma)$ is simple.

2) Follows by 1, and Proposition 2.1.

ACKNOWLEDGMENT

This paper forms a part of the authors dissertation. At this place he wants to thank his teacher Prof. G. Pilz for his constant encouragement and advice.

REFERENCES

[1] R. Baer, Linear algebra and projective geometry, New York, Academic Press (1965).

[2] C. C. Chang-H. J. Keisler, Model Theory, North-Holland, Amsterdam (1973).

[3] P. Fuchs, Ultraproducts of Ω-groups, Dissertation, Univ. Linz (1985).

[4] P. Fuchs-G. Pilz, Ultraproducts and ultralimits of near-rings, Mh. Math 100, 105-112, (1985).

[5] H. Lausch-W. Nöbauer, Algebra of polynomials, North Holland, Amsterdam (1973).

[6] G. Pilz, Near-rings, Revised edition, North-Holland, Amsterdam (1983).

[7] K. Wolfson, Two sided ideals of the affine near-ring, Americ. Math. Monthly 65 (1958) 29-30.

Department of Mathematics
Texas A&M University
College Station, TX 77843
U.S.A.

Near-rings and Near-fields, G. Betsch (editor)

RADICALS OF Ω-GROUPS DEFINED BY MEANS OF ELEMENTS

G.K. Gerber*

Mathematics Department, University of Port Elizabeth, P O Box 1600, Port Elizabeth 6000, South Africa.

ABSTRACT. In this paper radicals of Ω-groups are defined by means of elements. The following generalizations of radicals of rings are obtained: The nil radical, the E_6-radical, the λ-regular radical and the f-regular radical.

1980 Mathematics subject classification (Amer. Math. Soc.): primary 20 N 99; secondary 16 A 12, 16 A 22, 08 A 99.

1. INTRODUCTION

Wiegandt ([11]) published an article where he defined radicals of rings by means of elements. The definitions, theorems and proofs of Wiegandt can be generalised to Ω-groups. We apply these theorems to obtain the nil radical (Buys and Gerber, [2]) and generalizations of the E_6-radical for rings, the radical class of λ-regular rings as well as the radical class of f-regular rings for varieties of Ω-groups.

2. NOTATION AND DEFINITIONS

Throughout this paper we shall use the definitions and notation of Higgins ([7]). Whenever we refer to G it is meant to be an Ω-group. $P \triangleleft G$ denotes an ideal P of G. A^G will denote the ideal generated by $A \subseteq G$ in G. In particular we write a^G for $\{a\}^G$. Higgins ([7]) called words which involve only operations $\omega \in \Omega$ monomials. We shall call monomials Ω-words.

2.1 Definition

Let Ω be a fixed set of operations. $\omega \in \Omega$ is called a *trivial operation* in a variety V of Ω-groups if $\underline{a}\omega = 0$ for all $\underline{a} \in G \in V$. $\omega \in \Omega$ is a *non-trivial operation* if it is not trivial. An Ω-word involving only non-trivial operations will be called a *non-trivial Ω-word*.

If $f(x_1, x_2, \ldots, x_n) = f(\underline{x})$ is any word, then $f(x, x, \ldots, x)$ will be denoted by $f(x)$. Furthermore

$$A_1 A_2 \ldots A_n\omega = \{a_1 a_2 \ldots a_n\omega \mid a_i \in A_i,\ i = 1,2,\ldots n\},\ A_i \subseteq G,\ i = 1,2,\ldots,n.$$

The author gratefully acknowledges financial assistance from the University of Port Elizabeth.

2.2 Definition

An Ω-group G is called *zero-symmetric* if for each $\omega \in \Omega$ and $a_1, a_2, \ldots, a_n \in G$ it is true that $a_1 \ldots a_{i-1} 0 a_{i+1} \ldots a_n \omega = 0$, $i = 1,2,\ldots,n$.

2.3 Theorem (Buys and Gerber, [3])

G is a zero-symmetric Ω-group if and only if for each $\omega \in \Omega$ and $H_1, H_2, \ldots, H_n \triangleleft G$ is is true that

$$H_1 H_2 \ldots H_n \omega \subseteq \bigcap_{i=1}^{n} H_i.$$

2.4 Definition

Let $f(x_1, x_2, \ldots, x_n) = x_1$ (the so-called identity-word of Higgins ([7]) and $H_1, H_2, \ldots, H_n \triangleleft G$. Define the ideal $f'(H_1, H_2, \ldots, H_n)$ as follows:

$$f'(H_1, H_2, \ldots, H_n) = H_1.$$

For any Ω-word $f(x_1, x_2, \ldots, x_n)$ and $H_1, H_2, \ldots, H_n \triangleleft G$, the ideal $f'(H_1, H_2, \ldots, H_n)$ is defined inductively. Let ℓ be the number of operations which occurs in an Ω-word. If $\ell = 1$ then there exists an $\omega \in \Omega$ such that $f(\underline{x}) = \underline{x}\omega$. Then we define $f'(H_1, H_2, \ldots, H_n) = H_1 H_2 \ldots H_n \omega^G$.

Suppose that $f'(H_1, H_2, \ldots, H_n)$ has been defined for any Ω-word $f(\underline{x})$ with $\ell \leq k$. Let $f(\underline{x})$ be any Ω-word with $\ell = k+1$. Then $f(\underline{x}) = f_1(\underline{x}) f_2(\underline{x}) \ldots f_r(\underline{x}) \omega$ where some $f_j(\underline{x})$ are identity words and the remainder are Ω-words with $\ell \leq k$. Then we define

$$f'(H_1, H_2, \ldots, H_n) = f_1'(H_1, H_2, \ldots, H_n) f_2'(H_1, H_2, \ldots, H_n) \ldots f_r'(H_1, H_2, \ldots, H_n) \omega^G.$$

2.5 Definition (Buys and Gerber, [4])

$P \triangleleft G$ is called an *Ω-prime* (*Ω-semi-prime*) ideal if for every non-trivial $\omega \in \Omega$ and $A_1, A_2, \ldots, A_n \triangleleft G$ ($A \triangleleft G$) such that $A_1 A_2 \ldots A_n \omega \subseteq P$ ($AA \ldots A\omega \subseteq P$) it follows that $A_i \subseteq P$ ($A \subseteq P$) for an i.

The concepts accessible Ω-subgroup, subdirectly irreducible Ω-group, heart of a subdirectly irreducible Ω-group, regular class, radical class and upper radical of a regular class are defined as in the variety of rings.

3. RADICALS DEFINED BY MEANS OF ELEMENTS

Analogously to Wiegandt ([11]) we let P denote a property an element of an Ω-group may possess. Property P will be an abstract property and we assume that 0 always possesses property P for any P.

3.1 Definition

(a) An element a of an Ω-group is called a P-element if it has property P.

(b) An Ω-group G (Ω-subgroup A, $H \triangleleft G$) will be called a P-Ω-group (P-Ω-subgroup, P-ideal) if every element of G (A,H) is a P-element of G (A,H).

The following conditions will be used frequently. Let $H \triangleleft G$.

(A) If a is a P-element of G, then a+H is a P-element of G/H.

(B) If a is a P-element of G, then a is a P-element of H.

(C) If $a+H \neq H$ is a P-element of G/H, then a is a P-element of H.

(D) If a is a P-element of H, then a is a P-element of G.

(E) If a+H is a P-element of G/H and H is a P-ideal, then a is a P-element of of G.

(F) If a is a P-element of G and H has no non-zero P-elements, then a+H is a P-element of G/H.

Let ~ denote logical negation.

3.2 Lemma (Wiegandt, [11])

P satisfies (A), (B), (C), (D), (E), (F) if and only if ~P satisfies (C), (D), (A), (B), (F), (E) respectively.

Similar to Wiegandt ([11]) we can prove the following theorems for Ω-groups.

3.3 Theorem (Wiegandt, [11], Theorem 1)

Let property P satisfy conditions (A), (D) and (E). Then $R_1(P) = \{G \mid G$ is a P-Ω-group$\}$ is a radical class. If P also satisfies condition (B), then $R_1(P)$ is a hereditary radical class.

Furthermore,

$$SR_1(P) = \{G \mid \text{Each non-zero ideal of } G \text{ has a non-zero } {\sim}P\text{-element}\}$$

is the semisimple class of $R_1(P)$. If $T = {\sim}P$ and T satisfies conditions (B), (C) and (F), then

$$R_1^*(T) = \{G \mid G \text{ has no non-zero } T\text{-elements}\}$$

is a radical class which is hereditary if T satisfies condition (D).

3.4 Theorem (Wiegandt, [11], Theorem 2)

If property P satisfies conditions (A) and (B), then $R_2(P) = \{G \mid$ Every homomorphic image of G has no non-zero P-elements$\}$ is a radical class. The semisimple class of $R_2(P)$ is given by

$$SR_2(P) = \{G \mid \text{Every non-zero ideal of } G \text{ has a homomorphic image containing non-zero } P\text{-elements}\}.$$

If $T = {\sim}P$ and T satisfies conditions (C) and (D) then

$$R_2^*(T) = \{G \mid \text{Every homomorphic image of } G \text{ is a } T\text{-}\Omega\text{-group}\}$$

is a radical class.

3.5 Theorem (Wiegandt, [11], Theorem 3)

If property P satisfies condition (D), then

$$R_3(P) = \{G \mid \text{Every non-zero homomorphic image } \overline{G} \text{ of } G \text{ has a non-zero accessible } \Omega\text{-subgroup such that each element is a } P\text{-element of } \overline{G}\}$$

is a radical class. If $T = {\sim}P$ satisfies condition (B), then

$R_3^*(T)$ = {G | Every non-zero homomorphic image $\overline{G}$ of G has a non-zero accessible Ω-subgroup containing no non-zero T-elements of $\overline{G}$}

is a radical class.

3.6 Theorem (Wiegandt, [11], Proposition)

(a) $R_i(P) = R_i^*(\sim P)$, $i = 1,2,3$.

(b) $R_2^*(P) \subseteq R_1(P)$, $R_2(P) \subseteq R_3^*(P) \subseteq UR_1(P)$.

(c) $R_1(P) \cap R_2(P) = R_1(P) \cap R_3^*(P) = \{0\}$.

(d) If property P satisfies condition (A), then $R_1(P) = R_2^*(P) \subseteq R_3(P)$.

(e) If property P satisfies condition (B), then $R_1(P) \subseteq SR_2(P)$.

Note that the full analogue of Wiegandt is not proved (part (iii) of the proposition is omitted).

4. APPLICATIONS

Wiegandt ([11]) had many examples that he could fit into this mould. Ω-groups, on the other hand, have few known examples of radicals.

4.1 Definition (Buys and Gerber, [2])

$a \in G$ is called a *nilpotent element* of G if there exists a non-trivial Ω-word $f(\underline{x})$ such that $f(a) = 0$.

Let P_1 be the property nilpotent.

4.2 Theorem

P_1 satisfies conditions (A), (B), (D), (E) and (F).

Proof

Since $f(a+H) = f(a)+H$ for any Ω-word $f(\underline{x})$, it follows that P_1 satisfies condition (A). It is easy to see that (B), (D), and (F) are also satisfied.

Consider the nilpotent element a+H of G/H with H being a P_1-ideal (i.e. a nil ideal). Thus we can find a non-trivial Ω-word $f(\underline{x})$ such that $f(a+H) = f(a)+H = H$. Hence $f(a) \in H$. Since H is a P_1-ideal there is a non-trivial Ω-word $g(\underline{y})$ such that $g(f(a)) = 0$. Thus a is a nilpotent element of G since $g(f(\underline{x}))$ is a non-trivial Ω-word.

It follows that $R_1(P_1)$, $R_2(P_1)$, $R_3(P_1)$ and $R_3^*(P_1)$ are radical classes and that $R_1(P_1)$ is hereditary.

4.3 Definition

$a \in G$ is called a *central element* of G if $[a,G] = 0$.

From Higgins ([7], Lemma 4.3) it follows that a is a central element if and only if $[a^G,G] = 0$. Let P_2 be the property being central.

4.4 Theorem

P_2 satisfies conditions (A), (B) and (F).

Proof

Let $H \lhd G$, a a central element of G and $\theta : G \to G/H$ be the natural homomorphism. Then

$0 = [a,G]\theta = [a\theta, G\theta]$ (Higgins, [7], Lemma 3.1)

$= [a+H, G/H]$.

Hence a+H is a central element of G/H. Thus P_2 satisfies condition (A) and also (F). Since $[a,H] \subseteq [a,G] = 0$ we have that P_2 satisfies condition (B).

By 3.4 it follows that $R_2(P_2)$ is a radical class.

We now interpret this property for the varieties of associative rings, Γ-rings (Barnes, [1]) and cubic rings (Nobusawa, [8]) using results of Higgins ([7], Theorem 4B and Lemma 4.3).

It is easy to see that a is a central element of an associative ring R if and only if a ∈ Ann R (Ann R being the annihilator of R). Hence, by Wiegandt ([11], p.123) $R_2(P_2)$ coincides with the E_6-radical of Szasz ([10]).

If we define the annihilator of a Γ-ring M by Ann M = $\{a \in M \mid a\Gamma M = 0$ and $M\Gamma a = 0\}$ then it follows that a is a central element of M if and only if a ∈ Ann M.

In the variety of cubic rings it is easy to see that a is a central element of a cubic ring if and only if a = 0. Hence $R_2(P_2)$ is the class of all cubic rings.

4.5 Definition

a ∈ G possess property P_3 if for each non-trivial ω ∈ Ω it is true that $a \in aG\ldots G\omega$.

4.6 Theorem

P_3 satisfies conditions (A), (D) and (F).

The proof follows trivially.

By 3.5 it follows that $R_3(P_3)$ is a radical class. In the case of associative rings this is the class of λ-regular rings (Wiegandt, [11]).

4.7 Definition

a ∈ G is called *idempotent* if $aa\ldots a\omega = a$ for all non-trivial ω ∈ Ω.

4.8 Theorem

a is idempotent if and only if f(a) = a for each non-trivial Ω-word $f(\underline{x})$.

Proof

Trivial by induction on the number of operations performed in an Ω-word.

Let P_4 be the property idempotent.

4.9 Theorem

P_4 satisfies conditions (A), (B), (D) and (F).

Proof

Let a be a P_4-element of G and $f(\underline{x})$ any non-trivial Ω-word. Then f(a+H) = f(a)+H = a+H. Hence condition (A) is satisfied and thus also (F). (B) and (D) follows trivially.

By 3.4 and 3.5 $R_2(P_4)$, $R_3(P_4)$ and $R^*_3(P_4)$ are radical classes.

4.10 Definition

(a) G is called a *non-trivial Ω-group* if there exists a non-trivial operation $\omega \in \Omega$.

(b) $e \in G$ is called an *identity-element* of a non-trivial Ω-group G if for each non-trivial $\omega \in \Omega$ and $g \in G$ it is true that $e\ldots ege\ldots e\omega = g$ where g can be in any position.

Let P_5 be the property identity-element.

4.11 Theorem

P_5 satisfies conditions (A), (B) and (F).

Proof

The proof is trivial.

By 3.4 and 3.5 it follows that $R_2(P_5)$ and $R_3^*(P_5)$ are radical classes.

4.12 Theorem

Let G be a non-trivial Ω-group and e an identity-element of G.

(a) If for arbitrary non-trivial $\omega \in \Omega$ and $H \triangleleft G$ it is true that $G\ldots GHG\ldots G\omega \subseteq H$ (H in any position), then $e^G = G$.

(b) If $G \neq 0$ is a zero-symmetric Ω-group then $e \neq 0$ and $e^G = G$.

Proof

(a) Let $g \in G$. For any non-trivial $\omega \in \Omega$ we have $g = e\ldots ege\ldots e\omega \subseteq G\ldots e^G GG\ldots G\omega \subseteq e^G$. Hence $G = e^G$.

(b) By 2.3 we have $G\ldots GHG\ldots G\omega \subseteq H$.

Hence $e^G = G$ by (a). Let $0 \neq g \in G$. Since G is zero-symmetric, $0\ldots 0g0\ldots 0\omega = 0$ for any non-trivial $\omega \in \Omega$. Therefore, $e \neq 0$.

Zero-symmetric is a necessary condition in 4.12(b). Let $G = Z_2$ and define multiplication by

$\cdot$	$\bar{0}$	$\bar{1}$
$\bar{0}$	$\bar{0}$	$\bar{1}$
$\bar{1}$	$\bar{1}$	$\bar{0}$.

Then G is an Ω-group with $\bar{0}$ as identity-element and $\bar{0}^G = \bar{0} \neq G$.

It is also interesting to inquire into the uniqueness of identity-elements. The following example show that this is not so even if the Ω-group is distributive. Let $G = Z_3$. Define a ternary operation as follows: $\bar{a}\bar{b}\bar{c} = \overline{abc}$ mod 3. G is a distributive Ω-group (hence also zero-symmetric) and both $\bar{1}$ and $\bar{2}$ are identity-elements.

4.13 Definition

(a) $a \in G$ is called *f-regular* if for each non-trivial $\omega \in \Omega$ it holds that $a \in a^G a^G \ldots a^G \omega^G$.

(b) $H \triangleleft G$ is called *idempotent* if for each non-trivial $\omega \in \Omega$ it holds that $HH\ldots H\omega^G = H$.

(c) G is called *hereditarily idempotent* if each ideal of G is idempotent.

4.14 Theorem

$H \triangleleft G$ is idempotent if and only if $f'(H) = H$ for each non-trivial Ω-word $f(\underline{x})$

Proof

The sufficiency is trivial. The necessity is proved by induction on the number of operations performed in an Ω-word.

Let P_6 be the property f-regular.

4.15 Theorem

P_6 satisfies conditions (A), (D), (E) and (F).

Proof

Let a be an f-regular element of G and $H \lhd G$. By definition of a homomorphism and Lemma 3.1 (Higgins, [7]) it follows that for each non-trivial $\omega \in \Omega$

$(a+H)^{G/H}(a+H)^{G/H}\dots(a+H)^{G/H}\omega^{G/H} = (a^G a^G\dots a^G\omega^G + H)$.

Since $a \in a^G a^G\dots a^G\omega^G$ it follows that $a+H \in (a+H)^{G/H}(a+H)^{G/H}\dots(a+H)^{G/H}\omega^{G/H}$ for each non-trivial $\omega \in \Omega$. Thus P_6 satisfies condition (A) and hence also (F).

Condition (D) is satisfied trivially.

Let H be an f-regular ideal of G. Let $a+H \in G/H$ be an f-regular element of G/H and $\omega \in \Omega$ a non-trivial operation. Hence

$$a+H \in (a+H)^{G/H}(a+H)^{G/H}\dots(a+H)^{G/H}\omega^{G/H} = (a^G a^G\dots a^G\omega^G + H)/H$$

It follows that $a \in a^G a^G\dots a^G\omega^G + H$. Hence there exists $c \in a^G a^G\dots a^G\omega^G \subseteq a^G$ such that $-c+a = b \in H$. Since $c \in a^G$, we have $-c+a \in a^G$. Thus $b^G \subseteq a^G$. Since H is an f-regular ideal, it follows that for any non-trivial $\omega \in \Omega$

$$b \in b^H b^H\dots b^H\omega^H \subseteq b^G b^G\dots b^G\omega^G$$
$$\subseteq a^G a^G\dots a^G\omega^G.$$

Since $b,c \in a^G a^G\dots a^G\omega^G$ it follows that $a = c+b \in a^G a^G\dots a^G\omega^G$.

Hence P_6 also satisfies condition (E).

By 3.3 and 3.5 we have that $R_1(P_6)$ and $R_3(P_6)$ are radical classes. Furthermore, by 3.6, $R_1(P_6) = R_2^*(P_6)$.

4.16 Theorem

Let G be a subdirectly irreducible -group. The heart, H, of G is either nilpotent or idempotent. (An ideal H is called *nilpotent* if there exists a non-trivial Ω-word $f(\underline{x})$ such that $f'(H) = 0$.)

Proof

If H is nilpotent then we are finished. Suppose H is not nilpotent. Hence $f'(H) \neq 0$ for each non-trivial Ω-word $f(\underline{x})$. Since $f'(H) \subseteq H$ and $0 \neq f'(H) \lhd G$ we must have $H = f'(H)$ for any non-trivial Ω-word $f(\underline{x})$. By 4.14 it follows that H is idempotent.

The next theorem generalizes two theorems in ring theory (Szasz, [10], Proposition 13.4 and Courter, [5]).

4.17 Theorem

If G is a zero-symmetric Ω-group then the following statements are equivalent

(a) G is hereditarily idempotent.

(b) G is f-regular that is, every element of G is f-regular.

(c) For each non-trivial $\omega \in \Omega$ and $B_1, B_2, \ldots, B_n \lhd G$ we have $B_1 B_2 \ldots B_n \omega^G = \bigcap_{i=1}^{n} B_i$.

(d) G has no subdirectly irreducible homomorphic image with nilpotent heart.

(e) Every homomorphic image of G is a subdirect sum of subdirectly irreducible Ω-groups, each with an idempotent heart.

(f) Every $I \lhd G$ is the intersection of all Ω-prime ideals P of G such that $I \subseteq P$ and G/P has a minimal ideal.

(g) Every ideal of G is an Ω-semi-prime ideal.

Proof

(a) $\Rightarrow$ (b) Let $a \in G$ and ω be a non-trivial operation. Then $a \in a^G = a^G a^G \ldots a^G \omega^G$. Thus G is f-regular.

(b) $\Rightarrow$ (c) Let $\omega \in \Omega$ be a non-trivial operation and $B_1, B_2, \ldots, B_n \lhd G$. By 2.3, $B_1 B_2 \ldots B_n \omega^G \subseteq \bigcap_{i=1}^{n} B_i$. Let $b \in \bigcap_{i=1}^{n} B_i$. Since G is f-regular, we have $b \in b^G b^G \ldots b^G \omega^G \subseteq B_1 B_2 \ldots B_n \omega^G$. Hence equality follows.

(c) $\Rightarrow$ (d) If (c) holds for G, then it also holds for any homomorphic image of G. Let $\theta : G \to G'$ be a homomorphism onto G'. Let $\omega \in \Omega$ be a non-trivial operation and $B'_1, B'_2, \ldots, B'_n \lhd G'$. By Theorem 3B of Higgins ([7]) there exist $B_i \lhd G$, $i = 1,2,\ldots,n$ such that $\mathrm{Ker}\ \theta \subseteq B_i$ and $B_i\theta = B'_i$, $i = 1,2,\ldots,n$. Hence

$$\begin{aligned} B'_1 B'_2 \ldots B'_n \omega^G &= (B_1 B_2 \ldots B_n \omega^G)\theta \\ &= (\bigcap_{i=1}^{n} B_i)\theta \\ &\subseteq \bigcap_{i=1}^{n} B'_i. \end{aligned}$$

Let $b' \in \bigcap_{i=1}^{n} B'_i$ and $b_i \in B_i$ such that $b_i\theta = b'$, $i = 1,2,\ldots,n$. Hence $b_i\theta = b' = b_j\theta$, $i \neq j$. Therefore, there exists $k \in \mathrm{Ker}\ \theta$ such that $b_i = k + b_j$. But $\mathrm{Ker}\ \theta \subseteq B_j$, $j = 1,2,\ldots,n$ and hence $b_i = k + b_j \in B_j$ for $j \neq i$. Therefore,

$$b_i \in \bigcap_{i=1}^{n} B_i = B_1 B_2 \ldots B_n \omega^G, \quad i = 1,2,\ldots,n.$$

Thus $b' = b_i\theta \in (B_1 B_1 \ldots B_n \omega^G)\theta = B'_1, B'_2, \ldots, B'_n \omega^{G'}$ and hence $\bigcap_{i=1}^{n} B'_i = B'_1 B'_2 \ldots B'_n \omega^{G'}$.

Let $G\theta$, with nilpotent heart H, be a subdirectly irreducible homomorphic image of G. By what we have just proved, it follows that H is idempotent. This contradicts 4.16. Therefore, (d) follows.

(d) $\Rightarrow$ (e) By Birkhoff's Theorem (Gratzer [6], §20 Theorem 3) every homomorphic image $G\theta$ of G, is a subdirect sum of subdirectly irreducible Ω-groups. Since each subdirect summand is a homomorphic image of $G\theta$, and hence also of G, it follows that the heart of each subdirect summand is not nilpotent and thus idempotent by 4.16.

(e) $\Rightarrow$ (f) Let $I \triangleleft G$. G/I is a homomorphic image of G and hence a subdirect sum of subdirectly irreducible Ω-groups $(G/I)/(I_\alpha/I) \cong G/I_\alpha$ with idempotent hearts H_α/I_α and $\cap(I_\alpha/I) = 0$. Hence $\cap I_\alpha = I$.

Let $\omega \in \Omega$ be a non-trivial operation and $A_1/I_\alpha, A_2/I_\alpha, \ldots, A_n/I_\alpha \triangleleft G/I_\alpha$ such that $A_i/I_\alpha \neq 0$, $i = 1,2,\ldots,n$. Then $H_\alpha/I_\alpha \subseteq A_i/I_\alpha$, $i = 1,2,\ldots,n$. Hence

$$\begin{aligned} 0 \neq H_\alpha/I_\alpha &= (H_\alpha/I_\alpha)(H_\alpha/I_\alpha)\ldots(H_\alpha/I_\alpha)\omega^{G/I_\alpha} \\ &\subseteq (A_\alpha/I_\alpha)(A_\alpha/I_\alpha)\ldots(A_n/I_\alpha)\omega^{G/I_\alpha}. \end{aligned}$$

By Buys and Gerber ([4], Theorem 2.3) it follows that G/I_α is an Ω-prime Ω-group and hence I_α is an Ω-prime ideal of G for each α (Buys and Gerber, [4], Corollary 2.10). This $I = \cap I_\alpha$, I_α an Ω-prime ideal of G such that $I \subseteq I_\alpha$ and H_α/I_α being a minimal ideal of G/I_α. Now (f) follows easily.

(f) $\Rightarrow$ (a) Let $A \triangleleft G$ and $\omega \in \Omega$ be any non-trivial operation. Since $AA\ldots A\omega^G \triangleleft G$ it follows from the assumption that

$$AA\ldots A\omega^G = \cap\{P_\alpha \triangleleft G \mid P_\alpha \text{ an } \Omega\text{-prime ideal of } G \text{ such that } AA\ldots A\omega^G \subseteq P \text{ and } G/P_\alpha \text{ has a minimal ideal}\}.$$

Since $AA\ldots A\omega^G \subseteq P_\alpha$ and P_α is Ω-prime, it follows that $A \subseteq P_\alpha$ for each α. Hence $A \subseteq \cap P_\alpha = AA\ldots A\omega^G$. Equality follows since $A \triangleleft G$. Hence G is hereditarily idempotent.

(b) $\Rightarrow$ (g) Let $P \triangleleft G$, $\omega \in \Omega$ a non-trivial operation and $a \in G$ such that $a^G a^G \ldots a^G \omega^G \subseteq P$. Since G is f-regular, $a \in a^G a^G \ldots a^G \omega^G$ and thus $a \in P$. Therefore, P is an Ω-semi-prime ideal.

(g) $\Rightarrow$ (b) Let $a \in G$ and $\omega \in \Omega$ a non-trivial operation. $a^G a^G \ldots a^G \omega^G \triangleleft G$ and must be an Ω-semi-prime ideal by assumption. By definition of Ω-semi-prime and $a^G a^G \ldots a^G \omega^G \subseteq a^G a^G \ldots a^G \omega^G$ we have $a^G \subseteq a^G a^G \ldots a^G \omega^G$. Hence $a \in a^G \subseteq a^G a^G \ldots a^G \omega^G$. Thus the required result follows.

We also give the following relating result.

4.18 Theorem

The class of idempotent Ω-groups is a radical class.

Proof

We first note that $f'(G\theta) = f'(G)\theta$ easily follows by induction on the number of operations performed in any Ω-word $f(\underline{x})$.

Let G be an idempotent Ω-group and $I \triangleleft G$. Let $f(\underline{x})$ be any non-trivial Ω-word. Then

$$f'(G/I) = (f'(G) + I)/I = G/I \quad \text{using 4.14.}$$

Again, by 4.14, G/I is idempotent. Property R3 (Rjabuhin, [9]) are satisfied.

Let H_α be an idempotent ideal of G and $f(\underline{x})$ any non-trivial Ω-word. Then $f'(\Sigma H_\alpha) \subseteq \Sigma H_\alpha$ by definition. But $H_\alpha = f'(H_\alpha) \subseteq f'(\Sigma H_\alpha)$ for any α and hence $\Sigma H_\alpha = f'(\Sigma H_\alpha)$. Thus ΣH_α is an idempotent ideal. Property R5 (Rjabuhin, [9]) is satisfied.

Let $H = \Sigma H_\alpha$ be the sum of all the idempotent ideals. Let $J/H \triangleleft G/H$ be an idempotent ideal. For any non-trivial Ω-word $f(\underline{x})$, $f'(J/H) = J/H$, that is, $f'(J)/H = J/H$ since $H = f'(H) \subseteq f'(J)$ because $H \subseteq J$. Therefore, $f'(J) = J$ and so $J \subseteq H$. Property R6 (Rjabuhin, [9]) is satisfied. By Theorem 1.2 of Rjabuhin ([9]) the class of idempotent Ω-groups is a radical class.

REFERENCES

[1] Barnes, W.E., On the Γ-rings of Nobusawa, Pacific. J. Math., 18(1966), No.3, 411-422.
[2] Buys, A. and Gerber, G.K., Nil and s-prime Ω-groups, J. Austral. Math. Soc. (Series A), 38(1985), 222-229.
[3] Buys, A. and Gerber, G.K., Prime and k-prime ideals in Ω-groups, Quaestiones Mathematicae, 8(1985), No.1, 15-32.
[4] Buys, A. and Gerber, G.K., The prime radical for Ω-groups, Communications in Algebra, 10(1982), No.10, 1089-1099.
[5] Courter, R.C., Rings all whose factor rings are semi-prime, Canad. Math. Bull., 12(1969), No.4, 417-426.
[6] Grätzer, George, Universal Algebra (2nd Ed.), (Springer-Verlag, New York Inc., 1979).
[7] Higgins, P.J., Groups with multiple operators, London Math. Soc. Proc., (3), 6(1956), 366-416.
[8] Nobusawa, Nobuo, On a generalization of the ring theory, Osaka J. Math., 1(1964), 81-89.
[9] Rjabuhin, Ju. M., Radicals in Ω-groups I, General Theory, Mat. Issled, 3(1968), 123-160 (in Russian).
[10] Szasz, F.A., Radicals of Rings, (John Wiley & Sons, 1981).
[11] Wiegandt, Richard, Radicals of rings defined by means of elements, Österreich Akad. Wiss. Math. Naturwiss. SB II, 184(1975), No.1-4, 117-125.

Near-rings and Near-fields, G. Betsch (editor)

NOTE ON THE COMPLETELY PRIME RADICAL IN NEAR-RINGS

N.J. Groenewald

University of Port Elizabeth, P O Box 1600,
6000 Port Elizabeth

ABSTRACT: In this note a completely prime radical is defined for near-rings. It is then proved that for the class of all distributively generated near-rings the completely prime radical coincides with the generalized nil radical. We also show that in general the completely prime radical and also the prime radical is not hereditary.

1. INTRODUCTION

Let N be a near-ring. An ideal I of N is completely prime if $a,b \in N$, $ab \in I$ implies $a \in I$ or $b \in I$. If (0) is a completely prime ideal then N is called a completely prime near-ring. We define the completely prime radical of the near-ring N as the intersection of all the completely prime ideals and denote it by $\mathcal{C}(N)$.

For the notations about general radical theory in near-rings we refer to [7] Let M be a class of completely prime near-rings. Clearly M is hereditary, i.e. if $A \in M$ and $I \triangleleft A$ then $I \in \boldsymbol{M}$. Let $\mathcal{U}M$ be the upper radical class determined by M. We have $\mathcal{U}M$ = {A : A has no nonzero homomorphic image in $\boldsymbol{M}$}

= {A : every nonzero homomorphic image of A has a nonzero divisor of zero}.

As in the case of rings [2] and [3] we call $\mathcal{U}M$ the generalized nil radical and denote it by $\boldsymbol{N}_g$.

We have the following theorem:

Theorem 1. $\mathcal{U}M = \boldsymbol{N}' = \{A : A \text{ a near-ring such that } A = \mathcal{C}(A)\}$.

Let $\mathcal{D}$ denote the class of all distributively generated near-rings. If M is a class of prime near-rings, Kaarli [4] defined the class M to be $\mathcal{D}$-special if $M \cap \mathcal{D}$ is hereditary with respect to ideals and if $U \triangleleft S \triangleleft N \in \mathcal{D}$ and $S/U \in \boldsymbol{N}$ imply $U \triangleleft N$ and $N/(U:S) \in \boldsymbol{N}$, where $(U:S) = \{n \in N : nS \subseteq U\}$.

In what follows, let M be the class of all completely prime near-rings.

Theorem 2. The class M of all completely prime near-rings is $\mathcal{D}$-special.

Proof. $\boldsymbol{M} \cap \mathcal{D}$ is clearly hereditary. From [4] we have that the class of all prime near-rings is $\mathcal{D}$-special. Since a completely prime near-ring is also a prime near-ring, we have $U \triangleleft S \triangleleft N \in \mathcal{D}$ and $S/U \in M$ implies $U \triangleleft N$. We only have to show that $(U:S)$ is a completely prime ideal in N. Let $x \in N$ such that $x^2 \in (U:S)$. Suppose $x \notin (U:S)$, i.e. there exists $b \in S$ such that $xb \notin U$. We also have $bx \notin U$, for if $bx \in U$ then $xbxb \in U$ and since $S/U \in M$, i.e. U completely prime ideal of S, it follows that $xb \in U$. Let $a \in S$ be arbitrary. Since $x^2 \in (U:S)$ we have $x^2a \in U$ and therefore $bx^2a \in U$. Since $bx \notin U$ and U is a completely prime ideal, we have $xa \in U$ and consequently $x \in (U:S)$. Hence (U:S) is a completely semi-prime ideal. It is now easy to show that, since (U:S) is a prime ideal (c.f. [5]), (U:S) is also completely prime. Hence $N/(U:S) \in M$.

In the definition of $\mathcal{D}$-spesial classes the following condition plays an important role:

(F) $U \triangleleft S \triangleleft N$ and $S/U \in M$ imply $U \triangleleft S$ where M is a class of near-rings.

The following example shows that in general condition (F) is not satisfied if M is either the class of prime near-rings or the class of completely prime near-rings.

Example 1. Consider the dihedral group N with addition and multiplication defined as in Pilz [6], p.345, number 11. N is a near-ring which is not zero symmetric

·	0	1	2	3	4	5	6	7
0	0	0	0	0	0	0	0	0
1	0	1	0	1	0	1	1	0
2	0	2	0	2	0	2	2	0
3	0	3	0	3	0	3	3	0
4	4	4	4	4	4	4	4	4
5	4	5	4	5	4	5	5	4
6	4	6	4	6	4	6	6	4
7	4	7	4	7	4	7	7	4

It is easy to check that $I = \{0,1,2,3\}$ is the only nontrivial ideal of N. Furthermore, $K = \{0,2\} \triangleleft I$ but $\{0,2\}$ is not an ideal in N. Let M denote the class of all prime near-rings, i.e. $N \in M$ if for every two nonzero ideals A and B of N, we have $AB \neq 0$. Clearly the near-ring in the example is prime, i.e. 0 is a prime ideal. Hence $N \in M$. Furthermore, if $a,b \notin K$, we have $\langle a\rangle_I = \langle a\rangle_N = I$ and $\langle b\rangle_I = \langle b\rangle_N = I$. Hence $\langle a\rangle\langle b\rangle \not\subseteq K$. Therefore $I/K \in M$ and condition (F) is not satisfied since $K \not\triangleleft N$.

If M is the class of completely prime near-rings, we have $N \notin M$ since $2 \cdot 2 = 0$. I is a completely prime ideal in N for if $x,y \notin I$, then $xy \notin I$. If x,y are elements of I such that $x,y \notin K$, then $xy \notin K$. Hence K is a completely prime ideal in I and, therefore, $I/K \in M$. Condition (F) is not satisfied for the class of all completely prime near rings since $K \not\triangleleft N$.

The following theorem is proved in [1].

Theorem 3 ([1], Theorem 1). Let M be a regular and essentially closed class of near-rings satisfying condition (F). Then $\bar{M} = \{A : A = \sum_{\text{subdirect}} (A_\alpha : A_\alpha \in M)\}$ is a semisimple class, in fact $\bar{M} = \mathcal{SUM}$, moreover, the upper radical class

$\gamma = \mathcal{UM}$ is hereditary, i.e. $I \triangleleft A \in \gamma$ implies $I \in \gamma$.

Theorem 4. If M is the class of all completely prime near-rings, then for every $N \in \mathcal{D}$ and $I \triangleleft N$ we have $\mathcal{N}_g(N) \cap I = \mathcal{N}_g(I)$ and $\mathcal{N}_g(N) = \mathcal{C}(N)$.

Proof. We show that the conditions of Theorem 3 is satisfied. Since M is $\mathcal{D}$-special, we only have to show that M is essentially closed. Let N be a near-ring and A an essential ideal of N. We show that if $A \in M$, then $N \in M$. It is easy to show that $\ell(A) = \{n \in N : nA = 0\}$ is a two-sided ideal of N. But $(\ell(A) \cap A)^2 \subseteq \ell(A)\cdot A = 0$. Since $\ell(A) \cap A \triangleleft A$ and $A \in M$, we have $\ell(A) \cap A = 0$. From the fact that A is an essential ideal it follows that $\ell(A) = 0$. Let $x,y \in N$ such that $x,y \neq 0$. Since $\ell(A) = 0$, we can find $a,b \in A$ such that xa and $yb \neq 0$. Furthermore, we have $ax \neq 0$ for if $ax = 0$, then $xaxa = 0$ and since $A \in M$, it follows that $xa = 0$. Now $ax\cdot yb \neq 0$ and hence also $xy \neq 0$. Thus $N \in M$. From Theorem 3 we have $\mathcal{N}_g$ is hereditary and from [1], Proposition 3, we have $\mathcal{SUM}$ hereditary. Hence, for every $N \in \mathcal{D}$ and every $I \triangleleft N$ we have $\mathcal{N}_g(I) = N \cap \mathcal{N}_g(N)$. For $N \in \mathcal{D}$ we have that $\mathcal{C}(N) = \mathcal{N}_g(N)$. To show this, let $\{T_\alpha\}$ be the set of all ideals T of N for which $N/T_\alpha \in M$. We have $\mathcal{N}_g(N) \subseteq B = \cap T_\alpha$. Let us suppose $\mathcal{N}_g(N) \neq B$. In this case we have $\mathcal{N}_g(B) = B \cap \mathcal{N}_g(N) = \mathcal{N}_g(N) \neq B$. Hence there exists an ideal $C \triangleleft B$ with $0 \neq B/C \in M$. From Theorem 2 we have $N/(C:B) \in M$. Consequently $(C:B) \in \{T_\alpha\}$. $(C:B) \cap B = C$ for if $x \in (C:B) \cap B$, then $x \in B$ and $x^2 \in C$. Now, since $B/C \in M$, we have $x \in C$. Hence $(C:B) \cap B \subseteq C$. Inclusion in the other direction is clear. Thus $B = \cap T_\alpha = B \cap (C:B) = C \neq B$ a contradiction. Hence $B = \mathcal{N}_g(N) = \mathcal{C}(N)$.

Remark. From [4] it follows that if $N \in \mathcal{D}$, then $\mathcal{P}(N) \cap I = \mathcal{P}(I)$ where $I \triangleleft N$ and $\mathcal{P}(N)$ denotes the prime radical of N. In general, the prime and completely prime radicals are not hereditary. To show this we use the example above. Since N is not a completely prime near-ring, we have $\mathcal{C}(N) = I$ since I is the only completely prime ideal of N. I, considered as a near-ring, is not completely prime since $2\cdot 2 = 0$. Since K is the only completely prime ideal of I, we have $\mathcal{C}(I) = K \subsetneq I \cap \mathcal{C}(N) = I$.

Since $\mathcal{N}_g$ is a Kurosh-Amitsur radical (as the upper radical of a hereditary class) and $\mathcal{N}_g = \{N : N = \mathcal{C}(N)\}$ but $\mathcal{N}_g(N) \neq \mathcal{C}(N)$ in general, the completely prime radical is not a Kurosh-Amitsur radical for near-rings.

We note that N is a prime near-ring, hence $\mathcal{P}(N) = 0$. The near-ring $I = \{0,1,2,3\}$ is not prime since $0 \neq K = \{0,2\} \triangleleft I$ and $K^2 = 0$. Since K is the only prime ideal of the near-ring I, it follows that $\mathcal{P}(I) = K \supsetneq \mathcal{P}(N) \cap I = 0$. This shows that, in general, Theorem 5.62 of Pilz [6] is not satisfied if I is not a direct summand of N.

REFERENCES

[1] Anderson, T., Kaarli, K. and Wiegandt, R., Radicals and subdirect decompositions, Comm. in Algebra, 13(1985), 479-494.

[2] Andrunakievic, V.A. and Rhabuhin, Ju. M., Rings without nilpotent elements and completely simple ideals, Dokl. Akad. Nauk. SSSR, 180(1968, 9-1 (in Russian - Engelish translation in Soviet. Math. Dokl. 9(1968), 565-568).

[3] Divinsky, N., Rings and radicals (Math. Expositions No.14, University of Toronto Press, 1965).

[4] Kaarli, K., Special radicals in near-rings, Tartu Riikl. ÜL Toimetesed 610(1982), 53-68.

[5] McCoy, N.H., Completely prime and completely semi-prime ideals, Rings, Modules and Radicals, Coll. Math. Soc. S. Bolyai, 6, North Holland (1973), 147-152.

[6] Pilz, G., Near-rings (North-Holland Mathematics Studies, 1977).

[7] Wiegandt, R., Near-rings and Radical Theory, Conf. on near-rings and near-fields, San Benedetto del Tronto (Italy) (1981), 49-58.

Near-rings and Near-fields, G. Betsch (editor)

ON p-ADIC NEARFIELDS

Theo GRUNDHÖFER

Mathematisches Institut, Universität Tübingen
Auf der Morgenstelle 10
D-7400 Tübingen 1 , BRD

This is an announcement of a result concerning the classification of locally compact nearfields, namely the following: all p-adic nearfields with continuous multiplication are Dickson nearfields. We indicate a proof for some easier special cases, and give a rough sketch of a proof for the general case.

A topological nearfield is a nearfield with a topology such that addition, multiplication and inversion are continuous. The locally compact connected nearfields have been classified by Kalscheuer [3]: these are the fields of real and complex numbers and Dickson nearfields derived from the skew field of Hamilton's quaternions (cp. also Tits [7], Salzmann [6]7.26). The following is an extension of Kalscheuer's result.

<u>Theorem 1</u> Every non-discrete locally compact nearfield of characteristic zero is a Dickson nearfield derived from a non-discrete locally compact skew field.

The nearfields covered by this theorem are those of finite dimension over the reals $\mathbb{R}$ or the p-adics $\mathbb{Q}_p$ with continuous multiplication. Theorem 1 is a consequence of the following group-theoretic assertion.

<u>Theorem 2</u> Let G be a closed subgroup of $GL(n, \mathbb{Q}_p)$ or $GL(n, \mathbb{R})$ which acts transitively on the nonzero vectors. Assume that the stabilizer G_v is discrete for every nonzero vector v. Then G is a subgroup of $\Gamma L(1,D)$ for some skew field D of dimension n over $\mathbb{Q}_p$ resp. $\mathbb{R}$.

Such a group G is a p-adic resp. real linear Lie group, and, as a consequence of our assumption on the stabilizers G_v , its linear Lie algebra L has the special property that all nonzero elements of L are invertible (Kalscheuer [3] p.414). The Lie algebra L can be used to construct the skew field D, as the proof of the following proposition illustrates.

Proposition Let F be a non-discrete locally compact nearfield of characteristic zero. Then the following are equivalent:

a) The multiplicative group F* of F has an open abelian subgroup.

b) The Lie algebra L of F* is abelian.

c) F is a Dickson nearfield derived from a finite field extension of $\mathbb{R}$ or $\mathbb{Q}_p$.

Proof. a) and b) are equivalent by Lie theory, and c) implies a), as every finite extension of the reals or the p-adics has only finitely many continuous field automorphisms. Suppose b) holds. Then L is contained in its centralizer C, which is a skew field by Schur's Lemma. Comparing dimensions gives $L = C$, hence L is a field. Furthermore, F* acts on L via the adjoint representation, i.e. L is normalized by F*, which gives c).

The non-abelian Lie algebra of dimension 2 has no faithful linear representation with the special property mentioned above. For every topological nearfield F of dimension 2 over $\mathbb{Q}_p$ or $\mathbb{R}$, the Lie algebra of F* is therefore abelian, and the Proposition applies, thus proving Theorem 1 for this special case.

Finally we give a very sketchy description of a proof for Theorem 2 in the p-adic case. The Lie algebra L of F* is reductive, i.e. the direct sum of its center and its semisimple commutator algebra L'. The simple p-adic Lie algebras have been classified by Veisfeiler, Kneser [4], Bruhat-Tits [1], Tits [9]. This classification yields that L' is a direct sum of commutator algebras D_i' of skew fields D_i . The Proposition (or rather its group-theoretic analogue) allows us to assume that L' is not trivial. Then one of the D_i' acts irreducibly. Analysis of the representation of such an irreducible ideal D_i' shows that $D := D_i$ satisfies the conclusion of Theorem 2. Several steps of the proof make use of the representation theory developed in Tits[8].

It seems to be difficult to remove the characteristic zero assumption in Theorem 1. Note that Rink constructed non-discrete locally compact nearfields with finite kernel (see Hähl [2] p.279). The Dickson nearfields derived from unramified extensions of local fields have been classified in Rink[5].

REFERENCES

[1] F. Bruhat and J. Tits, Groupes rêductifs sur un corps local I,II , Publ. Math. I.H.E.S. 41 (1972) 5-251, 60 (1984) 5-184.

[2] H. Hähl, Kriterien für lokalkompakte topologische Quasikörper, Arch. Math. (Basel) 38 (1982) 273-279.

[3] F. Kalscheuer, Die Bestimmung aller stetigen Fastkörper über dem Körper der reellen Zahlen als Grundkörper, Abh. Math. Sem. Hamburg 13 (1940) 413-435.

[4] M. Kneser, Galois-Kohomologie halbeinfacher algebraischer Gruppen über $\mathfrak{p}$-adischen Körpern I,II , Math. Z. 88 (1965) 40-47, 89 (1965) 250-272.

[5] R. Rink, Zur Konstruktion lokal kompakter Dicksonscher Fastkörper, Geom. Dedicata 20 (1986).
[6] H. Salzmann, Topological planes, Advances Math. 2 (1967) 1-60.
[7] J. Tits, Sur les groupes doublement transitifs continus, Comment. Math. Helv. 26 (1952) 203-224. Correction et compléments, ibid. 30 (1956) 234-240.
[8] J. Tits, Représentations linéaires irréductibles d'un groupe réductif sur un corps quelconque, J. Reine Angew. Math. 247 (1971) 196-220.
[9] J. Tits, Reductive groups over local fields, in: A. Borel and W. Casselman (eds.), Proc. Sympos. Pure Math. 33 (1979), part 1, pp. 29-69, Amer. Math. Soc., Providence (R.I.), 1979.

Near-rings and Near-fields, G. Betsch (editor)

EUCLIDEAN SEMINEARRINGS AND NEARRINGS

Udo HEBISCH and Hanns Joachim WEINERT

1. INTRODUCTION

This paper is mainly based on the doctoral dissertation [6] of the first author, where the concept of a left Euclidean ring (cf. Def. 3.1) has been generalized to arbitrary (2,2)-algebras $(S,+,\cdot)$ (cf. § 2 and Def. 3.2). Several statements on left Euclidean rings or on commutative ones, e. g. a well known characterization of Euclidean integral domains due to Motzkin [11], could be generalized in [6] to those left Euclidean (2,2)-algebras without any further assumptions. Other results of this kind have been proved using merely the right distributive law for $(S,+,\cdot)$ and the associativity of $(S,\cdot)$, i. e. for (right distributive) seminearrings in the more general meaning of this term (cf. § 2). So [6] contains a first version of a theory of left Euclidean seminearrings and nearrings, which will be presented here in a self-contained and improved way.

The only paper in this context known to the authors is the unpublished manuscript [5]. It deals - in our notation - with left Euclidean nearrings S which are embeddable into a nearfield of right quotients of S. We also mention that there are some recent papers on Euclidean semirings. They consider rather specialized Euclidean functions on semirings $(S,+,\cdot)$ (cf. § 2) for which both operations are associative and commutative. For an extensive discussion of these publications we refer to [8]. In the latter paper we have dealt with semirings for which both operations are only assumed to be associative, and which are left Euclidean algebras according to Def. 3.2 given here.

Coming back to this paper, there are some results on left Euclidean seminearrings which are essential for our subject, but in fact true for arbitrary (2,2)-algebras $(S,+,\cdot)$ or those with associative multiplication. In order to see which assumptions are really needed and since the proofs cannot be simplified for seminearrings, we present those statements in their general version.

So, after some preliminaries in § 2, we define our basic concepts in § 3 for arbitrary (2,2)-algebras and show in particular, that a ring S is a left Euclidean (2,2)-algebra according to our

Def. 3.2 iff S is a left Euclidean ring. The next section deals with Φ-derivations of seminearrings in order to show that left Euclidean seminearrings can be obtained as Φ-derivations of other ones. In this way we get examples of left Euclidean nearrings as Φ-derivations of left Euclidean rings.

In § 5 we show that each homomorphic image of a left Euclidean (2,2)-algebra is again left Euclidean, and we give sufficient conditions such that the direct product of a finite number of left Euclidean (2,2)-algebras is also left Euclidean. The latter yields that such a direct product of left Euclidean nearrings all of which have left identities is a left Euclidean nearring. Sufficient conditions for the existence of (right) identities are given for left Euclidean (2,2)-algebras S such that $(S,\cdot)$ is associative.

A main point for each left Euclidean ring S (say with a left identity) is that the left Euclidean algorithm yields the existence of a greatest common right divisor for any two elements $a,b \in S$, and that each chain of proper right divisors in S is finite. According to our considerations in § 6, these results are also true for each left Euclidean seminearring satisfying one supplementary assumption which is always satisfied for nearrings. Further, each left Euclidean ring with a left identity is a left principal ideal ring. Also this result transfers to left Euclidean seminearrings and nearrings, but we emphasize that one needs in this context a concept of "left ideal" for seminearrings and nearrings which is in the latter case not the usual one.

In the last section, we consider the constant subnearring S_1 and the zero-symmetric one S_2 of a nearring S and the generalizations of these concepts for a seminearring S with a left absorbing zero as introduced in § 4 of [20]. Whereas a constant seminearring S_1 is always left Euclidean, we investigate conditions such that this property transfers from S to S_2 or conversely.

2. PRELIMINARIES ON (2,2)-ALGEBRAS AND SEMINEARRINGS

We consider (2,2)-algebras $S = (S,+,\cdot)$ with operations called addition and multiplication. The cardinality of S, also called the order of the algebra, is denoted by $|S|$. For subsets $A,B \subseteq S$ we introduce $A+B = \{a+b \mid a \in A,\ b \in B\}$ and $AB = \{ab \mid a \in A,\ b \in B\}$, and use e. g. aB for $\{a\}B$. If there exists a neutral element o of $(S,+)$ [e of $(S,\cdot)$], it is unique and called the *zero* [the *identity*] of the algebra S. Accordingly we speak about *left* (or *right*) *identities* of S. An element $a \in S$ is called *left absorbing* iff $aS = \{a\}$ holds. Corresponding left-right dual concepts and their two-sided versions are always taken for granted. We further introduce the notation S^* by $S^* = S \setminus \{o\}$ if S has a zero o, and by $S^* = S$ otherwise.

A (2,2)-algebra S is called *multiplicatively left cancellative* iff $|S| \geq 2$ holds and each $a \in S^*$ is left cancellable in $(S,\cdot)$. If $(S,\cdot)$ is associative, we have the following statement (cf. [18], Lemma 1): S is multiplicatively left cancellative iff either all elements of S, including the zero if there is one, are left cancellable in $(S,\cdot)$ or S has a (two-sided) absorbing zero o and $(S^*,\cdot)$ is a left cancellative subsemigroup of $(S,\cdot)$.

Our main interest is with (2,2)-algebras $(S,+,\cdot)$ such that $(S,\cdot)$ is a semigroup and $(a+b)c = ac + bc$ holds for all $a,b,c \in S$. According to Def. 1.1 of [20], we call such an algebra a *right seminearring*, briefly a *seminearring* in this paper. (The more restricted version of this concept includes also the associativity of $(S,+)$.) To avoid confusion, the term (*right*) *nearring* is used for a seminearring such that $(S,+)$ is a group. A seminearring which satisfies also $c(a+b) = ca + cb$ for all $a,b,c \in S$ is called a *distributive seminearring*. (The term *semiring*, in this paper mainly used in § 1, means sometimes such a distributive seminearring, whereas in its more frequent version the associativity of both operations is included.)

Finally, let $(S,+,\cdot)$ be a (2,2)-algebra which may have a zero or not. Then an absorbing zero $o \notin S$ can be adjoined to S extending the operations by $o+x = x+o = x$ and $xo = ox = o$ for all $x \in S \cup \{o\}$. Clearly, the resulting (2,2)-algebra $(S \cup \{o\},+,\cdot)$ is again a seminearring iff $(S,+,\cdot)$ is one.

3. LEFT EUCLIDEAN (2,2)-ALGEBRAS AND SEMINEARRINGS

The concept of a Euclidean ring, say S in our context, was originally defined for commutative rings without zero divisors, and with Euclidean functions γ from $S^* = S \setminus \{o\}$ into the set $\mathbb{N}_0$ of non-negative integers. Various generalizations (cf. [17], [12], [9], [11], [15], [14], [10], [16], [1]) have included non-commutative rings, or those with zero divisors, or replaced $\mathbb{N}_0$ by any well ordered set W_γ. The most general one, given in [2] for arbitrary (associative) rings, reads as follows:

<u>DEFINITION 3.1.</u> A *left Euclidean function* (briefly LEF) on a ring S

is defined as a mapping $\gamma: S^* \to W_\gamma$ of S^* into a well ordered set W_γ with the following property:

(3.1) For all $a \in S$, $b \in S^*$ there are $q,r \in S$ satisfying $a = qb+r$ and $r = o$ or $\gamma(r) < \gamma(b)$.

If such a LEF on S exists, S is called a *left Euclidean ring.*

Clearly, the zero o of S is used in (3.1) as an additive neutral to include $a = qb$ by $r = o$. Moreover, since o is left absorbing, $a = ob+a$ holds for all $a,b \in S$. Hence (3.1) is equivalent to

(3.2) For all $a,b \in S^*$ such that $\gamma(a) \geqq \gamma(b)$ there are $q,r \in S$ satisfying $a = qb+r$ and $r = o$ or $\gamma(r) < \gamma(b)$.

It is near by hand to apply Def. 3.1 also to any (2,2)-algebra $(S,+,\cdot)$ with a left absorbing zero, hence in particular to each nearring, and to look for a suitable generalization of the resulting concept for arbitrary (2,2)-algebras. The following definition presents such a generalization as we shall prove in Prop. 3.3.

DEFINITION 3.2. A *left Euclidean function* on a (2,2)-algebra S is defined as a mapping $\delta: S \to W_\delta$ of S into a well ordered set W_δ with the following property:

(3.3) For all $a,b \in S$ such that $\delta(a) \geqq \delta(b)$ there are $q,r \in S$ satisfying $a = qb$ or $a = qb+r$ and $\delta(r) < \delta(b)$.

If such a LEF on S exists, we call S a *left Euclidean (2,2)-algebra,* and e. g. a *left Euclidean seminearring* if S is a seminearring.

PROPOSITION 3.3. Let S be a (2,2)-algebra with a left absorbing zero o. If $\delta: S \to W_\delta$ is a LEF on S according to Def. 3.2, its restriction to S^* is a mapping $\gamma: S^* \to W_\gamma \subseteq W_\delta$ which satisfies (3.1) and, equivalently, (3.2). Conversely, each mapping $\gamma: S^* \to W_\gamma$ of this kind can by extended to a LEF δ' on S satisfying (3.3) by adding a value $\delta'(o) \notin W_\gamma$ to W_γ as a greatest element.

In particular, a left Euclidean ring according to Def. 3.1 is just a left Euclidean (2,2)-algebra which is also a ring.

Proof. If $\delta: S \to W_\delta$ satisfies (3.3), its restriction γ on S^* satisfies (3.2), which is clearly equivalent to (3.1) for each (2,2)-algebra S with a left absorbing zero. Conversely, let $\gamma: S^* \to W_\gamma$ satisfy (3.2). Then we extend W_γ by one element, say $\infty \notin W_\gamma$, to the well ordered set $W_{\delta'} = W_\gamma \cup \{\infty\}$ such that $w < \infty$ for all $w \in W_\gamma$, and define $\delta': S \to W_{\delta'}$ by $\delta'(s) = \gamma(s)$ for all $s \in S^*$ and $\delta'(o) = \infty$. Since $\delta'(o) > \delta'(s)$ holds for all $s \in S^*$, (3.1) for γ

implies (3.3) for δ' and all $a \in S$, $b \in S^*$. In the remaining case $a = b = o$ one has $a = ob$.

In the following, we use *"left Euclidean function"* (LEF) only with respect to Def. 3.2. For a (2,2)-algebra S with a left absorbing zero, a mapping $\gamma: S^* \to W_\gamma$ according to Prop. 3.3 will be called a *restricted* LEF henceforth. In this context we note that a similar proof as the above one provides

PROPOSITION 3.4. Let S be a (2,2)-algebra which may have a zero or not and let $T = S \cup \{o\}$ be the (2,2)-algebra obtained from S by adjoining an absorbing zero $o \notin S$. Then S is left Euclidean iff the same holds for T.

Various examples of distributive left Euclidean seminearrings (in fact those with associative addition) have been given in [8]. It is also clear that each (zero-symmetric) nearfield and each seminearfield S such that $|S^*| \geq 2$ (cf. [19]) is left Euclidean. So we present here in particular proper nearrings which are left Euclidean (Expls. 4.4 and 4.5) in the more general context of § 4. Moreover, the first four statements of § 5 can be used to obtain further examples of left Euclidean seminearrings and nearrings. So we close this section considering two special properties of left Euclidean functions needed later:

DEFINITION 3.5. Let S be a left Euclidean (2,2)-algebra and $\delta: S \to W_\delta$ a LEF on S. Then δ is called *pseudo-isotone* iff

(3.4) $\delta(b) \leq \delta(qb)$ holds for all $q, b \in S$,

and *alternative* iff for all $a, b, q_1, q_2, r \in S$

(3.5) $a = q_1 b$ excludes $a = q_2 b + r$ for $r \in S^*$ and $\delta(r) < \delta(b)$.

If $(S, \cdot)$ is associative, it is possible (and often convenient) to use only pseudo-isotone left Euclidean functions according to

PROPOSITION 3.6. Let S be a multiplicatively associative left Euclidean (2,2)-algebra and $\delta: S \to W_\delta$ any LEF on S. Then $\tilde{\delta}: S \to W_\delta$ defined by

(3.6) $\tilde{\delta}(b) = \min\{\delta(x) \mid x \in Sb \cup \{b\}\}$ for all $b \in S$

is also a LEF on S which is pseudo-isotone.

Proof. Note that $\tilde{\delta}(b) \leq \delta(b)$ holds for all $b \in S$ and that $\tilde{\delta}$ is pseudo-isotone since $Sqb \cup \{qb\} \subseteq Sb \cup \{b\}$ implies $\tilde{\delta}(b) \leq \tilde{\delta}(qb)$. So it remains to show that $\tilde{\delta}$ satisfies (3.3) and we assume $\tilde{\delta}(a) \geq \tilde{\delta}(b)$. Then we have $\delta(a) \geq \tilde{\delta}(a) \geq \tilde{\delta}(b)$ and either $\tilde{\delta}(b) = \delta(sb)$ for some

$s \in S$ or $\tilde{\delta}(b) = \delta(b)$. In the first case, we apply (3.3) for δ to a and sb and obtain $a = qsb$ or $a = qsb+r$ and $\delta(r) < \delta(sb)$. Because of $\tilde{\delta}(r) \leqq \delta(r) < \delta(sb) = \tilde{\delta}(b)$, this is (3.3) for a and b with respect to $\tilde{\delta}$. The second case follows in the same way.

REMARK 3.7. Let S be a multiplicatively associative left Euclidean (2,2)-algebra with a left absorbing zero. Then a restricted LEF γ on S is called *pseudo-isotone* iff

(3.4') $\gamma(b) \leqq \gamma(qb)$ holds for all $q, b \in S^*$ such that $qb \in S^*$.

If $\gamma\colon S^* \to W_\gamma$ is any restricted LEF on S, one obtains a pseudo-isotone restricted LEF $\tilde{\gamma}\colon S^* \to W_\gamma$ on S similarly as above by

(3.6') $\tilde{\gamma}(b) = \min\{\gamma(x) \mid x \in (Sb \cup \{b\}) \cap S^*\}$ for all $b \in S^*$.

Note that (3.4') is often included in the definition of a (left) Euclidean ring, in particular for those without zero divisors for which $qb \in S^*$ is superfluous. By the above considerations, this supplementary assumption does not restrict the concept of a (left) Euclidean ring itself.

Finally, each pseudo-isotone LEF δ on a nearring S is also alternative since $\delta(b) \leqq \delta((q_1-q_2)b) = \delta(r)$ contradicts $\delta(r) < \delta(b)$.

4. Φ-DERIVATIONS OF (LEFT EUCLIDEAN) SEMINEARRINGS

We generalize a well known procedure for nearrings as follows:

DEFINITION 4.1. Let $(S,+,\cdot)$ be a seminearring. We write $a \to a^{\eta_i}$ for each endomorphism $\eta_i \in End(S,+,\cdot)$ of $(S,+,\cdot)$ and $\eta_1 * \eta_2$ for the endomorphism given by $a \to a^{\eta_1 * \eta_2} = (a^{\eta_1})^{\eta_2}$. A mapping $\Phi\colon S \to End(S,+,\cdot)$ is called a *coupling map* of $(S,+,\cdot)$ iff

(4.1) $\Phi(a) * \Phi(b) = \Phi(a^{\Phi(b)} \cdot b)$ holds for all $a, b \in S$.

The latter and $\Phi(S) \subseteq End(S,+,\cdot)$ provide that one defines by

(4.2) $a \circ b = a^{\Phi(b)} \cdot b$ for all $a, b \in S$

again a seminearring $(S,+,\circ)$, called the *Φ-derivation* of $(S,+,\cdot)$.

One also checks that each seminearring $(S,+,\circ)$ is the Φ-derivation of a seminearring $(S,+,\cdot)$ with the trivial multiplication $a \cdot b = a$ for all $a, b \in S$, given by $a^{\Phi(b)} = a \circ b$. Further, let Φ be a coupling map of a seminearring $(S,+,\cdot)$ with a zero o. Then $o^{\Phi(a)} = o$ holds if $\Phi(a)$ is surjective, and it holds for all $\Phi(a)$ if $(S,+)$ is left or right cancellative. Moreover, $o^{\Phi(a)} = o$ for all $a \in S^*$ yields the equivalence $o \cdot S = \{o\} \Leftrightarrow o \circ S = \{o\}$.

Now let $(S,+,\circ)$ be a Φ-derivation of any seminearring $(S,+,\cdot)$. If $\Phi(a)$ is injective for all $a \in S^*$ and $(S,+,\cdot)$ is multiplicatively right cancellative, the same holds for $(S,+,\circ)$. The inverse implication is true if $\Phi(a)$ is surjective for all $a \in S^*$. Finally, let $\Phi(a)$ be bijective for all $a \in S^*$. Then $(S,+,\circ)$ is left cancellative or a seminearfield if the same holds for $(S,+,\cdot)$. (The converse implications fail to be true by the first remark behind Def. 4.1.) The proofs for these statements are more or less straightforward, but we need Thm. 4.1 (f) of [19] to show that $(S,+,\circ)$ is a seminearfield. We omit these proofs here since our main interest is with the following

THEOREM 4.2. Let $(S,+,\cdot)$ be a left Euclidean seminearring.
a) Regardless whether there is a zero, for each coupling map Φ of $(S,+,\cdot)$ such that $\Phi(b)$ is surjective for all $b \in S$, the Φ-derivation $(S,+,\circ)$ is also left Euclidean, and each LEF δ on $(S,+,\cdot)$ is also a LEF on $(S,+,\circ)$.
b) If there is a zero satisfying $S\cdot o = \{o\}$, the same statement as above holds for each coupling map Φ such that $\Phi(b)$ is surjective for all $b \in S^*$.

Proof. For all $a,b \in S$ such that $\delta(a) \geq \delta(b)$, by (3.3) there are $q,r \in S$ satisfying $a = q\cdot b$ or $a = q\cdot b + r$ and $\delta(r) < \delta(b)$. If $\Phi(b)$ is surjective, $q = q'^{\Phi(b)}$ holds for some $q' \in S$, and $q\cdot b = q'^{\Phi(b)}\cdot b = q'\circ b$ by (4.2) shows that δ is also a LEF on $(S,+,\circ)$. For b) we only have to exclude $b = o$ in these considerations, which is trivial since $q\cdot o = o$ holds by assumption and implies $q\circ o = o$.

We will use b) to give examples of left Euclidean nearrings as Φ-derivations of left Euclidean rings $(S,+,\cdot)$, since it is convenient to use the zero endomorphism for $\Phi(o)$. We further restrict ourselves to one simple procedure according to

LEMMA 4.3. Let $(S,+,\cdot)$ be a ring without zero divisors, η an automorphism of $(S,+,\cdot)$ and n a homomorphism of $(S^*,\cdot)$ into $(\mathbb{N}_0,+)$ satisfying $n(b^\eta) = n(b)$ for all $b \in S^*$. Then a coupling map of $(S,+,\cdot)$ is defined by

(4.3) $\quad a^{\Phi(b)} = a^{\eta^{n(b)}}$ if $b \in S^*$ and $a^{\Phi(o)} = o$ for all $a \in S$.

The proof is straightforward and we shall see that there are very natural realizations for those η and n.

Example 4.4. Let $S = K[x]$ be a polynomial ring over a skew field K. Then K[x] is left (and right) Euclidean (cf. [15], § 86) with respect to the restricted LEF γ given by the degree homomorphism

$n: (S^*,\cdot) \to (\mathbb{N}_0,+)$. Each linear polynomial $cx+d$ with coefficients in the centre of K provides an automorphism η of $(S,+,\cdot)$ by $k^\eta = k$ for all $k \in K$ and $f(x)^\eta = (\Sigma k_i x^i)^\eta = \Sigma k_i (cx+d)^i$, and we clearly have $n(b^\eta) = n(b)$ for all $b \in S^*$. So we can apply Lemma 4.3 and Thm. 4.2 b) and get for each automorphism η of this kind a coupling map Φ of $(S,+,\cdot)$ by (4.3) and a zero-symmetric left Euclidean nearring $(S,+,\circ)$. One easily checks that $(S,+,\circ)$ is not left distributive (provided that $cx+d \neq x$), but multiplicatively cancellative by the above remarks. We note that in a similar way $S = K[x]$ over a commutative field K was used in [5] to obtain examples for "Euclidean near domains" as considered in that paper (cf. § 1).

Example 4.5. Let $(S,+,\cdot)$ be a commutative Euclidean ring without zero divisors and with an automorphism η which is not the identical mapping, e. g. the ring of Gaussian integers $\mathbb{Z}+\mathbb{Z}i$. Each ring $(S,+,\cdot)$ of this kind has an identity (cf. Thm. 5.7) and is hence a Gaussian ring in the meaning that each $b \in S^*$ is either a unit or has an essentially unique factorization into prime elements. Denote by $n(b)$ the number of the latter in this factorization, including $n(b) = 0$ if b is a unit. This clearly defines a homomorphism n of $(S^*,\cdot)$ into $(\mathbb{N}_0,+)$ such that $n(b^\eta) = n(b)$ holds for each automorphism η of $(S,+,\cdot)$. So η yields a coupling map Φ of $(S,+,\cdot)$ by (4.3). Hence each Euclidean ring as above provides for each (non-trivial) automorphism η a left Euclidean nearring $(S,+,\circ)$ as a Φ-derivation by Thm. 4.2 b). Again $(S,+,\circ)$ is zero-symmetric and multiplicatively cancellative, but in general not left distributive. For instance, if η^2 is the identical mapping as for $S = \mathbb{Z}+\mathbb{Z}i$, the left distributive law yields by $p = p\circ((p+1)-p) = p\circ(p+1) + p\circ(-p)$ the contradiction $p^\eta = p$ for each prime element p.

5. SOME RESULTS ON LEFT EUCLIDEAN (2,2)-ALGEBRAS

We first consider the question, whether the property "left Euclidean" transfers from given (2,2)-algebras to other ones. Clearly, subalgebras of a left Euclidean (2,2)-algebra S are in general not left Euclidean (otherwise each subring of a field would be Euclidean). As a contrast we have the following transfer without any further assumptions:

THEOREM 5.1. Each homomorphic image T of a left Euclidean (2,2)-algebra S is also left Euclidean.

Proof. Let $f: S \to T$ be a homomorphism of $(S,+,\cdot)$ onto $(T,+,\cdot)$ and $\delta: S \to W_\delta$ a LEF on S. Then $\tau: T \to W_\delta$ defined by

(5.1) $\tau(f(a)) = \min\{\delta(a') \mid f(a') = f(a)\}$ for all $a,a' \in S$

is a LEF on T. To show this we consider $f(a), f(b) \in T$ satisfying $\tau(f(a)) \geqq \tau(f(b))$. Then $\tau(f(a)) = \delta(a')$ and $\tau(f(b)) = \delta(b')$ hold for suitable elements $a', b' \in S$ by (5.1) such that $f(a') = f(a)$ and $f(b') = f(b)$. From $\delta(a') \geqq \delta(b')$ it follows $a' = qb'$ or $a' = qb'+r$ and $\delta(r) < \delta(b')$ for some $q,r \in S$. This yields in T $f(a) = f(q)f(b)$ or $f(a) = f(q)f(b) + f(r)$, which together with $\tau(f(r)) \leqq \delta(r) < \delta(b') = \tau(f(b))$ proves that τ is a LEF on T.

Next we consider the direct product $T = S\times\ldots\times S'$ of a finite number of left Euclidean (2,2)-algebras $S,\ldots,S'$, also called their direct sum. We shall see that T is again left Euclidean if one of its components, say S, satisfies that

(5.2) for all $q,b \in S$ there is some $s \in S$ such that $qb = sb+b$,

another one, say S', has a left absorbing zero, and the remaining components satisfy both conditions. Note that (5.2) holds e. g. for a nearring S with a left identity e_ℓ by $s = q-e_\ell$ (or one which satisfies (6.1), cf. also Thm. 5.7). Since both, (5.2) and the existence of a left absorbing zero, transfer from any two (2,2)-algebras to their direct product, the above statement is a consequence of the following

THEOREM 5.2. Let S and S' be left Euclidean (2,2)-algebras such that S satisfies (5.2) and S' has a left absorbing zero o'. Then their direct product $T = S\times S'$ is a left Euclidean (2,2)-algebra.

Proof. Let $\delta: S \to W$ be a LEF on S, $\delta': S' \to W'$ a LEF on S' and $V = W\times W'$ the well ordered set obtained by the lexicographic order, i. e. $(u,u') < (w,w')$ iff $u < w$ in W or $u = w$ and $u' < w'$ in W'. We define

(5.3) $\tau: S\times S' \to V$ by $\tau(s,s') = (\delta(s), \delta'(s'))$.

To show that τ is a LEF on $T = S\times S'$, assume $\tau(a,a') \geqq \tau(b,b')$ for elements $(a,a'), (b,b') \in T$. Then we have $\delta(a) \geqq \delta(b)$ by (5.3), hence $a = qb$ or $a = qb+r$ and $\delta(r) < \delta(b)$ for some $q,r \in S$. If $\delta'(a') \geqq \delta'(b')$ holds, there are $q',r' \in S'$ satisfying $a'= q'b'+r'$ and $\delta(r') < \delta(b')$ if $r' \neq o'$. But the same holds for $\delta'(a') < \delta'(b')$ according to $a' = o'b'+a'$ with $r' = a'$. Now assume $a = qb$. Then we either have $(a,a') = (q,q')(b,b')$ or, using $a = qb = sb + b$ by (5.2), $(a,a') = (s,q')(b,b')+(b,r')$, where $\tau(b,r') < \tau(b,b')$ follows from $\delta'(r') < \delta'(b')$. For $a = qb + r$ we get $(a,a') = (q,q')(b,b')+(r,r')$, and $\delta(r) < \delta(b)$ implies $\tau(r,r') < \tau(b,b')$.

COROLLARY 5.3 Let $S, \ldots, S'$ be a finite number of left Euclidean nearrings such that all of them except at most one have a left identity. Then their direct product $T = S\times\ldots\times S'$ is again a left Euclidean nearring.

This follows from Thm. 5.2 and the preceding remarks. We note that these statements for commutative Euclidean rings with identities (cf. Thm. 5.7) are due to [16], but with a construction of a LEF on $T = S\times S'$ which is more complicated than (5.3).

To obtain other examples of left Euclidean algebras in this context, note at first that each (2,2)-algebra $(S,+,\cdot)$ satisfying

(5.4) $\quad S \subseteq Sb$ for all $b \in S$

is left Euclidean with any mapping $\delta: S \to W_\delta$ as a LEF. There are plenty of those algebras, e. g. each constant seminearring (cf. § 7) or each seminearfield without a zero (cf. [19]), and each left simple semigroup $(S,\cdot)$ (i. e. (5.4)) provides such a seminearring if one defines an addition e. g. by $a+b = a$ for all $a,b \in S$. Now we state:

PROPOSITION 5.4. Let S be any left Euclidean (2,2)-algebra and S' one which satisfies (5.4) and $S' \subseteq S'+S'$. Then their direct product $T = S\times S'$ is a left Euclidean (2,2)-algebra.

Proof. Let δ be a LEF on S. We show that $\tau(a,a') = \delta(a)$ defines a LEF τ on T and assume $\tau(a,a') \geqq \tau(b,b')$. Then we have $a = qb$ or $a = qb+r$ and $\delta(r) < \delta(b)$ in S. The first case is clear by (5.4), and for the second one we only need that $a' = q'b'+y'$ holds for all $a',b' \in S'$ with suitable $q',y' \in S'$. The latter clearly follows from $a' = x'+y' \in S'+S'$ and $x' = q'b'$ by (5.4).

In the second part of this section we use Prop. 3.6 to give sufficient conditions for the existence of (right) identities in left Euclidean (2,2)-algebras. We prepare them by the following

LEMMA 5.5. Let S be a multiplicatively associative left Euclidean (2,2)-algebra. Then one has

(5.5) $\quad M = \{c \in S \mid \delta(c) = \min \delta(S) \in W_\delta\} = \{c \in S \mid S \subseteq Sc\} \neq \emptyset$

for each pseudo-isotone LEF $\delta: S \to W_\delta$ of S and $(M,\cdot)$ is a subsemigroup of $(S,\cdot)$. Thus $(M,\cdot)$ contains idempotents if M is finite.

Proof. From $a = qc$ for all $a \in S$ it follows $\delta(c) \leqq \delta(a)$ by (3.4). Conversely, each $c \in S$ such that $\delta(c)$ is minimal in $\delta(S)$ satisfies $a = qc$ for all $a \in S$ by (3.3) since $\delta(r) < \delta(c)$ is impossible.

Finally, assume $c_1, c_2 \in M$. Then $c_2 = q_2c_2$ and $q_2 = q_1c_1$ for some $q_i \in S$ implies $c_2 = q_1c_1c_2$ and hence $\delta(c_1c_2) \leq \delta(c_2)$.

The proofs for the following lemma and theorem are the same as those for corresponding statements on additively associative left Euclidean semirings given in [8] (Lemma 5.1 and Thm. 5.3).

LEMMA 5.6. Let S be a multiplicatively associative left Euclidean (2,2)-algebra and M as in (5.5). Then the following statements are equivalent and satisfied if M is finite:

(i) S has a right identity,

(ii) M contains an idempotent of $(S,\cdot)$

(iii) M contains a right cancellable element of $(S,\cdot)$.

Moreover, $c \in S$ is a right identity of S iff $c = c^2 \in M$ holds, and $c \in S$ is an identity of S iff c is a central idempotent of $(S,\cdot)$ contained in M.

Note in this context that, by the remarks preceding Prop. 5.4, there are algebras S of this kind such that each element of S is a right identity (e. g. any constant nearring) as well as those without a left or right identity. The latter holds since there are left simple semigroups $(S,\cdot)$ without idempotents (cf. [3], Chap. 8).

THEOREM 5.7. Let S be a multiplicatively associative left Euclidean (2,2)-algebra which is multiplicatively cancellative (cf. § 2) or multiplicatively commutative. Then S has an identity e, and the group of units of $(S,\cdot)$ coincides with M in (5.5).

6. DIVISIBILITY IN LEFT EUCLIDEAN SEMINEARRINGS

Let S be any seminearring and $a,b \in S$. We say that b is a *right divisor* of a iff $a \in Sb$ holds. For simplicity, we denote this by $b|a$ instead of $b|_r a$. This relation on S is reflexive iff

(6.1) $s \in Ss$ holds for all $s \in S$.

We will use (6.1) as an assumption in some statements of this section. It is weaker than that to demand a left identity (cf. Thm. 5.7), since various types of semigroups $(S,\cdot)$, e. g. idempotent or regular or inverse ones (cf. [3]), satisfy (6.1). We further call $d \in S$ a *greatest common right divisor*, abbreviated by g. c. r. d., of a subset $A \subseteq S$ iff $d|a$ holds for all $a \in A$ and $t|a$ for all $a \in A$ implies $t|d$. A main point for a (left) Euclidean ring S is that the *(left) Euclidean algorithm*

$$(6.2)\quad \begin{cases} a = q_1 b + r_1 & \delta(r_1) < \delta(b) \\ b = q_2 r_1 + r_2 & \delta(r_2) < \delta(r_1) \\ \vdots & \vdots \\ r_{n-2} = q_n r_{n-1} + r_n & \delta(r_n) < \delta(r_{n-1}) \\ r_{n-1} = q_{n-1} r_n & \\ (\delta \text{ a LEF on } S \text{ and all } r_\nu \in S^*) & \end{cases}$$

yields some $r_n \in S$ as a greatest common (right) divisor of $a, b \in S$.

Now let $\delta\colon S \to W_\delta$ be a LEF on a left Euclidean seminearring S which satisfies (6.1). Then (6.2) is applicable to each pair of elements $a, b \in S$ as we may assume $\delta(a) \geqq \delta(b)$, and since each chain $\delta(b) > \delta(r_1) > \delta(r_2) > \ldots$ in a well ordered set W_δ is finite, after n steps we obtain a last $r_n \in S^*$. (If $a = q_1 b$ holds, clearly b is a g. c. r. d. of a and b.) Using (6.1) for r_n, we have

$$r_n | r_{n-1}, \quad r_n | r_{n-2} \quad \text{by} \quad r_{n-2} = (q_n q_{n-1} + x) r_n, \ldots$$

which leads to $r_n | b$ and $r_n | a$. Now we assume $t|a$ and $t|b$ for some $t \in S$. Then we get $t|r_1$, $t|r_2, \ldots, t|r_n$ if S satisfies the following *right divisor property*

$$(6.3)\qquad xt = yt + z \;\Rightarrow\; z \in St \quad \text{for all} \quad t, x, y \in S, \quad z \in S^*.$$

Since (6.3) clearly holds for each nearring S, we have shown the following

THEOREM 6.1. Let S be a left Euclidean seminearring which satisfies (6.1) and (6.3), in particular a left Euclidean nearring with a left identity. Then for each LEF δ on S and all $a, b \in S$ such that $\delta(a) \geqq \delta(b)$, any performance of the left Euclidean algorithm (6.2) provides an element $r_n \in S$ which is a g. c. r. d. of a and b. Consequently, each pair of elements $a, b \in S$ and thus each finite subset $A \subseteq S$ has a g. c. r. d. in S.

Note that in general the elements r_ν and q_ν in (6.2) as well as the number n of steps are not uniquely determined, not even in the case of rings and for a suitable choice of a LEF δ. We further state with respect to the condition (6.3):

LEMMA 6.2. Let S be an additively associative left Euclidean seminearring. Then S satisfies the right divisor property (6.3) iff there is at least one alternative LEF δ on S, and in this case each pseudo-isotone LEF on S is alternative.

Proof. Assume that δ is an alternative LEF on S and $xt = yt+z$ for some $t, x, y \in S$ and $z \in S^*$. Then $\delta(z) < \delta(t)$ is excluded by (3.5).

Hence $\delta(z) \geq \delta(t)$ implies $z = qt$ or $z = qt+r$ and $\delta(r) < \delta(t)$ for some $q \in S$ and $r \in S^*$. Now we use that $(S,+)$ is associative - for the only time in this section - to obtain $xt = yt + qt + r = (y+q)t + r$ in the second case. Since this is again excluded by (3.5), it remains $z = qt \in St$ and we have proved (6.3). For the converse we show that each pseudo-isotone LEF δ on S is alternative if (6.3) holds. By way of contradiction, assume $a = q_1 b$ and $a = q_2 b+r$, $r \in S^*$ and $\delta(r) < \delta(b)$ for suitable elements of S. Then (6.3) and (3.4) for δ yield $r = xb \in Sb$ and $\delta(b) \leq \delta(r)$.

Let $(S,+,\cdot)$ be a seminearring and $b \in S$. It is reasonable to consider the subset $A \subseteq S$ of all elements which have b as right divisor, i. e. $A = Sb$. Such a subset A obviously satisfies

(6.4) $A+A \subseteq A$ and $SA \subseteq A$.

If S is a ring, a subset A of S satisfying (6.4) is called a left ideal of S. The same notion is common if S is a semiring, and $SA \subseteq A$ is also the usual definition of a left ideal of a semigroup $(S,\cdot)$. This justifies the following

DEFINITION 6.3. A subset A of a seminearring S is called a *left ideal* of S iff A satisfies (6.4). If such a left ideal A is generated by one of its elements $b \in A$ (i. e. if A is the smallest subset of S with (6.4) which contains b), A is called a *principal left ideal* of S.

In the following, we use "left ideal" always in this meaning, in particular if S happens to be a nearring. In the latter case we speak about a *nearring left ideal* A of S iff A is a normal subgroup of $(S,+)$ and satisfies $s(t+a)-st \in A$ for all $s,t \in S$ and $a \in A$. Note that for a zero-symmetric nearring S each nearring left ideal is also a left ideal according to Def. 6.3.

REMARK 6.4. If S is a seminearring which satisfies (6.1), the principal left ideals of S are just given by $A = Sa$ for all $a \in S$, and one has $Sb \supseteq Sa$ iff $b|a$ for all $a,b \in S$.

The near by hand concept of a left principal ideal seminearring would be a seminearring S for which all left ideals are principal ones. But such a concept would be much to strong and inpracticable.

DEFINITION 6.5. Let S be a seminearring. A left ideal A of S is called *left k*-closed* iff

(6.5) $a_1+x = a_2$ for some $a_i \in A$, $x \in S^*$ implies $x \in A$,

and S is called a *left principal ideal seminearring* iff each left k*-closed left ideal A of S is a principal left ideal of S.

REMARK 6.6. Let S be a seminearring satisfying (6.1). Then, obviously, all principal left ideals $A = Sa$ of S are left k*-closed iff S satisfies the right divisor property (6.3).

THEOREM 6.7. Let S be a left Euclidean seminearring. Then S is a left principal ideal seminearring, and each left k*-closed left ideal A of S satisfies $A = Sb$ for some $b \in A$.

Proof. Let A be any left ideal of S which is left k*-closed, and $\delta: S \to W_\delta$ a LEF on S. The subset $\delta(A)$ of W_δ has a minimal element, say $\delta(b)$ for some $b \in A$. Then one has $\delta(a) \geqq \delta(b)$ for each $a \in A$, hence $a = qb$ or $a = qb+r$ and $\delta(r) < \delta(b)$ for some $q \in S$, $r \in S^*$. Since $r \in A$ holds by (6.5), the latter case is impossible. This yields $A \subseteq Sb \subseteq A$.

THEOREM 6.8. Let S be a left principal ideal seminearring satisfying (6.1) and (6.3). Then each proper chain $A_1 \subset A_2 \subset A_3 \subset \ldots$ of left k*-closed left ideals of S is finite. Hence S satisfies the chain condition for right divisors, i. e. each chain $a_1, a_2, a_3, \ldots$ in S such that $a_\nu | a_{\nu-1}$ and $a_{\nu-1} \nmid a_\nu$ hold is finite.

Proof. Due to Remarks 6.4 and 6.6, both statements are equivalent by $A_\nu = Sa_\nu$ and $a_\nu | a_{\nu-1} \Leftrightarrow Sa_\nu \supseteq Sa_{\nu-1}$. Now we assume that there are infinite chains of this kind. Then the union $A = \cup\{A_\nu \mid \nu \in \mathbb{N}\}$ is clearly again a left k*-closed left ideal of S. Hence $A = Sa$ holds for some $a \in A$, which is contained in one of the A_ν, say A_n. This yields $a = xa_n$ and $a_{n+1} = ya$ for some $x, y \in S$, hence the contradiction $a_n | a_{n+1}$.

As a supplement to our statement on nearrings in Thm. 6.1 we obtain from the last both theorems:

COROLLARY 6.9. Let S be a left Euclidean nearring which satisfies (6.1), e. g. one with a left identity. Then each proper chain of right divisors in S as described above is finite.

7. CONSTANT AND ZERO-SYMMETRIC SUBSEMINEARRINGS OF LEFT EUCLIDEAN SEMINEARRINGS AND NEARRINGS

In this section we only consider seminearrings $(S,+,\cdot)$ such that $(S,+)$ is associative and has a zero.

LEMMA 7.1. Let S be an additively associative left Euclidean seminearring with a zero o. Then $o^2 = o$ holds and for each pseudo-isotone LEF on S we have the equivalence

(7.1) $\delta(b) \leqq \delta(o) \Leftrightarrow ob = o$ for all $b \in S$.

Proof. If $\delta(b) \leqq \delta(o)$ holds, by (3.3) there are $q, r \in S$ satisfying $o = qb+r$, where r may be the zero or not. This yields $ob = o$ and thus $o^2 = o$ by $o = (o+q)b + r = ob + qb + r = ob + o = ob$. The other implication of (7.1) is immediate by (3.4).

The following theorem applies in particular to each left Euclidean nearring S with a left identity and its constant and zero-symmetric subnearrings S_1 and S_2. We use the generalization of the latter concepts for a seminearring S as introduced in § 4 of [20].

THEOREM 7.2. Let S be an additively associative left Euclidean seminearring with a left absorbing zero o, S_1 its constant and S_2 its zero-symmetric subseminearring. Assume that S_2 (or S) satisfies (6.1), and that $S = S_1+S_2$ holds or that $(S_1,+)$ does not contain a bicyclic subsemigroup (cf. [3], 1.12). Then both, S_1 and S_2, are left Euclidean with respect to the restrictions of any pseudo-isotone LEF δ on S.

Proof. Each pseudo-isotone LEF δ on S satisfies

(7.2) $\delta(s) \leqq \delta(o) = \delta(t_1)$ for all $s \in S$ and $t_1 \in S_1$,

due to (7.1) and $t_1 o = t_1$, which yields $\delta(o) \leqq \delta(t_1)$. Moreover, each constant seminearring S_1 is always left Euclidean as already stated as a consequence of (5.4). By assumption, we have $a_2 \in S_2 a_2$ or $a_2 \in Sa_2$, say $a_2 = xa_2$, for each $a_2 \in S_2$. Then

(7.3) $\delta(a_2) \leqq \delta(s_1+a_2+t_1)$ holds for all $a_2 \in S_2$ and $s_1, t_1 \in S_1$

because of $(s_1+x+t_1)a_2 = s_1a_2 + xa_2 + t_1a_2 = s_1 + a_2 + t_1$. To show that S_2 is left Euclidean, assume $\delta(a_2) \geqq \delta(b_2)$ for some $a_2, b_2 \in S_2$. Due to $S = S_1+S_2$, by (3.3) for δ and S there are q_1+q_2 and r_1+r_2 which satisfy $a_2 = (q_1+q_2)b_2$ or $a_2 = (q_1+q_2)b_2 + r_1 + r_2$ and $\delta(r_2) \leqq \delta(r_1+r_2) < \delta(b_2)$, the latter by (7.3). In the first case, we multiply $a_2 = q_1b_2 + q_2b_2$ from the right by o and get $q_1 = o$, hence $q_1+q_2 \in S_2$. In the second case we have $a_2 = q_1b_2 + q_2b_2 + r_1 + r_2$. Again by $S = S_1+S_2$ we have $q_2+r_1 = s_1+s_2$ for some $s_i \in S_i$, which yields multiplied by o from the right $r_1 = s_1$. We apply this to $q_2b_2 + r_1 = q_2b_2 + r_1b_2$ and obtain $q_2b_2 + r_1 = r_1 + s_2b_2$. As above, from $a_2 = q_1b_2 + r_1 + s_2b_2 + r_2$ it follows $q_1b_2 + r_1 = q_1 + r_1 = o$. So it remains $a_2 = s_2b_2 + r_2$ as we were to show.

Without the assumption $S = S_1+S_2$, we merely have $a_2 = qb_2$ or $a_2 = qb_2 + r$ and $\delta(r) < \delta(b_2)$ for suitable $q, r \in S$. The first case yields $o = a_2o = qb_2o = qo$ and hence $q \in S_2$. In the second one we obtain $qo+ro=o$. This yields either $ro+qo=o$, or qo and ro generate a bicyclic subsemigroup (which contains an infinite set of idempotents) of $(S,+)$ according to [3], Lemma 1.31. If the latter is excluded by assumption, we go on with

$$a_2 = qb_2 + ro + qo + r = (q+ro)b_2 + qo + r.$$

From $(q+ro)o=o$ it follows $q+ro \in S_2$, and likewise $qo+r \in S_2$. So it remains to prove $\delta(qo+r) < \delta(b_2)$, which follows again by (7.3) from $\delta(qo+r) \leqq \delta(ro+qo+r) = \delta(r)$ and $\delta(r) < \delta(b_2)$.

<u>THEOREM 7.3.</u> Let S be a nearring with an identity, S_1 its constant and S_2 its zero-symmetric subnearring. Then S is left Euclidean iff S_2 is and S satisfies that

(7.4) for all $b_2, q_2 \in S_2$ and $b_1 \in S_1$ there exists an $x_2 \in S_2$ such that $x_2(b_1+b_2) = x_2b_1 + q_2b_2$.

In particular, (7.4) holds if S_1 is a nearring ideal of S.

<u>Proof.</u> Assume that S is left Euclidean. Since S has an identity e, we can apply Thm. 7.2 and obtain that S_2 is left Euclidean. Further,

(7.5) $\delta(a) = \delta(s_1+a+t_1)$ holds for all $a \in S$ and $s_1, t_1 \in S_1$,

which follows from $a \in Sa$ as (7.3) since $(S,+)$ is a group. It is well known that $e=e_2$ belongs to S_2, and $\delta(e_2)$ is minimal in $\delta(S)$ for each pseudo-isotone LEF δ on S due to Lemma 5.6. The same holds for $\delta(b_1+e_2)$ by (7.5). So for each $a_1+a_2 \in S = S_1+S_2$ there is some $x_1+x_2 \in S$ satisfying $a_1+a_2 = (x_1+x_2)(b_1+e_2) = x_1 + x_2(b_1+e_2)$. Since we have $x_2(b_1+e_2) = s_1+s_2 \in S_1+S_2$, we obtain $s_1 = x_2b_1$ and $s_2 = a_2$, in particular for $a_2 = q_2$

$$x_2(b_1+e_2) = x_2b_1+q_2, \text{ hence } x_2(b_1+b_2) = x_2(b_1+e_2)b_2 = x_2b_1 + q_2b_2.$$

This shows (7.4). Conversely, let $\delta: S_2 \to W_\delta$ be any LEF on S_2. We extend δ to $\tau: S \to W_\delta$ by $\tau(a_1+a_2) = \delta(a_2)$ for all $a_1+a_2 \in S_1+S_2 = S$, and assume $\tau(a_1+a_2) \geqq \tau(b_1+b_2) = \delta(b_2)$. Then there are $q_2, r_2 \in S_2$ satisfying $a_2 = q_2b_2+r_2$ and $\delta(r_2) < \delta(b_2)$ if $r_2 \neq o$. We choose $x_2 \in S$ according to (7.4) and $x_1 \in S_1$ such that $a_1 = x_1 + x_2b_1$ holds. Then we obtain $(x_1+x_2)(b_1+b_2)+r_2 = x_1 + x_2b_1 + q_2b_2 + r_2 = a_1 + a_2$ and $\tau(r_2) = \delta(r_2) < \delta(b_2) = \tau(b_1+b_2)$ if $r_2 \neq o$. This proves that τ is a LEF on S. Now let S_1 be a nearring (left) ideal of S and b_2, q_2, b_1 as in (7.4). Then $q_2(b_1+b_2) - q_2b_2 \in S_1$ yields $q_2(b_1+b_2) = s_1 + q_2b_2$ for some $s_1 \in S$. Multiplying by o from the right we get $q_2b_1 = s_1$. So we even have shown $q_2(b_1+b_2) = q_2b_1 + q_2b_2$, which clearly is more than (7.4).

REFERENCES

[1] Albis-Gonzáles, V. S. and R. K. Markanda, Rings of fractions of euclidean rings, Comm. Algebra 6 (1978), 353 - 360.

[2] Brungs, H. H., Left Euclidean Rings, Pacific J. Math. 45 (1973), 27 - 33.

[3] Clifford, A. H. and G. B. Preston, The algebraic theory of semigroups, Amer. Math. Soc. Vol. I (1961), Vol. II (1967).

[4] Graves, J. A. and J. J. Malone, Embedding near domains, Bull. Austral. Math. Soc. 9 (1973), 33 - 42.

[5] Graves, J. A. and J. J. Malone, Euclidean near domains, Manuscript 1975.

[6] Hebisch,U., (2,2)-Algebren mit euklidischen Algorithmen, Diss. TU Clausthal, Germany 1984.

[7] Hebisch,U., Left euclidean functions on (2,2)-algebras, to appear.

[8] Hebisch,U. and H. J. Weinert, On Euclidean semirings, to appear.

[9] Jacobson, N., A note on non-commutative polynomials, Ann. of Math. 35 (1934), 209 - 210.

[10] Jategaonkar, A. V., Left Principal Ideal Rings, Lecture Notes in Mathematics 123, Springer, Berlin-Heidelberg-New York 1970.

[11] Motzkin, Th., The Euclidean algorithm, Bull. Amer. Math. Soc. 55 (1949), 1142 - 1146.

[12] Ore, O., Theory of non commutative polynomials, Ann. of Math. 34 (1933), 480 - 508.

[13] Pilz, G., Nearrings, North-Holland Publ. Comp. 1977.

[14] Pollák, G., Mengentheoretische Betrachtung der euklidischen und Hauptidealringe, Magyar Tud. Akad. Mat. Kutató Int. Közl. 7 (1962), 323 - 333.

[15] Rédei, L., Algebra I, Akadémiai Kiadó, Budapest 1954, Leipzig 1959, London 1967.

[16] Samuel, P., About Euclidean Rings, J. Algebra 19 (1971), 282 - 301.

[17] Wedderburn, J. H. M., Non-commutative domains of integrity, J. Math. 167 (1932), 129 - 141.

[18] Weinert, H. J., Multiplicative cancellativity of semirings and semigroups, Acta Math. Acad. Sci. Hungar. 35 (1980), 335 - 338.

[19] Weinert, H. J., Seminearrings, seminearfields and their semigrouptheoretical background, Semigroup Forum 24 (1982), 231 - 254.

[20] Weinert, H. J., Partially and fully ordered seminearrings and nearrings, this proceedings.

Institut für Mathematik
Technische Universität Clausthal
Erzstraße 1
D-3392 Clausthal-Zellerfeld

Near-rings and Near-fields, G. Betsch (editor)

IDEALS AND REACHABILITY IN MACHINES

Gerhard HOFER

Institut für Mathematik, Johannes-Kepler-Universität Linz
A-4040 Linz, Austria

ABSTRACT

It is well-known that semigroups are very useful in the study of automata. These semigroups consist of mappings from the state set Q of the automaton into itself. If, as it is often the case, Q bears the structure of a module, we are studying mappings of a module into itself. The natural algebraic tool for that is the structure of a near-ring, called the syntactic near-ring of the automaton. In this paper we try to show how ideals of the zerosymmetric part of syntactic near-rings can be used for determining reachability in automata.

We start with the definition of (semi-) automata. See e.g. [2], [7] or [10] for the theory of these creatures.

Definition 1. A *semiautomaton* is a triple $\mathcal{A} = (Q,A,F)$ where Q and A are sets (called the *state* and *input set*) and F is a function from $Q \times A$ into Q (called the *state-transition function*). If Q and A are R-modules (with the same R) and F is an R-homomorphism we call $\mathcal{A}$ a module-semiautomaton and abbreviate this by MSA and $\mathcal{A} = (Q,A,F)_R$.

For any semiautomaton $\mathcal{A} = (Q,A,F)$ we get a collection of mappings $F_a : Q \to Q$, one for each $a \in A$, which are given by

$$F_a(q) := F(q,a)$$

If the input $a_1 \in A$ is followed by the input a_2, the semiautomaton "moves" from the state $q \in Q$ first into $F_{a_1}(q)$ and then into $F_{a_2}(F_{a_1}(q))$. If we extend (as usual) A to the free monoid A^* over A (consisting of all finite sequences of elements of A, including the empty sequence Λ), we therefore get

$$F_{a_1a_2} = F_{a_2}F_{a_1}$$

moreover $M(\mathcal{A}) := \{F_{\bar{a}} \mid \bar{a} \in A^*\}$ (with composition of maps) is a monoid with identity $F_\Lambda = id_Q$ (called the *syntactic monoid* of $\mathcal{A}$). In the case of MSA's we are also able to study the superposition $F_{a_1} + F_{a_2}$ (defined pointwisely) of two "simultaneous" inputs $a_1, a_2 \in A$. Hence it is natural to consider $\{F_a \mid a \in A\} \cup \{F_\Lambda\}$ and all its sums and products (= composition of maps). The obvious framework for that is the structure of a near-ring:

Definition 2. A (right) *near-ring* is a triple (N,+,.), where (N,+) is a group (not necessarily abelian), (N,.) is a semigroup and $(a+b)c = ac + bc$ for all $a,b,c \in N$.

Prominent examples of near-rings are (w.r.t. pointwise addition and composition of mappings) the following

Examples. Let G be an additive group with identity 0 and A an R-module. Near-rings are e.g.

$M(G) := \{f:G \to G\}$

$M_0(G) := \{f \in M(G) \mid f(0) = 0\}$

$M_c(G) := \{f \in M(G) \mid f \text{ constant}\}$

$M_{aff}(A) := \{f:A \to A \mid f \text{ affine map}\}$

Thus near-rings are generalized rings and most suitable to study mappings on groups. See [12] for the theory of near-rings. Since near-rings are Ω-groups, there is no need for defining homomorphisms, ideals, etc..

We need a few more concepts for near-rings. Let N be a near-ring. $n \in N$ is called *zero-symmetric* if $n0 = 0$, *constant* if $n0 = n$ and *distributive* if $n(a+b) = na + nb$ for all $a,b \in N$. Let N_0, N_c, N_d denote the sets of all zero-symmetric (constant, distributive, respectively) elements. Note that $N_d \subseteq N_0$. It is well-known that $N_0 \cup N_c$ generates (N,+). If (N,+) is abelian and $N_0 = N_d$ then N is called *abstract affine near-ring* (abbr.: *a.a.n.r.*). $M_{aff}(A)$ is of this type.

The linear case. Let $A = (Q,A,F)_R$ be a MSA. Because of

$$F_a(q) = F(q,a) = F((q,0) + (0,a)) = F(q,0) + F(0,a) = F_0(q) + F_a(0)$$

we get $F_a = F_0 + \bar{F}_a$, where F_0 is an R-homomorphism, while $\bar{F}_a$ is the map with constant value $F_a(0)$. By induction one sees that

$$F_{a_1 a_2 \ldots a_n} = F_0^n + \sum_{i=1}^{n} F_0^{n-i}(F_{a_i}(0)) \qquad \text{(see [5] or [6])}$$

where the sum denotes a constant map. Each power F_0^n is an R-homomorphism, hence M(A) is a subset of $M_{aff}(Q)$ and the next definition comes in a natural way (see [5] or [6]).

Definition 3. Let $A = (Q,A,F)_R$ be a MSA. The subnear-ring N(A) of $M_{aff}(Q)$ generated by M(A) is called the *syntactic near-ring* of A.

Thus N(A) is an a.a.n.r. with identity. In the following, <X> indicates the group generated by X. The zerosymmetric part, which we need in the following, is given by

$N_0(A) := (N(A))_o = \langle id_Q, F_o \rangle = \{z_0 id_Q + z_1 F_0^1 + \ldots + z_n F_0^n \mid z_i \in \mathbb{Z}, n \in \mathbb{N}_0\}$

hence $N_0(A)$ is a commutative ring with identity id, which is generated by $\{id, F_0\}$ (see [5]).

We motivate the next definition.

Example. Let $G = (g_{ij}) \in (\mathbb{Z}_n)_2^2$ (= the collection of all 2×2-matrices over $\mathbb{Z}_n$), where at least one g_{ij} $(i \neq j)$ is invertible. By Cayley-Hamilton and our assumption on G it follows that $\langle E,G \rangle = \{z_0 E + z_1 G \mid z_0, z_1 \in \mathbb{Z}_n\}$ has cardinality n^2 (E is the identity matrix). Let $A = (Q,A,F)_R$ be a MSA where R is commutative and Q a finite dimensional unitary free R-module, then F_0 corresponds to a matrix A_{F_0} (see [1], [9] or [13]), thus $N_0(A) = \langle E, A_{F_o} \rangle$. Since there seems to be no hope to get explicite results about reachability in general, we restrict our considerations to the following special situations.

Definition 4. Let R be a ring with identity 1, char $R = n \in \mathbb{N}$. If $R = \langle 1,r \rangle$ where $r^2 = \alpha 1 + \beta r$ for some $\alpha, \beta \in \mathbb{Z}$, ord $r = n$ and $\langle 1 \rangle \cap \langle r \rangle = \{0\}$ then R is called a *syntactic-ring* (abbr.: *s.r.*) with *generator* r and *syntactic triple* (abbr.: *s.tr.*) $(\alpha,\beta;n) \in \mathbb{Z} \times \mathbb{Z} \times \mathbb{N}$. We abbreviate this situation by $R(\alpha,\beta;n)$.

In the following, we write z instead of z1 (for $z \in \mathbb{Z}$).

Remark 1. Let R be a syntactic ring with generator r.
$R = \{z_0 + z_1 r \mid z_0, z_1 \in \mathbb{Z}\}$ has cardinality n^2 and R is commutative. The map $z_0 + z_1 r \to (z_0, z_1)$ is a group isomorphism between $(R,+)$ and $(\mathbb{Z}_n \times \mathbb{Z}_n, +)$.

In the following, we study the structure of syntactic rings and try to apply the results in order to describe "reachability" in machines.

Theorem 1. Let $R(\alpha,\beta;p)$ be a syntactic ring with generator r and p prime. Consider $x^2 + \beta x - \alpha = 0$, $y^2 = 4\alpha + \beta^2$ as equations over $\mathbb{Z}_p$ (abbr. by (*),(+)), then the following hold:

a) $\langle 1 \rangle$ is never an ideal of R; $\langle z+r \rangle$ is an ideal iff z is a solution of (*).

b) $(R,+)$ has p+1 nontrivial subgroups, namely $\langle 1 \rangle$ and $\langle z+r \rangle$ for $z \in \{0,\ldots,p-1\}$.

c) (*) has

no solution iff R is isomorphic to $GF(p^2)$

one solution z iff $\langle z+r \rangle$ is the unique nontrivial ideal in R

two solutions z_1, z_2 iff $\langle z_1+r \rangle$, $\langle z_2+r \rangle$ are the only nontrivial ideals in R.

d) If $p \geq 3$ then we get: (+) has
no solution iff R is a field, isomorphic to $GF(p^2)$
one solution, namely 0, iff $<(\frac{p-1}{2}\beta)+r>$ is the unique nontrivial ideal in R
two solutions $\pm\tilde{y}$ iff $<(\frac{p-1}{2}(\beta+\tilde{y}))+r>$, $<(\frac{p-1}{2}(\beta-\tilde{y}))+r>$ are the only nontrivial ideals in R.
e) R has at most two nontrivial ideals.

Proof. a) $<z+r>$ is an ideal in $R \Leftrightarrow \forall a,b \in \mathbb{Z}$:
$(a+br)(z+r) = (za+b\alpha)+(bz+b\beta+a)r \in <z+r> \Leftrightarrow \forall a,b \in \mathbb{Z}$:
$za+b\alpha = zb(z+\beta)+za \Leftrightarrow \alpha = z(z+\beta) \Leftrightarrow z^2+\beta z-\alpha = 0$
$<1>$ is not in ideal in R since $r.1 = r \notin <1>$.
b) $(R,+)$ is isomorphic to $(\mathbb{Z}_p \times \mathbb{Z}_p,+)$ /Remark 1), hence a 2-dimensional vector space over $\mathbb{Z}_p$. This proves b).
c) and e) are trivial consequences of a)
d) follows by c) and elementary transformations of (*) (see [3] and [4]).

Remark 2. Later we will see that, with the notation of Theorem 1, $<z+r>$, $z \in \mathbb{Z}$, are the "essential" subgroups of R for the applications. Hence the following definitions are explicable.

Definition 5. Let p prime. $<(z,1)>$ for each integer $z \in \{1,\dots,p\}$, and the trivial subgroups of $\mathbb{Z}_p \times \mathbb{Z}_p$ are called *important subgroups*.

For a generalization we need the following

Notation. Let $R(\alpha,\beta;n)$ be a syntactic ring, where $n = \prod_{i \in A} p_i$ is squarefree. Then

$T_1(\alpha,\beta;n) := \{i \in A \mid x^2+\beta x-\alpha = 0$ has no solution over $\mathbb{Z}_{p_i}\}$,
$T_2(\alpha,\beta;n) := \{i \in A \mid x^2+\beta x-\alpha = 0$ has exactly one solution z_i over $\mathbb{Z}_{p_i}\}$,
$T_3(\alpha,\beta;n) := \{i \in A \mid x^2+\beta x-\alpha = 0$ has two solutions z_{1i}, z_{2i} over $\mathbb{Z}_{p_i}\}$.

We abbreviate, if no confusions can arise, $T_1(\alpha,\beta;n)$ by T_1, etc..
$h_1: (z_0+z_1 r) \to (z_0,z_1)$ denotes the group isomorphism between R and $(\mathbb{Z}_n)^2$,
$h_2: (z_0,z_1) \to ((z_0,z_1),\dots,(z_0,z_1))$ the group isomorphism between $(\mathbb{Z}_n)^2$ and $\bigoplus_{i \in A} (\mathbb{Z}_{p_i})^2$ (Chinese-Remainder-Theorem).

Definition 6. Let $R(\alpha,\beta;n)$ be a syntactic ring, where $n = \prod_{i \in A} p_i$ is squarefree and $U \leq (R,+)$. U is called a *useful subgroup* if all H_i's of $\bigoplus_{i \in A} H_i := h_2(h_1(U))$ $(\leq \bigoplus_{i \in A} (\mathbb{Z}_{p_i})^2)$ are important subgroups.

Now we have the ingredients for the next Theorem.

Theorem 2. Let $R(\alpha,\beta;n)$ be a syntactic ring with generator r, where $n = \prod_{i\in A} p_i$ is squarefree and $U \leqq (R,+)$. The the following hold:

a) U is cyclic if and only if $|U|$ divides n.

b) All useful subgroups of R are given by $<\prod_{i\in T} p_i> \dotplus <\prod_{j\in S} p_j(z+r)>$ where $S \subseteq T \subseteq A$, $z \in \{1,\ldots \prod_{l\in T\setminus S} p_l\}$.

Proof. $h_1(U)$ is a subgroup of $(\mathbb{Z}_n)^2$ and isomorphic (by h_2) to $\bigoplus_{i\in A} H_i$, where H_i is a subgroup of $(\mathbb{Z}_{p_i})^2$. U and $h_1(U)$ are cyclic iff each H_i is cyclic. H_i is cyclic iff $|H_i| \leqq p_i$, hence a) is proved. With the abbreviations $t := \prod_{i\in T} p_i$, $s := \prod_{j\in S} p_j$ for $T \subseteq A$ and $S \subseteq T$, we get for $z \in \mathbb{Z}$:

$$h_1(<t> + <s(z+r)>) = <(t,o)> + <s(z,1)> =: H.$$

By the last expression one can easily see that the sum is direct.

If $h_2(H) =: \bigoplus_{i\in A} H_i$ then

$H_k := <(0,0)> \dotplus <(0,0)>$ for $k \in S$,

$H_l := <(0,0)> \dotplus <s(z,1)> = <(z,1)>$ for $l \in T\ S$

$H_m := <(t,0)> \dotplus <s(z,1)> = <(1,0)> \dotplus <(0,1)> = \mathbb{Z}_{p_m} \times \mathbb{Z}_{p_m}$ for $m \notin T$.

Hence $<t> \dotplus <s(z+r)>$ is a useful subgroup.

If z runs through all integers of $\{1,\ldots,a\}$ where $a := \prod_{l\in T\setminus S} p_l$, then, for $l \in T\setminus S$, the H_l's run through all nontrivial important subgroups of $(\mathbb{Z}_{p_l})^2$. The reason is the isomorphism $(z,1) \to ((z,\ldots,z),(1,\ldots,1))$ between $(\mathbb{Z}_a)^2$ and $(\bigoplus_{l\in T\setminus S} \mathbb{Z}_{p_l})^2$ (Chinese-Remainder-Theorem). Hence we get all useful subgroups of R if T runs through all subsets of A, S through all subsets of T and z through all integers of $\{1,\ldots,\prod_{l\in T\setminus S} p_l\}$.

Corollaries. Let $R(\alpha,\beta;n)$ be a syntactic ring with generator r, where $n = \prod_{i\in A} p_i$ is squarefree. U is an ideal in R iff each H_i of $\bigoplus_{i\in A} H_i := h_2(h_1(U))$ is an ideal in $(\mathbb{Z}_{p_i})^2$. Hence, by Theorem 1 and b) of Theorem 2, all ideals of R are now explicitely known (for that, there are only to solve some quadratic equations $x^2 + \beta x - \alpha = 0$ over $\mathbb{Z}_{p_i}$, $i \in A$). Note that $A = T_1 \dot\cup T_2 \dot\cup T_3$. In particular we get:

a) All cyclic subgroups of R which are ideals are given by $<\prod_{i\in S} p_i(z+r)>$, where $T_1 \subseteq S \subseteq T = A$ and $z \equiv \begin{cases} z_j \pmod{p_j} & \text{for } j \in T_2\ S \\ z_{1k} \text{ or } z_{2k} \pmod{p_k} & \text{for } k \in T_3\ S \end{cases}$

b) If $T_1 = A$, then all ideals of R are given by $<\prod_{i\in T} p_i> \dotplus <(\prod_{j\in T} p_j)r>$ where $T \subseteq T_1$.

c) If $T_2 = A$, then all ideals of R are given by $<\prod_{i\in T} p_i> \dot{+} <\prod_{j\in S} p_j(z+r)>$ where $S \subseteq T \subseteq T_2$ and z is an integer such that $z \equiv z_k \pmod{p_k}$ for $k \in T_2$.

Now we give some applications for both theorems. First of all (compare [11]) let $A = (Q,A,F)_R$ be a MSA. If $M_c(Q) \subseteq N(A)$ (which is often the case) then clearly for all $q,q' \in Q$ there is some $F_\alpha \in N(A)$, $\alpha = a_1 a_2 \dots a_n \in A^*$, such that $F_\alpha(q) = q'$. But this says very little about A, because this F_α can be chosen to be the constant map with value q'. In some sense $N_0(A)$ represents the "autonomous part" of A, which is defined and studied for linear automata (cf. [11]).

Definition 7. Let $A = (Q,A,F)_R$ be a MSA, U a subset of Q. U is called *reachable from* q if U is a subset of $N_0(A)q$. U is *reachable* if U is a subset of the intersection of all $N_0(A)q$, $q \in Q \setminus \{0\}$. A is called *strictly connected* if Q is reachable.

Notation. Let (G,+) any (abelian) group, $g \in G$, $n \in \mathbb{N}$, then $\mathbb{Z}_n g := \{0,g,\dots,(n-1)g\}$.

Theorem 3. Let $A = (Q,A,F)_R$ be a MSA where $N_0 := N_0(A)$ is a s.r. with generator F_0 and s.tr. $(\alpha,\beta;p_1)$, p_1 prime. Then the following hold:
a) If $T_1 = \{1\}$ then $N_0 q = <q> \dot{+} <F_o(q)>$ for all $q \in Q$.
b) If $T_2 = \{1\}$ then $N_0 q = <q>$ for all $q \in \mathrm{Ker}(z_1 \mathrm{id} + F_0)$, $N_0 q = <q> \dot{+} <F_o(q)>$ otherwise.
c) If $T_3 = \{1\}$ then $N_0 q = <q>$ for all $q \in \mathrm{Ker}(z_{11}\mathrm{id} + F_0) \cup \mathrm{Ker}(z_{21}\mathrm{id} + F_0)$, $N_0 q = <q> \dot{+} <F_o(q)>$ otherwise.

Proof. Consider the map $h_q: N_0 \to N_0 q$, $n_0 \to n_0 q$. h_q is an N_0-epimorphism (note that Q is also a N_0-module). Furthermore $n_0 \in \mathrm{Ann}\, q$ if and only if $q \in \mathrm{Ker}\, n_0$. Clearly $N_0 q = <q> + <F_o(q)>$. Now we use mainly b) of Theorem 1.

If $T_1 = \{1\}$ then N_0 is a field and a) is proved.

If $T_2 = \{1\}$ then $<z_1 \mathrm{id} + F_o>$ is the unique proper ideal in N_0, hence N_0 is a local ring. Thus exactly the elements of $<z_1\mathrm{id} + F_o>$ are not bijective; moreover, they have the same kernel $(z(z_1\mathrm{id} + F_0)(q) = 0$ where $z \in \mathbb{Z}$, $z \neq 0$, implies $(z_1\mathrm{id} + F_0)(q) = 0$ since z^{-1} exists in $(\mathbb{Z}_{p_1}, .)$, the converse direction is trivial). If $q \in \mathrm{Ker}(z_1\mathrm{id} + F_0)$ then $F_0(q) = -z_1 q$, hence $N_0 q \subseteq <q>$. But $<q> = \mathbb{Z}_{p_1} q$ and p_1 is prime, thus $N_0 q = <q>$. If $q \notin \mathrm{Ker}(z_1\mathrm{id} + F_0)$ then Ann q is $\{0\}$ and b) is proved.

If $T_3 = \{1\}$ then $<z_{11}\mathrm{id} + F_o>$, $<z_{21}\mathrm{id} + F_o>$ are the unique proper ideals in N_0. The isomorphic images $h_1(<z_{11}\mathrm{id} + F_o>)$, $h_1(<z_{21}\mathrm{id} + F_o>)$ are given by $<(z_{11},1)>$, $<(z_{21},1)>$. Since $z_{11} \neq z_{21}$, we get $\mathbb{Z}_{p_1} \times \mathbb{Z}_{p_1} = <(z_{11},1)> \dot{+} <(z_{21},1)>$, hence

$N_o = \langle z_{11}\mathrm{id} + F_o\rangle \dot{+} \langle z_{21}\mathrm{id} + F_o\rangle$. The ideal generated by an element $\neq 0$ of N_o is either $\langle z_{11}\mathrm{id} + F_o\rangle$ or $\langle z_{21}\mathrm{id} + F_o\rangle$ or N_o itself. Since N_o is a commutative ring with identity, exactly the elements of $\langle z_{11}\mathrm{id} + F_o\rangle \cup \langle z_{21}\mathrm{id} + F_o\rangle$ are not invertible, that means not bijective. The rest follows similar to b).

Remark 3. With the same assumption and notation as in Theorem 3, the following is also valid. If $T_3 = \{1\}$, then $N_o = \langle z_{11}\mathrm{id} + F_o\rangle \dot{+} \langle z_{21}\mathrm{id} + F_o\rangle$ hence $(z_{11}\mathrm{id} + F_o) \circ (z_{21}\mathrm{id} + F_o) = (z_{21}\mathrm{id} + F_o) \circ (z_{11}\mathrm{id} + F_o) = 0$. Thus $\mathrm{Im}(z_{11}\mathrm{id} + F_o)$ is a submodule of $\mathrm{Ker}(z_{21}\mathrm{id} + F_o)$ and $\mathrm{Im}(z_{21}\mathrm{id} + F_o)$ is one of $\mathrm{Ker}(z_{11}\mathrm{id} + F_o)$. Since $\mathrm{Ann}\, q \neq N_o$ for all $q \neq 0$, the intersection of $\mathrm{Ker}(z_{11}\mathrm{id} + F_o)$ with $\mathrm{Ker}(z_{21}\mathrm{id} + F_o)$ is trivial. If Q is finite or a finite-dimensional vector space, then $\mathrm{Ker}(z_{11}\mathrm{id} + F_o) = \mathrm{Im}(z_{21}\mathrm{id} + F_o)$, $\mathrm{Ker}(z_{21}\mathrm{id} + F_o) = \mathrm{Im}(z_{11}\mathrm{id} + F_o)$, respectively. The last assertion needs an explanation. If $|Q| = n \in \mathbb{N}$, then $\mathrm{Ker}(z_{21}\mathrm{id} + F_o) = \mathrm{Im}(z_{11}\mathrm{id} + F_o)$ follows by
$n = |\mathrm{Ker}(z_{11}\mathrm{id} + F_o)|\ |\mathrm{Im}(z_{11}\mathrm{id} + F_o)| \leq |\mathrm{Ker}(z_{11}\mathrm{id} + F_o)|\ |\mathrm{Ker}(z_{21}\mathrm{id} + F_o)| \leq n$.
The equality holds because of the homomorphism theorem, the first inequality because of $\mathrm{Im}(z_{11}\mathrm{id} + F_o) \leq \mathrm{Ker}(z_{21}\mathrm{id} + F_o)$, the second because the sum of both kernels is direct, together with the homomorphism theorem. If Q is a finite-dimensional vector space then the assertion follows similarly by
$\dim Q = \dim \mathrm{Ker}(z_{11}\mathrm{id} + F_o) + \dim \mathrm{Im}(z_{11}\mathrm{id} + F_o) \leq \dim \mathrm{Ker}(z_{11}\mathrm{id} + F_o) +$
$+ \dim \mathrm{Ker}(z_{21}\mathrm{id} + F_o) \leq \dim Q$.

Now we present two methods for determining reachability in A if char $N_o(A)$ is not prime. Let $A = (Q,A,F)_R$ be a MSA where $N_o := N_o(A)$ is a s.r. with generator F_o, s.tr. $(\alpha,\beta;n)$, and let $q \in Q^*$:

First method. Determine $a := \mathrm{ord}\, q$ and $b := \min\{m \in \mathbb{N} \mid mF_o(q) \in \langle q\rangle\}$. Then the following is easy to see (cf. [4]): $\mathbb{Z}_a\,\mathrm{id} + \mathbb{Z}_b\,F_o$ is a system of representatives of $N_{o/\mathrm{Ann}\,q}$ and, by considering $h_q : N_o \to N_oq$, $N_oq = \mathbb{Z}_a q + \mathbb{Z}_b F_o(q)$ is the "smallest" representation of N_oq.

If $n = \prod_{i\in A} p_i$ is squarefree, then we can use also the following

Second method.

1) Determine all ideals in N_o (that means determine T_1, T_2, T_3)
2) Determine Ann q
3) There are some $T \subseteq A$, $S \subseteq T$, $z \in \mathbb{Z}$ (because of Th. 2 and its Cor.) such that $\mathrm{Ann}\, q = \langle \prod_{i\in T} p_i\rangle \dot{+} \langle \prod_{j\in S} p_j(z+r)\rangle$. Then $\mathbb{Z}_a\,\mathrm{id} + \mathbb{Z}_b\,F_o$, where $a := \prod_{i\in T} p_i$, $b := \prod_{j\in S} p_j$, is a system of representatives of $N_{o/\mathrm{Ann}\,q}$ (see [4]). Hence $N_oq = \mathbb{Z}_a q + \mathbb{Z}_b F_o(q)$ is the "smallest" representation of N_oq.

The first method works more generally. But the second method works sometimes faster and offers the opportunity to determine $N_0 q$ for a lot of states q simultaneously:

Proposition 1. Let $A = (Q,A,F)_R$ be a MSA where $N_0 := N_0(A)$ is a s.r. with generator F_0 and s.tr. $(\alpha,\beta;n)$ where $n = \prod_{i\in A} p_i$ is squarefree and $T_1 = A$. If $a := \operatorname{ord} \bar{q}$, $\bar{q} \in Q$, then the following holds:

a) $\operatorname{Ann} \bar{q} = \langle a\ id\rangle \dot{+} \langle a\ F_0\rangle$

b) $N_0 \bar{q} = \mathbb{Z}_a \bar{q} + \mathbb{Z}_a F_0(\bar{q})$

c) If B is the set of all proper divisors of a, $M_b := \operatorname{Ker}(b\ id)$ for $b \in B$, $\bar{M}_b$ denotes the complement of M_b, then we get: $N_0 q = \mathbb{Z}_a q + \mathbb{Z}_a F_0(q)$ for all $q \in \bigcap_{b\in B} M_a \cap \bar{M}_b$.

Proof. $nq = n\,id(q) = 0$ for all $q \in Q$ because N_0 is a s.r., hence ord q divides n for all $q \in Q$. Note that for $tq \in \mathbb{Z}_n q$ we get:
$tq = 0 \Leftrightarrow q \in \operatorname{Ker} t\,id \Leftrightarrow t\,id \in \operatorname{Ann} q$. The ideal generated by tid is given by $\langle tid\rangle \dot{+} \langle t\ F_0\rangle$ and also contained in Ann q. Since $T_1 = A$, $\operatorname{Ann} \bar{q}$ must be of the form $\langle tid\rangle \dot{+} \langle t\ F_0\rangle$ (see b) of the Corollary). Now $a = \operatorname{ord} \bar{q}$ yields the desired minimal t. This proves a), and b) follows.
$q \in \bigcap_{b\in B} M_a \cap \bar{M}_b \Leftrightarrow q \in \operatorname{Ker} a\ id \wedge \bigwedge_{b\in B} q \notin \operatorname{Ker} b\ id \Leftrightarrow \operatorname{ord} q = a$. This proves c). □

Hence, for $T_1 = A$, all $q \in \bigcap_{b\in B} M_a \cap \bar{M}_b$ have the same order, consequently the same annihilator, which in turn gives instant knowledge about $N_0 q$ for all these q. The situation in the case $T_2 = A$ is similar.

Proposition 2. Let $A = (Q,A,F)_R$ be a MSA where $N_0 := N_0(A)$ is a s.r. with generator F_0 and s.tr. $(\alpha,\beta;n)$ where $n = \prod_{i\in A} p_i$ is squarefree and $T_2 = A$.
Let $z \in \mathbb{Z}$ with $z \equiv z_i \pmod{p_i}$ for $i \in A$, $q \in Q$ and $a := \operatorname{ord} q$, $b := \operatorname{ord} \bar{q}$, where $\bar{q} := (z\,id + F_0)(q)$, then the following hold:

a) $\operatorname{Ann} q = \langle a\,id\rangle \dot{+} \langle b(z\,id + F_0)\rangle$

b) $N_0 q = \mathbb{Z}_a q + \mathbb{Z}_b F_0(q)$

Proof. On the one hand we have $\operatorname{Ann} q = \langle c\ id\rangle \dot{+} \langle d(z\ id + F_0)\rangle$ by c) of the Corollary for some $c,d \leq n$. Hence $cq = 0$ and $d\bar{q} = 0$, that means $a|c$ and $b|d$. But then $\langle c\ id\rangle \dot{+} \langle d(z\ id + F_0)\rangle \subseteq \langle a\ id\rangle \dot{+} \langle b(z\ id + F_0)\rangle$. On the other hand, $aq = 0$ and $b\bar{q} = 0$, so $\langle a\ id\rangle \dot{+} \langle b(z\ id + F_0)\rangle \subseteq \operatorname{Ann} q$, and we are done. □

For "mixed" cases (i.e. $T_1 \neq A$, $T_2 \neq A$) and for more information about properties of ideals in syntactic rings, the Jacobson-radical of a syntactic ring and so on, see [4].

In private communications, G. Betsch (Tübingen) suggested the following point of view. If a state $q \in Q$ is reachable from another state $q' \in Q$ (i.e. if $q \in Nq'$ or $q \in N_0 q'$), we can say that q' *divides* q ($q'|q$). The relation | is then a preorder on Q. The corresponding equivalence classes consist of states which can mutually be reached from another. These and related subjects will be studied in a subsequent paper.

REFERENCES

[1] Blyth, T.S., Module theory, Clarendon Press, Oxford, 1977.

[2] Eilenberg, S., Automata, languages and machines, Academic Press, New York, 1974.

[3] Hlawka, E. and Schoißengeier, J., Zahlentheorie, Manzsche Verlags- und Universitätsbuchhandlung, Wien, 1979.

[4] Hofer, G., Near-rings and group-automata, Doctoral dissertation, Univ. Linz, Austria, 1986.

[5] Hofer, G. and Pilz, G., Group-automata and near-rings, Contrib. Gen. Algebra 2, Klagenfurt, Austria, 1983.

[6] Holcombe, W.M.L., Algebraic automata theory, Cambridge University Press, Cambridge, 1982.

[7] Holcombe, W.M.L., The syntactic near-ring of a linear sequential machine, Proc. Edinbg. Math. Soc. 26 (1983), 15-24.

[8] Kalman, R.E., Algebraic theory of linear systems, in: Topics in math. systems theory, McGraw-Hill, New York, 1969.

[9] Lang, S., Algebra, Addison-Wesley Publishing Company, Reading, Mass., 1984.

[10] Lidl, R. and Pilz, G., Applied abstract algebra, Springer-Verlag, New York-Heidelberg-Berlin, 1984.

[11] Pilz, G., Near-rings, 2nd ed., North-Holland/American Elsevier, Amsterdam-New York, 1983.

[12] Pilz, G., Algebra, Universitätsverlag R. Trauner, Linz, Austria, 1984.

[13] Pilz, G., Strictly connected group automata, submitted.

Near-rings and Near-fields, G. Betsch (editor)

COUPLINGS AND DERIVED STRUCTURES

Helmut KARZEL

Mathematisches Institut, Technische Universität München,
Arcisstr. 21, D-8000 München 2

Near vector spaces which can be derived from usual vector spaces by the method of couplings will be discussed. The main results are stated in the theorems (8) and (16).

1. INTRODUCTION

A <u>structure group</u> $(P,\Sigma,\cdot)$ (cf.[2], §2) is a group $(P,\cdot)$ where the set P is provided with a structure Σ such that for any $a \in P$ the map $a_\ell : P \to P;\ x \to ax$ is an automorphism of the structure (P,Σ).

If $\mathrm{Aut}(P,\Sigma)$ denotes the group of all automorphisms of the structure (P,Σ) then a map $\varphi : P \to \mathrm{Aut}(P,\Sigma,\cdot);\ a \to a_\varphi$ is called a <u>coupling</u> if for all $a,b \in P$ the functional equation

(C) $a_\varphi b_\varphi = (a\cdot a_\varphi(b))_\varphi$ holds.

Then we have the statement:

(1) Let $(P,\Sigma,\cdot)$ be a structure group, let $\varphi : P \to \mathrm{Aut}(P,\Sigma)$ be a coupling and let $a \circ b := a\cdot a_\varphi(b)$, then $(P,\Sigma,\circ) =: (P,\Sigma,\cdot)^\varphi$ is again a structure group called the <u>φ-derivation</u> of $(P,\Sigma,\cdot)$.

For a class $\mathfrak{C}$ of structure groups we denote by $\mathfrak{D}(\mathfrak{C})$ the class of all structure groups which are φ-derivations of structure groups of the class $\mathfrak{C}$. Since the constant mapping $\varphi : x \to \mathrm{id}$ is a coupling for any structure group we have $\mathfrak{C} \subset \mathfrak{D}(\mathfrak{C})$. All members of $\mathfrak{D}(\mathfrak{C})$ are called <u>Dickson structure groups</u> with respect to $\mathfrak{C}$.

Now we can formulate the following Dickson problems:

P1 Let $\mathfrak{C}$ be a class of structure groups. Find couplings and determine $\mathfrak{D}(\mathfrak{C})$.

P2 Let $\mathfrak{C}$ be a class of structure groups and $\mathfrak{S}$ a subclass of $\mathfrak{C}$ characterized by additional properties. When do we have $\mathfrak{D}(\mathfrak{S}) = \mathfrak{C}$?

Remarks. 1. Any (left) nearfield $(N,+,\cdot)$ can be considered as a structure group if we set $P := N^* = N\setminus\{0\}$, $(P,\cdot) := (N^*,\cdot)$ and if Σ denotes the additive structure $+$ on $N = P \cup \{0\}$. Because of the left distribution law for any $a \in P$ the map $a_\ell : x \to a\cdot x$ is an automorphism of (P,Σ).

2. If $\mathfrak{N}$ denotes the class of all left nearfields and $\mathfrak{F}$ the subclass of all fields, then the members of the class $\mathfrak{D}(\mathfrak{F})$ are called Dickson nearfields. For the classes $\mathfrak{N}_f$ of finite nearfields and $\mathfrak{F}_f$ of finite fields we know by the theorem of Zassenhaus that $\mathfrak{N}_f \setminus \mathfrak{D}(\mathfrak{F}_f)$ consists of exactly seven nearfields. Till now all known examples of infinite nearfields belong to $\mathfrak{D}(\mathfrak{F})$. (Cf. [11])

3. Let $\mathfrak{A}$ be the class of all affine incidence groups, i.e. the class of all structure groups $(P,\mathfrak{L},\cdot)$ where $(P,\mathfrak{L})$ is either an affine space of $\dim(P,\mathfrak{L}) \geq 3$ or a translation plane. Then $\mathfrak{A}$ contains the subclass $\mathfrak{T}$ of all translation geometries $(P,\mathfrak{L},+)$ defined by the property, that for each $a \in P$ the map $a^+: P \to P; x \to a+x$ is a translation. (Here $(P,+)$ is always a commutative group.) By [2] we have here the result $\mathfrak{A} = \mathfrak{D}(\mathfrak{T})$, which tells us that any affine incidence group is Dicksonian with respect to a translation geometry.

Any translation geometry $(P,\mathfrak{L},+)$ has the nice property that the set $\mathfrak{K}$ of all lines through 0 obeys the following conditions:

(F1) For each $A \in \mathfrak{K}$, $A \leq (P,+)$ and $A \neq \{0\}, P$.

(F2) $\bigcup \mathfrak{K} = P$.

(F3) For each $A, B \in \mathfrak{K}$, $A = B$ or $A \cap B = \{0\}$.

Now let $(G,+)$ be any group (not necessarily commutative) and $\mathfrak{K}$ a collection of subgroups of G satisfying (F1,2,3). Then $(G,\mathfrak{K})$ is called a fibered group and $\mathfrak{K}$ a fibration (or partition). $\mathfrak{K}$ is called a kinematic fibration if further

(FK) For each $a \in G$, for each $A \in \mathfrak{K}$, $a + A - a \in \mathfrak{K}$ is satisfied.

For a fibered group $(G,\mathfrak{K})$ let End G be the set of all endomorphisms of the group $(G,+)$ and $E(G,\mathfrak{K}) := \{f \in \text{End}\, G \mid \forall X \in \mathfrak{K}, f(X) \subset X\}$. Then $E(G,\mathfrak{K})$ is a semigroup with identity. But if $(G,+)$ is not abelian in general the sum of two endomorphisms is not an endomorphism and the sets
$\text{d.g.}(\text{End}\, G) := \{\sum \delta_i f_i \mid \delta_i \in \{1,-1\}, f_i \in \text{End}\, G\}$ and
$\text{d.g.}(E(G,\mathfrak{K})) := \{\sum \delta_i f_i \mid \delta_i \in \{1,-1\}, f_i \in E(G,\mathfrak{K})\}$ are near-rings distributively generated by End G and $E(G,\mathfrak{K})$ resp. using pointwise addition and composition (cf.[8] Part III, §6). For $g \in \text{d.g.}(E(G,\mathfrak{K}))$ we have still $g(X) \subset X$ if $X \in \mathfrak{K}$.

At the Harrisonburg Conference on near-rings and near-fields in 1983 it was suggested to study near-rings of the type $\text{d.g.}(E(G,\mathfrak{K}))$. By [5],[7] we have the result:

(2) If $(G,\mathfrak{K})$ is a kinematic fibration with $|\mathfrak{K}| > 1$ and $E(G,\mathfrak{K}) \neq \{o, id\}$ then G is abelian, $E(G,\mathfrak{K})$ an integral domain and $(G,\mathfrak{K})$ can be embedded into

a vector space $(\hat{G},L)$ where $\hat{G} = G_\Delta$ and $L = Z(E(G,\mathfrak{K}))_\Delta$ with $\Delta = Z(E(G,K))\setminus\{0\}$ are the quotient structures. Therefore if $(G,\mathfrak{K})$ is kinematic and $(G,+)$ not commutative then $E(G,\mathfrak{K}) = \{o,id\}$ and hence d.g.$(E(G,\mathfrak{K}))$ is isomorphic to one of the rings $(\mathbb{Z},+,\cdot)$ or $(\mathbb{Z}_n,+,\cdot)$ with $n \in \mathbb{N}$. To avoid this trivial case and also the case where d.g.$(E(G,\mathfrak{K})) = E(G,\mathfrak{K})$, we have to consider non abelian groups $(G,+)$ with a fibration $\mathfrak{K}$ which is not kinematic such that $E(G,\mathfrak{K}) \neq \{0,id\}$.

Here we will discuss the problem how to obtain examples of fibered groups using the method of couplings. As the simplest class $\mathfrak{B}$ of fibered groups we consider all examples $(V,\mathfrak{K})$ where $(V,+)$ is the additive group of a left vector space (V,K) over a skew field K and where $\mathfrak{K}$ is a fibration consisting of vector subspaces of the same dimension. For instance the set $\mathfrak{K} := \{Ka \mid a \in V\setminus\{o\}\}$ of all one dimensional vector subspaces is such a fibration.

If $(V,\mathfrak{K})$ is any fibered group of the class $\mathfrak{B}$ then $L := E(V,\mathfrak{K})$ is a skew field containing K and any $X \in \mathfrak{K}$ is a L-vector subspace over L. Therefore we will always assume $K = L = E(V,\mathfrak{K})$. $\mathfrak{K}$ contains a subset $\mathfrak{B}$ such that $V = \Sigma\{B \mid B \in \mathfrak{B}\}$ is the direct sum of the subspaces of $\mathfrak{B}$; we call $|\mathfrak{B}|$ the dimension of the fibered group $(V,\mathfrak{K})$. If we set $\mathfrak{L} := \{a+X \mid a \in V, X \in \mathfrak{K}\}$ then it is easy to prove that $(V,\mathfrak{L})$ is an affine space of dimension $|\mathfrak{B}|$ where V is the set of points and $\mathfrak{L}$ the set of lines, and $(V,\mathfrak{L},+)$ is a translation geometry. If $|\mathfrak{B}| \geq 3$, then $(V,\mathfrak{L})$ is a desarguesian affine space (cf.p.e.[6] §10) and hence $\mathfrak{K}=\{Lx \mid x \in V^*\}$. If $|\mathfrak{B}| = 2$, this result is true if and only if $(V,\mathfrak{L})$ is desarguesian; in all other cases $(V,\mathfrak{L})$ is a proper translation plane.

In [4] p. 207 and [3] the so called <u>near vector spaces</u> $(G,\oplus,F,*)$ are studied, i.e. $(G,\oplus)$ is a group, $(F,+,\cdot)$ is a near field and $* : F \times G \longrightarrow G$ is a operation such that

(V1) $\forall \lambda,\mu \in F, \quad \forall x \in G : (\lambda+\mu) * x = \lambda * x \oplus \mu * x$

(V2) $\forall \lambda,\mu \in F, \quad \forall x \in G : (\lambda \cdot \mu) * x = \lambda * (\mu * x)$

(V3) $\forall x \in G \quad : 1 * x = x$

To any near vector space there belongs also a fibration $\mathfrak{K} := \{F * a \mid a \in G\setminus\{0\}\}$ such that each fiber $F * a$ is isomorphic to the additive group $(F,+)$ of the near field F.
So we have the questions:

P3 If $\hat{\mathfrak{B}}$ denotes the class of all additive groups $(V,+)$ belongings to a vector space (V,K) does then $\mathfrak{D}(\hat{\mathfrak{B}})$ contain examples which can be made to near vector spaces?

P4 Do the examples of near vector spaces given in [3] and [4] belong to $\mathfrak{D}(\hat{\mathfrak{B}})$?

2. CONSTRUCTION OF NEAR VECTOR SPACES, SEMILINEAR CASE

To deal with the problems P3 and P4 let (V,K) be a (left) vector space. By $M(V)$ we denote the near-ring of all maps from V into V.

Let $\varphi : V \to M(V);\ a \to a_\varphi$ and $\psi : V \to M(V);\ a \to a_\psi := a_\varphi - \mathrm{id}$.

Then φ fulfills the coupling functional equation

(C) $\forall\, a,b \in V:\ a_\varphi b_\varphi = (a + a_\varphi(b))_\varphi$

if and only if ψ fulfills

(C') $\forall\, a,b \in V:\ (a + b + a_\psi(b))_\psi = a_\psi + b_\psi + a_\psi b_\psi$.

(3) For the operation $a \oplus b := a + a_\varphi(b) = a + b + a_\psi(b)$ we have

a) $x \oplus 0 = x \Leftrightarrow x_\psi(0) = 0$

b) $0 \oplus x = x \Leftrightarrow 0_\psi = 0$ (zero map)

c) $(V,\oplus)$ is associative $\Leftrightarrow$ (C') and $\psi(V) \subset \mathrm{End}\,(V,+)$.

d) Let $a \in V$, then for each $b \in V$ the equation $a \oplus x = b$ has a solution if and only if $a_\varphi = a_\psi + \mathrm{id}$ is a permutation.

From now on we assume for the map $\psi : V \longrightarrow M(V)$ the conditions (C'), $0_\psi = 0$ and $\psi(V) \subset \mathrm{End}(V,+)$. Then by (3) $(V,\oplus)$ is a semigroup with 0 as neutral element. By induction we have

(4) Let $n,m \in \mathbb{N}$, $a \in V$, $n * a := \underbrace{a \oplus a \oplus \ldots \oplus a}_{n\text{-times}}$ then

a) $n * a = \sum_{i=0}^{n-1} \binom{n}{i+1} (a_\psi)^i(a) = \sum_{i=0}^{n-1} (a_\varphi)^i(a)$, if $a_\psi^0 := \mathrm{id}$.

b) $(n * a)_\psi = \sum_{i=1}^{n} \binom{n}{i} (a_\psi)^i$

c) $n * a \oplus m * a = (n + m) * a$

d) $(n \cdot m) * a = n * (m * a)$

e) If there is a $n \in \mathbb{N}$ with $n * a = 0$ then a has a negative with respect to $\oplus$ and $\mathbb{N} * a$ is a finite cyclic subgroup of $(V,\oplus)$.

f) If a has a negative $\ominus a$ with respect to $\oplus$ let $(-n) * a := n * (\ominus a)$. Then c) and d) are valid for all $m,n \in \mathbb{Z}$ and hence $\mathbb{Z} * a$ is a cyclic subgroup of $(V,\oplus)$.

g) If a_ψ is nilpotent then $\ominus a$ exists and

$$\ominus a = (-1) * a = (-1)a + \sum_{i \in \mathbb{N}} \binom{-1}{i+1} (a_\psi)^i(a) .$$

(5) If $\psi : V \longrightarrow M(V)$ has the additional property that for each $a \in V$ the map a_ψ is semilinear then we have either

a) for each $a \in V$, a_ψ is linear, or

b) there is an $a \in V$ such that a_ψ is not linear, then $u_\psi v_\psi = 0$ for all $u,v \in V$ and there is an automorphism $\bar{\ }: K \longrightarrow K; \lambda \longrightarrow \bar{\lambda}$ such that $b_\psi(\lambda x) = \bar{\lambda} b_\psi(x)$ for all $b,x \in V$ and $\lambda \in K$.

Proof. For $a,b \in V$ let $a_\psi, b_\psi \neq 0$ and let α,β be the automorphisms corresponding to the semilinear maps a_ψ, b_ψ respectively. By (C') we have for $\lambda \in K$, $x \in V$:
$(a \oplus b)_\psi(\lambda x) = \alpha(\lambda) \cdot a_\psi(x) + \beta(\lambda) \cdot b_\psi(x) + \alpha\beta(\lambda) a_\psi b_\psi(x)$. Since $(a \oplus b)_\psi$ is semilinear we have $\alpha = \beta$ and if $\alpha = \beta \neq \mathrm{id}$ then $a_\psi b_\psi = 0$.

(6) If there is an exterior operation $* : K \times V \longrightarrow V$ such that (V1) is valid then for all $\lambda,\mu \in K$ and all $a \in V$:

a) $(\lambda * a)_\psi(\mu * a) = (\mu * a)_\psi(\lambda * a)$

b) $(\lambda * a)_\psi(\mu * a)_\psi = (\mu * a)_\psi(\lambda * a)_\psi$.

Proof. a) is a consequence of (V1). By (C') we have $(a \oplus b)_\psi = a_\psi + b_\psi + a_\psi b_\psi$ hence "$a_\psi b_\psi = b_\psi a_\psi \Leftrightarrow a \oplus b = b \oplus a$". Now by (V1) we obtain $\lambda * a \oplus \mu * a = (\lambda + \mu) * a = \mu * a \oplus \lambda * a$ which gives us b).

Next we will show that the examples of near vector spaces given in [4] belong to $\mathfrak{D}(\mathfrak{B})$. Let (V,K) be a 3-dimensional vector space over a commutative field K with char $K \neq 2$, let $\bar{\ }: K \longrightarrow K$ be an automorphism of the field K and let e_1, e_2, e_3 be a basis of (V,K). For $x \in V$ let $x_1, x_2, x_3 \in K$ be defined by $x = x_1 e_1 + x_2 e_2 + x_3 e_3$ and for $a \in V$ let $a_\psi(x) := \bar{a}_3 \bar{x}_1 e_2$. Then $a_\psi \in \mathrm{End}(V,+)$ and since $a_\psi(\lambda x) =: \bar{a}_3 \overline{\lambda x}_1 e_2 = \bar{\lambda}\, \bar{a}_3 \bar{x}_1 e_2 = \bar{\lambda}\, a_\psi(x)$ for $\lambda \in K$, the map a_ψ is semilinear. If $a_3 \neq 0$ then $a_\psi(V) = Ke_2$ and $a_\psi^{-1}(0) = Ke_2 + Ke_3$, and if $a_3 = 0$ then a_ψ is the zero map. This shows us $a_\psi b_\psi = 0$ and $(a_\psi(b))_\psi = 0$ for all $a,b \in V$. Since $(a+b)_\psi(x) = (\overline{a_3 + b_3}) \bar{x}_1 e_2 = \bar{a}_3 \bar{x}_1 e_2 + \bar{b}_3 \bar{x}_1 e_2$, $(a+b)_\psi = a_\psi + b_\psi$ and so (C'):
$(a + b + a_\psi(b))_\psi = a_\psi + b_\psi + (a_\psi(b))_\psi = a_\psi + b_\psi = a_\psi + b_\psi + a_\psi\, b_\psi$.
By (3) $(V,\oplus)$ with $a \oplus b := a + b + a_\psi(b)$ is a group which is not

commutative because $e_1 \oplus e_3 = e_1 + e_3 \neq e_1 + e_3 + e_2 = e_3 \oplus e_1$ and since $a_\psi(b) = \bar{a}_3\bar{b}_1 e_2$, $\oplus$ is the addition defined in [4], Sec. V. Since $(a_\psi)^2 = 0$ the formula (4)a) is reduced to $n * a = n \cdot a + \binom{n}{2} a_\psi(a) = n \cdot a + \frac{1}{2} n(n-1)\bar{a}_3\bar{a}_1 e_2$. Since $(\lambda e_2)_\psi = 0$, each element $a \in V$ has the same order with respect to $+$ and $\oplus$. So we have:

(7) The map $\psi : V \longrightarrow \mathrm{End}(V,+)$ defined above has the following properties a),b),c) and (C') is a consequence of a),b),c).

a) For each $a \in V$, a_ψ is semilinear and if $\lambda \in K$, then $(\lambda a)_\psi = \bar{\lambda} a_\psi$

b) $a_\psi b_\psi = 0$, $(a_\psi(b))_\psi = 0$

c) $(a+b)_\psi = a_\psi + b_\psi$

Now let us assume that there is an $a \in V$ such that the semilinear map a_ψ is not linear. Then by (5)b) there is an automorphism $\lambda \longrightarrow \bar{\lambda}$ of $(K,+,\cdot)$ different from the identity, such that $u_\psi(\lambda v) = \bar{\lambda} u_\psi(v)$ and $u_\psi v_\psi = 0$ for all $u,v \in V$. If $x_\psi(x) = 0$ for all $x \in V$ then by (4)a) $n * x = n \cdot x$ for $n \in \mathbb{N}$ and $\lambda a \oplus \mu a = (\lambda+\mu)a + (\lambda a)_\psi(\mu a) = (\lambda+\mu)a + \overline{\mu\lambda^{-1}}(\lambda a)_\psi(\lambda a) = (\lambda+\mu)a$ and $\lambda(\mu a) = (\lambda\mu)a$ so that $(V,\oplus,K,\cdot)$ is a near vector space. If there is an $a \in V$ with $a_\psi(a) \neq 0$ then $a, a_\psi(a)$ are linearly independent.

Trying to extend the operation $*$ of (4)a) onto $K \times V$ such that $(V,\oplus,K,*)$ becomes a near vector space we make the following approach: Let $f,g,h\colon K \longrightarrow K$ be functions such that for each $x \in V$ we have $\lambda * x := h(\lambda)\cdot x + g(\lambda)x_\psi(x)$ and $(\lambda * x)_\psi = f(\lambda)x_\psi$. Then $0 = 0 * a$ and $a = 1 * a$ implies $h(0) = g(0) = g(1) = 0$ and $h(1) = f(1) = 1$. We set $\alpha := a_\psi$ and recall that $\alpha^2 = 0$.

Since $\lambda * a \oplus \mu * a = (h(\lambda) + h(\mu))a + (g(\lambda) + g(\mu))\alpha(a) + f(\lambda)\alpha(h(\mu)a + g(\mu)\alpha(a)) = (h(\lambda)+h(\mu))a + (g(\lambda) + g(\mu) + f(\lambda)\bar{h}(\mu))\alpha(a)$,

$(\lambda+\mu) * a = h(\lambda+\mu)a + g(\lambda+\mu)\alpha(a)$

(V1) is equivalent with

(α) $h(\lambda+\mu) = h(\lambda) + h(\mu)$

(β) $g(\lambda+\mu) = g(\lambda) + g(\mu) + f(\lambda)\cdot\bar{h}(\mu)$

and since $\lambda * (\mu * a) = h(\lambda)\cdot \mu * a + g(\lambda)\cdot(\mu * a)_\psi(\mu * a) = h(\lambda)h(\mu)a + h(\lambda)g(\mu)\alpha(a) + g(\lambda)\cdot f(\mu)\,\alpha(h(\mu)a + g(\mu)\alpha(a)) = h(\lambda)h(\mu)a + (h(\lambda)g(\mu) + g(\lambda)\cdot f(\mu)\bar{h}(\mu))\alpha(a)$

$(\lambda\cdot\mu) * a = h(\lambda\,\mu)a + g(\lambda\,\mu)\alpha(a)$

(V2) is equivalent with

(γ) $h(\lambda\cdot\mu) = h(\lambda)\cdot h(\mu)$

(δ) $g(\lambda\cdot\mu) = h(\lambda)\cdot g(\mu) + g(\lambda)\cdot f(\mu)\,\bar{h}(\mu)$

Since $h(1)=1$ the equations (α) and (γ) tell us that $h: K \longrightarrow K$ is a monomorphism, and from (β) and $f(1)=1$ we obtain $f(\lambda)=g(\lambda+1)-g(\lambda)-g(1)=\overline{h}(\lambda)$ so that (δ) gets the form

(δ') $g(\lambda\cdot\mu)=h(\lambda)\cdot g(\mu)+g(\lambda)\overline{h}(\mu)\cdot\overline{h}(\mu)$.

By $g(1)=0$, $h(1)=f(1)=1$, (α) and (β) we have $h(n)=nh(1)=n$ for $n\in\mathbb{N}$, $g(2)=1$ and (δ') gives us $g(2\cdot\lambda)=h(\lambda)+g(\lambda)\cdot 4=$ $=2\cdot g(\lambda)+\overline{h}(\lambda)\cdot\overline{h}(\lambda)$ thus $g(\lambda)=\frac{1}{2}(\overline{h(\lambda)}^{2}-h(\lambda))$. Furthermore $\overline{h}(\lambda\cdot\mu)\overset{(\gamma)}{=}\overline{h}(\lambda)\cdot\overline{h}(\mu)=f(\lambda)\,\overline{h}(\mu)\overset{(\beta)}{=}g(\lambda+\mu)-g(\lambda)-g(\mu)=f(\mu)\,\overline{h}(\lambda)=$ $=\overline{h}(\mu)\cdot\overline{h}(\lambda)=\overline{h}(\mu\cdot\lambda)$ shows that K has to be commutative.

As in [1] and [2] we consider the subgroups $U:=\{x\in V\mid x_\psi=0\}$ and $V_\Gamma:=\{x\in V\mid V_\psi(x)=0\}$ of $(V,+)$. Then V_Γ is also a subgroup of $(V,\oplus)$, and if $V_\psi V_\psi=0$ then also U is a subgroup of $(V,\oplus)$ and $\{x_\psi(y)\mid x,y\in V\}\subset V_\Gamma$. In any case $(u+v)_\psi=u_\psi+v_\psi$ if $u\in U\cup V_\Gamma$ or $v\in U\cup V_\Gamma$ and V_Γ is a vector subspace of (V,K) because V_ψ consists of semilinear maps. Therefore $(\lambda * x)_\psi=(h(\lambda)x)_\psi+$ $+(g(\lambda)x_\psi(x))_\psi$ and (6) implies $(h(\lambda)x)_\psi(x)+(g(\lambda)\cdot x_\psi(x))_\psi(x)=$ $=\overline{h}(\lambda)x_\psi(x)$, hence $\overline{\mu}\,x_\psi(x)=(\mu x)_\psi(x)+(\frac{1}{2}(\overline{\mu}^{2}-\mu)x_\psi(x))_\psi(x)$ if $\mu\in h(K)$.

On the other hand one shows by calculation that also the part 2. of c) of the following theorem is valid:

(8) Theorem. Let (V,K) be a vector space, let $\varphi: V\longrightarrow \mathrm{Aut}(V,+)$ be a coupling such that for each $x\in V$ the map $x_\psi:=x_\varphi-\mathrm{id}$ is semilinear and there is an $a\in V$ where a_ψ is not linear, and let $^{-}: K\longrightarrow K;\ \lambda\longrightarrow\overline{\lambda}$ denote the automorphism of a_ψ. Further let $x\oplus y=x+x_\varphi(y)=x+y+x_\psi(y)$. Then

a) For all $u,v\in V$, $\lambda\in K$, $u_\psi(\lambda v)=\overline{\lambda}\,u_\psi(v)$ and $u_\psi\, v_\psi=0$ which means that φ is a <u>strong coupling</u> i.e. $(u+v)_\varphi=u_\varphi\, v_\varphi$.

b) If $x_\psi(x)=0$ for all $x\in V$ then $(V,\oplus,K,\cdot)$ is a near vector space.

c) Let $a\in V$ with $a_\psi(a)\neq 0$. 1. If there are functions $f,g,h: K\longrightarrow K$ with $(h(\lambda)\cdot x+g(\lambda)\cdot x_\psi(x))_\psi=f(\lambda)\cdot x_\psi$ such that $(V,\oplus,K,*)$ with $\lambda * x:=h(\lambda)\cdot x+g(\lambda)\cdot x_\psi(x)$ becomes a near vector space; then K is commutative, $\mathrm{Char}\,K\neq 2$, h is a monomorphism of the field K, $f(\lambda)=\overline{h(\lambda)}$, $g(\lambda)=\frac{1}{2}(\overline{h(\lambda)}^{2}-h(\lambda))$, and

(N) For each $x\in V$, $\mu\in h(K)$, $\overline{\mu}\,x_\psi(x)=(\mu x)_\psi(x)+(\frac{1}{2}(\overline{\mu}^{2}-\mu)x_\psi(x))_\psi(x)$.

2. If the field K is commutative with Char $K \neq 2$ and if there is a monomorphism h such that (N) is valid then $(V,\oplus)$ becomes a near vector space $(V,\oplus,K,*)$ for $\lambda * x := h(\lambda)x + \frac{1}{2}(\overline{h(\lambda)}^2 - h(\lambda))x_\psi(x)$.

Remark. 4. If $(\mu x)_\psi = \overline{\mu} x_\psi$ and $(x_\psi(x))_\psi = 0$ for each $x \in V$, $\mu \in K$ then (N) is valid. In this case $h = id$ determines exactly the exterior multiplication for the examples given in [4] p. 206. Here $(\mu a)_\psi(x) = \overline{\mu a_3} \overline{x}_1 e_2 = \overline{\mu}\, a_\psi(x)$ and $(a_\psi(a))_\psi(x) = (\overline{a}_3 \overline{a}_1 e_2)_\psi(x) = \overline{0}\, e_2 = 0$.

3. CONSTRUCTION OF NEAR VECTOR SPACES, LINEAR CASE

The near vector spaces considered in [3] one obtains in this way:

(9) Let (A,K) be an associative nilpotent algebra and let $m \in \mathbb{N}$ such that $x^m = 0$ for all $x \in A$ but $y^{m-1} \neq 0$ for at least one $y \in A$. If char $K = 0$ or $m \leq p :=$ char K then $(A,\oplus,K,*)$ with $x \oplus y := x+y+x\cdot y$ and $\lambda * x = (1+x)^\lambda - 1 = \sum_{i=1}^{m-1} \binom{\lambda}{i} x^i$ is a near vector space.

For an associative nilpotent algebra (A,K) the map $\psi : A \longrightarrow M(A)$ with $a_\psi(x) := a\cdot x$ has the properties $(a+b)_\psi = a_\psi + b_\psi$ and $(a_\psi(b))_\psi = a_\psi b_\psi$ hence $(a + b + a_\psi(b))_\psi = a_\psi + b_\psi + a_\psi b_\psi$ and for each $a \in A$ the map a_ψ is K-linear and nilpotent. Therefore by (3)c) and (4)g) $(A,\oplus)$ with $a \oplus b = a + b + a_\psi(b) = a+b+a\cdot b$ is a group and so the examples of [3] are also members of $\mathfrak{D}(\hat{\mathfrak{B}})$. Hence P4 is answered positively.

We remark that $(A,\oplus,K,*)$ is not a near vector space if in (9) $m > p :=$ char K; then there is an $a \in A$ with $a^p \neq 0$ but $a^{p+1} = 0$ and hence $p * a = \sum \binom{p}{i} a^i = a^p \neq 0$. Further if $x \in A$ then $(x_\psi)^m = 0$ and there is an $y \in A$ with $(y_\psi)^{m-2} \neq 0$.

The result of (9) can be reversed.

(10) If $\psi : V \longrightarrow M(V)$ has the following properties

a) for each $a \in V$, a_ψ is linear,

b) there is a $m \in \mathbb{N}$, $m > 1$ with $(x_\psi)^{m-1}(x) = 0$ for all $x \in V$ and there is an $y \in V$ with $(y_\psi)^{m-2}(y) \neq 0$ if $m > 2$,

c) for all $a,b \in V$, $(a+b)_\psi = a_\psi + b_\psi$ and $(a_\psi(b))_\psi = a_\psi b_\psi$,

then $(V,\oplus)$ with $a \oplus b := a + b + a_\psi(b)$ is a group and $(V,+,\cdot)$ with $a\cdot b = a_\psi(b)$ is an associative nilpotent ring with $x^{m+1} = 0$ for all $x \in V$ and there are $y,z \in V$ with $y^{m-1}z \neq 0$. $(V,+,\cdot)$ is a K-algebra if and only if $(\lambda a)_\psi = \lambda a_\psi$ for all $\lambda \in K$ and all $a \in V$.

Finally we discuss the problem P3 for the case of a coupling $\varphi : V \longrightarrow \mathrm{Aut}(V,+)$ where the corresponding map $\psi : a \longrightarrow a_\psi := a_\varphi - \mathrm{id}$ has the properties a) and b) of (10).

If $m = 2$, i.e. $x_\psi(x) = 0$ for all $x \in V$ then $(V,\oplus,K,\cdot)$ is a near vector space (cf.(7)b)). Therefore let $m > 2$. Let $a \in V$ be such that $(a_\psi)^{m-2}(a) \neq 0$, $\alpha := a_\psi$ and $A := \langle a, \alpha a, \ldots, \alpha^{m-2}(a)\rangle$. Then $\dim A = m-1$ and $\alpha(A) \subset A$. Again we ask if there is a function $g : K \times \mathbb{N} \longrightarrow K$ such that for $\lambda * x := \sum_{i=0}^{m-2} g(\lambda,i)(x_\psi)^i(x)$, $(V,\oplus,K,*)$ is a near vector space. Then by (6) $(\lambda * a)_\psi(a) = \sum_{i=0}^{m-3} g(\lambda,i)\alpha^{i+1}(a)$. If we further follow (4)b) and make the assumption that the linear map $(\lambda * a)_\psi$ is a linear combination of the α^i then

(11) $(\lambda * a)_\psi = \sum_{i=1}^{m-2} g(\lambda,i-1)\alpha^i$.

Now (V1) gives us

$$\Sigma\, g(\lambda+\mu,i)\,\alpha^i(a) = \Sigma(g(\lambda,i)+g(\mu,i))\alpha^i(a) + \Sigma\Sigma\, g(\lambda,j)\, g(\mu,i)\, \alpha^{j+i+1}(a)$$

and by comparing the coefficients we obtain for all i with $\alpha^i(a) \neq 0$:

(12) $g(\lambda+\mu,i) = g(\lambda,i) + g(\mu,i) + \sum_{j=1}^{i} g(\lambda,j-1)\, g(\mu,i-j)$.

By (V2) we have

$\Sigma\, g(\lambda\mu,i)\alpha^i(a) = \Sigma\, g(\lambda,i)(\mu * a)_\psi^i \Sigma\, g(\mu,j)\alpha^j(a)$ and using (11) the coefficients of $\alpha^0(a) = a$, $\alpha(a)$ and $\alpha^2(a)$ (if $\alpha^2(a) \neq 0$) yield

(13) $g(\lambda\cdot\mu,0) = g(\lambda,0)\cdot g(\mu,0)$

(14) $g(\lambda\cdot\mu,1) = g(\lambda,0)\cdot g(\mu,1) + g(\lambda,1)(g(\mu,0))^2$

(15) $g(\lambda\cdot\mu,2) = g(\lambda,0)\, g(\mu,2) + g(\lambda,2)(g(\mu,2))^3 + 2g(\mu,0)g(\lambda,1)g(\mu,1)$ if $\alpha^2(a) \neq 0$.

(12) and (13) shows that $\lambda \longrightarrow \overline{\lambda} := g(\lambda,0)$ is a monomorphism, then (12) and (14) yields $g(\lambda,1) = \binom{\overline{\lambda}}{2}$ and (12) and (15) $g(\lambda,2) = \binom{\overline{\lambda}}{3}$ if $\alpha^2(a) \neq 0$. With this method one obtains $g(\lambda,i) = \binom{\overline{\lambda}}{i+1}$ if $\alpha^i(a) \neq 0$ and from (11) we derive the following condition for the map ψ :

(N') $\forall x \in V$, $\forall \mu \in g(K,0) : \left(\sum_{i=0}^{m-2}\binom{\mu}{i+1}(x_\psi)^i(x)\right)_\psi = \sum_{i=1}^{m-2}\binom{\mu}{i} x_\psi^i$

The part 2. of b) of the following theorem one obtains by calculation.

(16) Theorem. Let $\varphi : V \longrightarrow \mathrm{Aut}(V,+)$ be a coupling such that for each $x \in V$ the map $x_\psi := x_\varphi - \mathrm{id}$ is linear, there is a $m \in \mathbb{N}$, $m > 1$ with $(x_\psi)^{m-1}(x) = 0$ for all $x \in V$ and $(y_\psi)^{m-2}(y) \neq 0$ for at least one $y \in V$.

a) If $m = 2$ then $(V, \oplus, K, \cdot)$ is a near vector space.

b) Let $m > 2$. 1. If there is a function $g : K \times \mathbb{N} \longrightarrow \mathbb{N}$ with $\left(\sum_{i=0}^{m-2} g(\lambda,i)(x_\psi)^i(x)\right)_\psi \in \langle x_\psi, x_\psi^2, \ldots, x_\psi^{m-2}\rangle$ for all $x \in V$ such that $(V, \oplus, K, *)$ with $\lambda * x := \sum_{i=0}^{m-2} g(\lambda,i)(x_\psi)^i(x)$ becomes a near vector space then K is a commutative field with $\mathrm{char}\, K = 0$ or $\geq m$, $\lambda \longrightarrow \overline{\lambda} = g(\lambda,0)$ is a monomorphism of the field K, $g(\lambda,i) = \binom{\overline{\lambda}}{i+1}$ for all $i \in \{0,1,\ldots,m-2\}$ and ψ fulfills (N').
2. If K is a commutative field with $\mathrm{char}\, K = 0$ or $\geq m$, $\lambda \longrightarrow \overline{\lambda}$ a monomorphism of K such that (N') holds then $(V, \oplus, K, *)$ with $\lambda * x := \sum_{i=0}^{m-2} \binom{\overline{\lambda}}{i+1} (x_\psi)^i(x)$ is a near vector space.

Remark. 5. If the map $\psi : V \longrightarrow \mathrm{End}(V,+)$ is linear (i.e. $(x+y)_\psi = x_\psi + y_\psi$ and $(\lambda x)_\lambda = \lambda x_\psi$ for $\lambda \in K$) then (N') is valid because then $a_\psi + b_\psi + (a_\psi(b))_\psi = (a + b + a_\psi(b))_\psi \overset{(C')}{=} a_\psi + b_\psi + a_\psi b_\psi$ hence $(a_\psi(b))_\psi = a_\psi b_\psi$ in particular $((x_\psi)^i(x))_\psi = x_\psi^{i+1}$.

Remark. 6. By [3],(4.10) the fibrations belonging to the near vector spaces of theorem (9) are kinematic so that by (2) we have either $\mathrm{d.g.}(E(V,\mathfrak{R})) = E(V,\mathfrak{R})$ or $\mathrm{d.g.}(E(V,\mathfrak{R}))$ is trivial. It remains the question whether there are other couplings in the sense of theorem (16) leading to near vector spaces whose fibrations are not kinematic.

In the theorems (8)b) and (16)a) we considered couplings with $x_\psi(x) = 0$ for all $x \in V$. For these maps we have:

(17) Let $\varphi : V \longrightarrow \mathrm{Aut}(V,+)$ be a coupling such that for all $x,y \in V$, $x_\psi := x_\varphi - \mathrm{id}$ is semilinear, $x_\psi(x) = 0$ and $x_\psi y_\psi = 0$. Then

a) $x_\psi(y) = -y_\psi(x)$, $x \oplus y \ominus x = y + 2x_\psi(y)$

b) $(x+y)_\psi = x_\psi + y_\psi$, $(\lambda x)_\psi = \overline{\lambda} x_\psi$ i.e. $\psi : (V,K) \longrightarrow \mathrm{End}(V,K)$ is a semilinear map.

c) $V_\Gamma = U$

d) The fibration $\mathfrak{R} := \{K \cdot x \mid x \in V \setminus \{0\}$ of the near vector space $(V, \oplus, K, \cdot)$ is kinematic if and only if $\mathrm{char}\, K = 2$ or x_ψ is linear for each $x \in V$.

e) If $U \neq V$ then $E((V,\oplus),\mathfrak{K}) = \{\lambda \in K \mid \overline{\lambda}^2 = \lambda\}$.

Proof. a) By $x_\psi(x) = 0$, $x_\psi y_\psi = 0$ and (C'), $0 = (x \oplus y)_\psi (x \oplus y) = $
$= (x_\psi + y_\psi)(x + y + x_\psi(y)) = x_\psi(y) + y_\psi(x)$ hence $x_\psi(y) = -y_\psi(x)$.
b) By a), $(x+y)_\psi(z) = -z_\psi(x+y) = -z_\psi(x) - z_\psi(y) = x_\psi(z) + y_\psi(z)$ and by (5) $(\lambda x)_\psi(y) = -y_\psi(\lambda x) = -\overline{\lambda} y_\psi(x) = \overline{\lambda} x_\psi(y)$.
c) Let $u \in U$ and $x \in V$ then $0 = u_\psi(x) = -x_\psi(u)$ hence $u \in V_\Gamma$. Now $v \in V_\Gamma$ and $x \in V$ implies $0 = x_\psi(v) = -v_\psi(x)$ thus $v \in U$.
d) is a consequence of a).
e) Since $U \neq V$ there are $x,y \in V \setminus U$ with $x_\psi(y) \neq 0$. Then $\lambda(x \oplus y) = \lambda x + \lambda y + \lambda x_\psi(y)$ and by b) $\lambda x \oplus \lambda y = \lambda x + \lambda y + \overline{\lambda}^2 x_\psi(y)$ hence $\lambda \in E((V,\oplus),\mathfrak{K})$ if $\overline{\lambda}^2 = \lambda$.

(18) (Examples for (17). Let K be a commutative field and $^-: K \to K$ an automorphism of K then the map $\psi: K^3 \to \mathrm{End}(K^3,K)$ defined by $(a_1,a_2,a_3)_\psi(x_1,x_2,x_3) = (0,0,\overline{a_1x_2} - \overline{a_2x_1})$ has the properties of (17). If $K = \mathbb{Z}_5(\tau)$ is the quadratic extension field of $\mathbb{Z}_5$ with $\tau^2 = 2$ then $E((V,\oplus),\mathfrak{K}) = \{0,1,2+2\tau,2+3\tau\}$.

Remark. 7. If $(V,\oplus,K,*)$ is a near vector space of the type of theorem (8)c) and if $(V,\oplus)$ is not commutative then $E((V,\oplus),\mathfrak{K}) = \{\lambda \in K \mid \overline{\lambda}^2 = \lambda\}$ (cf.[4], p. 206).

REFERENCES

[1] Karzel, H., Unendliche Dicksonsche Fastkörper. Arch.Math. 16 (1965), 247-256
[2] ——, Affine incidence groups. Rend. Sem. Mat. Brescia Vol. 7 (1984), 409-425
[3] ——, Fastvektorräume, unvollständige Fastkörper und ihre Abgeleiteten Strukturen. Mitt. Math. Sem. Giessen, Coxeter-Festschrift, Teil IV (1984), 127-139
[4] ——, and C.J. Maxson, Fibered groups with non-trivial centers. Res. Math. 7 (1984), 192-208
[5] ——, and C.J. Maxson, Kinematic spaces with dilatations. J. Geometry 22 (1984), 196-201
[6] ——, K. Sörensen and D. Windelberg, Einführung in die Geometrie. Göttingen 1973
[7] Marchi, M. and C. Perelli Cippo, Su una perticolare classe di S-spazi. Rend. Sem. Mat. Brescia 4 (1979), 3-42
[8] Pilz, G., Near-Rings. North-Holland Math. Studies 23, ^{1}1977, ^{2}1983.
[9] Pokropp, F., Gekoppelte Abbildungen auf Gruppen. Abh. Math. Sem. Univ. Hamburg 32 (1968), 147-159
[10] Wähling, H., Bericht über Fastkörper. Jber. Deutsch. Math. Verein. 76 (1974), 41-103
[11] Wähling, H., Theorie der Fastkörper, Thales Verlag, Essen 1987.

Near-rings and Near-fields, G. Betsch (editor)

MAXIMAL IDEALS IN NEAR-RINGS

Hermann KAUTSCHITSCH

Institut für Mathematik
Universität Klagenfurt
9022 Klagenfurt

The intention of this paper is, to relate the ideal structure of a special near-ring N to that of some homomorphic image N' of N or to its constant part N_c. In particular one gets information about maximal ideals. Such a relationship is well known in polynomial - or formal power series rings and it turns out, that one is successful in finding such a relationship for those near-rings, which have a little bit of a "polynomial-" or "power series-structure".

1. IDEALS AND HOMOMORPHISMS

Throughout this paper we assume that $(N,+,\circ)$ is a right near-ring with identity 1 and R is a commutative ring with identity. R[[x]] denotes the set of all formal power series over R (in one indeterminate x), $R_\circ[[x]]$ the set of all formal power series of positive order.

First we shall try to describe maximal ideals by certain homomorphisms.

We study the case, that $(N,+,\circ)$ is a near ring with identity 1, which has an epimorphism $h: N \rightarrow N'$, such that $h^{-1}(1')$ consists only of invertible elements in $(N,\circ)$. i.e., $h^{-1}(1') \subseteq I(N)$.

Examples: 1. Any finite dgnr N with identity: By Pilz [4], 6.31 we have $h(I(N))=I(h(N))$.

2. For $N=(R[[x]],+,.,)$ we set $h(p_\circ+p_1x+p_2x^2+...):=p_\circ$; then h is an epimorphism from N onto $(R,.)$, such that $h^{-1}(1)=\{1+p_1x+p_2x^2...\}$ consists only of invertible elements of $(N,.)$.

3. For $N=(R_\circ[[x]]+,\circ)$, where $\circ$ denotes the composition of power series, we set $h(r_1x+r_2x^2+...)=r_1$; then h is an epimorphism from N onto $(R,.)$, such that $h^{-1}(1)=\{x+r_2x^2...\}$ consists only of invertible elements of $(N,\circ)$ (see [1]).

THEOREM 1: If $\{M'_i\}$ is the set of all maximal ideals of the epimorphic image N', then $\{h^{-1}(M'_i)\}$ is the set of *all* maximal ideals of N.

Proof: 1. $h^{-1}(M'_i)$ is a maximal ideal of $N: h^{-1}(M'_i)\neq N$, otherwise $1\in h^{-1}(M'_i)$ and then $1':=h(1)\in M'_i$ and $M'_i=N'$. If $B \trianglelefteq N$ with

$B \supset h^{-1}(M'_i)$, then $h(B) \supset M'_i$, hence $h(B)=N'$ and $1' \in h(B)$. Since $b:=h^{-1}(1')$ is invertible in $(N,\circ)$, we get $1=b \circ v \in B$ and $B=N$. (The notation $b:=h^{-1}(1')$ means, that we choose an element $b \in N$ with $h(b)=1'$.)

2. If M is any maximal ideal of N, then $h(M)$ is a maximal ideal in $N':h(M)\neq N'$, otherwise $1' \in h(M)$ and M contains the invertible element $m:=h^{-1}(1')$, hence $1=m \circ u \in M$ and $M=N$ (again the notation $m:=h^{-1}(1')$ means, that we choose an element $m \in N$ with $h(m)=1'$).

Therefore there exists an element $a \in N' \setminus h(M)$. Hence there exists an element $b \in N \setminus M$ with $h(b)=a$. By the maximality of M we get $M+(b)=N$. By [5], p. 111, there are $m \in M$ and $p \in P_0(N) := (P(N))_0$ with $1=m+p(b)$. Here $P(N)$ is the set of polynomial functions on N (i.e. the near-ring generated by id_N and the constant maps on N). Since $1'=h(1)=h(m)+p'(h(b))= h(m)+p'(a)$ with $p' \in P_0(N')$, we get $1' \in h(M)+(a)$ and therefore $h(M)+(a)=N'$, so $h(M)$ is maximal in N.

COROLLARY 1: (i) All maximal ideals of the formal power series ring $(R[[x]],+,.)$ over a commutative ring R with 1 are given by the ideals $M_i+(x)$, where M_i is any maximal ideal of R.

(ii) All maximal ideals of the zero symmetric near-ring of formal power series $(R_0[[x]],+,\circ)$ over a commutative ring R with 1 are given by the ideals $\{m_1x+m_2x^2+\dots \mid m_1 \in M_i\}$, where M_i is any maximal ideal of R.

Proof: (i) By Theorem 1 we describe all maximal ideals M of $(R[[x]],+,.)$ with the epimorphism of Example 2, hence $M=h^{-1}(M_i)= \{m_0+m_1x+m_2x^2+\dots \mid m_i \in M_i\}$, where M_i is any maximal ideal of $(R,+,.)$, so $M=M_i+(x)$.

(ii) Similarly we describe all maximal ideals M of $(R_0[[x]],+,\circ)$ with the epimorphism of Example 3, hence $M=h^{-1}(M_i)= \{m_1x+m_2x^2+\dots \mid m_1 \in M_i\}$, where M_i is any maximal ideal of $(R,+,.)$.

COROLLARY 2: (i) A non-simple finite dgnr N with 1 is never a subdirect product of simple near-rings.

(ii) The ring $R[[x]]$ of formal power series over a commutative ring R with 1 is never a subdirect product of fields, even if R is such a ring.

(iii) The zero symmetric near-ring $R_0[[x]]$ of formal power series over a commutative R with 1 is never a subdirect product of simple near-rings, even if R is a subdirect product of fields.

Proof: Let Rad N (or RadN', respectively) denote the intersection of all maximal ideals of N (or N') and $J(R)$ the Jacobson radical of R. N (or R, respectively) is a subdirect product of simple near-rings (fields) iff Rad N=0 (J(R)=0). By Theorem 1 we get:

$$\mathrm{Rad}\, N = \bigcap_{\substack{M \text{ maximal} \\ \text{in } N}} M = \bigcap_{\substack{M' \text{ maximal} \\ \text{in } N'}} h^{-1}(M') = h^{-1}(\bigcap M') = h^{-1}(\mathrm{Rad}N')$$

(i) If N is not simple, then there exists a non-injective epimorphism $h:N \to N/I$, $I \trianglelefteq N$ and $h^{-1}(\mathrm{Rad}(N/I)) \neq 0$, even if $\mathrm{Rad}(N/I)=0$.

(ii) By Corollary 1(i) $\mathrm{Rad}(R[[x]],+,.)=J(R)+(x) \neq 0$, even if $J(R)=0$.

(iii) By Corollary 1(ii) $\mathrm{Rad}(R_0[[x]],+,\circ)=$ $\{m_1x+m_2x^2+\ldots / m_1 \in J(R)\} \neq 0$, even if $J(R)=0$.

To get some information about maximal left-ideals of N, we introduce an additional assumption on N:
I(N) is commutative and $e_1 \circ e_2 = m \circ (n+e_1) - m \circ n$ with suitable $m,n \in N$, $\forall\, e_1, e_2 \in I(N)$.

Examples: 1. Any near-ring, in which the invertible elements are zero symmetric and permutable: $e_1 \circ e_2 = e_2 \circ (0+e_1) - e_2 \circ 0$.

2. The zero symmetric part $R_0[x]$ of the polynomial near-ring $(R[x],+,\circ)$. In $R_0[x],+,\circ)$ all invertible elements are given by the polynomials ex, e a unit in (R,.), which are permutable with respect to $\circ$.

THEOREM 2: If $\{L'_i\}$ is the set of all maximal left ideals of the epimorphic image N' of N, then $\{h^{-1}(L'_i)\}$ is the set of all maximal left ideals of N.

Proof: 1. If L'_i is maximal in N', then $h^{-1}(L'_i)$ is maximal in $N: h^{-1}(L'_i) \neq N$, otherwise $1 \in h^{-1}(L'_i)$ and $1'=h(1) \in L'_i$. By the above assumption we get:
$n'=1' \circ n' = p' \circ (q'+1') - p' \circ q' \in L'_i\ \forall\, n' \in N'$, so $L'_i = N'$.
If $B \trianglelefteq_1 N$ and $B \supset h^{-1}(L'_i)$, then $h(B)=N'$ and $1' \in h(B)$, i.e., $1'=h(b)$, $b \in B$, so B contains the invertible element $b:=h^{-1}(1')$. Then $1 = b \circ u = m \circ (n+b) - m \circ n \in B$ and $B=N$.

2. If L is any maximal left ideal of N, then h(L) is a maximal left ideal of N':
$h(L) \neq N'$, otherwise $1' \in h(L)$, hence $1':=h(l)$ for some $l \in L$ and $l:=h^{-1}(1')$ is invertible in L, so L=N. Now let $a' \in N' \setminus h(L)$, then $a:=h^{-1}(a') \in N \setminus L$. But then $L+(a)=N$. So $1=l+p(a)$ with some zero symmetric polynomial $p \in P_0(N)$. Then
$1'=h(1)=h(l)+p'(h(a))=l'+p'(a')$ with $p' \in P_0(N')$, $l' \in h(L)$, hence $1' \in h(L)+(a')$ and then $h(L)+(a')=N'$, so h(L) is maximal in N'.

COROLLARY 3: If N is a finite and distributively generated near-ring with commutative I(N), then all maximal left ideals are given by the ideals $\{h^{-1}(L'_i)\}$, where L'_i is any maximal left ideal of an epimorphic image h(N) of N.

2. IDEALS AND THE CONSTANT PART N_c

Now we consider the case that N does not have an epimorphism of the type given in section 1. In order to describe the ideal structure of N by that of its constant part N_c, we assume, that in N an additional multiplication $\circ$ is defined, such that

(i) $(N,+,.,\circ)$ is a composition-ring

(ii) $\forall n,m,i \in N \exists s \in N: n\circ(m+i)=n\circ m+i.s$

This means, in particular, that (N,+,.) is a ring.

Examples: $(R[x],+,.,\circ)$ and $(R_N[[x]]+,.,\circ)$, where $R_N[[x]]$ denotes the set of formal power series with nilpotent initial coefficient (see [2]).

A nonempty set I is an ideal of a composition-ring N with property (ii) iff

(i) $i-j \in I$

(ii) $i\circ n \in I$

(iii) $i.n \in I$ and $n.i \in I$ $\forall i,j \in I$, $n \in N$.

First we introduce two special ideals (see also [3]):

1. If $I \trianglelefteq (N,+,.,\circ)$, then we set $I_c := I \cap N_c \trianglelefteq (N_c,+,.)$ and we get $I_c \trianglelefteq (N_c,+,.,\circ,)$: $i-j \in I\cap N_c$, $i.c \in I\cap N_c$, $c.i \in I\cap N_c$, $i\circ c \in I\cap N_c$, $\forall i,j \in I_c, c \in N_c$.

2. If $I' \trianglelefteq (N_c+,.)$, then (I') denotes the ideal of N generated by I': $(I') = \{\sum_{\text{finite}} p_i\circ j'_i \mid p_i \in P_\circ(N), j'_i \in I'\}$, where $P_\circ(N)$ denotes the zero symmetric polynomials over N (see [5]).

Remark: The ring-ideals I' of N_c coincide with the composition ideals of N_c, because $i\circ c=i$ for all $i \in I'$ and $c \in N_c$.

3. If $I' \trianglelefteq (N_c,+,.)$, then $[I'] := \{n \in N \mid n\circ c \in I' \ \forall c \in N_c\} \trianglelefteq N$: For all $i,j \in [I']$, $n \in N$ we get: $(i-j)\circ c = i\circ c - j\circ c \in I$, hence $i-j \in [I']$, $(i\circ n)\circ c = i\circ(n\circ c) = i\circ c' \in I'$ with $c' \in N_c$, hence $i\circ n \in [I']$, $(i.n)\circ c = (i\circ c).(n\circ c) = (i\circ c).c' \in I'$, hence $i.n \in [I']$ and similarly $n.i \in [I']$. We list some properties of these two ideals: Let I', J' denote ideals of N_c:

4. $[I']_c = I'$

If $i \in [I']_c$, then $i \in [I'] \cap N_c$, so $i = i\circ c \in I'$.

If $i \in I' \trianglelefteq N_c$, then $i\circ c = i \in I'$ for all $c \in N_c$, hence $i \in [I']$.

5. If $I'\subset J'$, then $[I']\subset[J']$ and $(I')\subset(J')$ and vice versa.
Let $i\in[I']$, then $i\circ c\in I'\subset J'\ \forall\ c\in N_c$, hence $i\in[J']$.
If $i\in(I')$, then $i=\sum p_i\circ j_i$ with $j_i\equiv 0 \bmod I'\subset J'$, so $j_i\equiv 0 \bmod J'$ and $p_i\circ j\equiv p_i\circ 0=0 \bmod J'$ and $i\in(J')$.

6. $[I'\cap J']=[I']\cap[J']$ and $(I'\cap J')=(I')\cap(J')$.

THEOREM 3: For all ideals $I\trianglelefteq N$ there exists an uniquely determinded ideal $I_c\trianglelefteq(N_c,+,.)$, such that $(I_c)\subseteq I\subseteq[I_c]$.
I_c is called the *enclosing ideal* of I.

Proof: We set $I_c=I\cap N_c$.
If $i\in(I_c)$, then $i=\sum_{\text{finite}} p_i\circ j_i$ with $j_i\in I_c$ and $p_i\in P_\circ(N)$.
Since $I_c\subseteq I$ we get from $j_i\equiv 0 \bmod I$ also $p_i\circ j_i\equiv p_i\circ 0=0 \bmod I$, hence $i\in I$.
If $i\in I$, then $i\circ c\in I\cap N_c=I_c$ for all $c\in N_c$, hence $i\in[I_c]$.
If K is another such an ideal with $(K)\subseteq I\subseteq[K]$, then $(K)\subseteq[I_c]$ and for all $k\in K$ we get $k=k\circ c\in I_c,\ \forall\ \in N_c$, hence $K\subseteq I_c$.
From $(I_c)\subseteq[K]$ we get $I_c\subseteq K$ in a similar way, hence $K=I_c$.

COROLLARY 4: The ideal lattice of N is a homomorphic image of the ideal lattice of N.

Proof: We set: $h(I)=I_c$ for $I\trianglelefteq N$.

a) $h(I)$ is uniquely determined by Theorem 3.

b) $h(I\cap J)=(I\cap J)_c=(I\cap J)\cap N_c=(I\cap N_c)\cap(J\cap N_c)=I_c\cap J_c=h(I)\cap h(J)$

c) $h(I+J)=(I+J)\cap N_c=(I\cap N_c)+(J\cap N_c)=I_c+J_c=h(I)+h(J)$.

To get all maximal ideals, we assume for N:

(i) $(N,.)$ has an identity $1_c\in N_c$.

or

(ii) N_c generates N, i.e. $(N_c)=N$.

THEOREM 4: All maximal ideals M of $(N,+,.,\circ)$ are given by the ideals $M=[M']$, where M' is a maximal ideal of $(N_c,+,.)$.

Proof: a) If M' is maximal in N_c, then $[M']$ is maximal in N: First $[M']\neq N$, otherwise $[M']_c=M'=N_c$. If there exists a proper ideal $B\trianglelefteq N$ with $N\supset B\supset[M']$, then $B_c\supset M'$, otherwise $B\subseteq[B_c]\subseteq[M']$, so $B_c=N$.
By property (i): $1_c\in B_c\subseteq B$ and $n=1_c.n\in B\ \forall\ n\in N$, so $B=N$.
By property (ii): If $B_c=N_c$, then $N=(B_c)\subseteq B$, so $B=N$.

b) If M is maximal in N, then M_c is maximal in N_c: $M_c\neq N$, otherwise we get by similar arguments as in a): $M=N$. If there exists a proper ideal $A\triangleleft N_c$ with $M_c\subseteq A\subset N_c$, then $M\subseteq[M_c]\subset[A]\subset[N_c]=N$ in contradiction to the maximality of M in N.

COROLLARY 5: All maximal ideals M of the polynomial composition-ring $(R[x],+,.,\circ)$ over a commutative ring with identity are given by the ideals $M=(M':R)$, where M' is a maximal ring ideal of R.

COROLLARY 6: If N has only the trivial annihilator with respect to $\circ$ and generates N or possesses an identity with respect to multiplication, then the composition ring N is a subdirect product of simple composition-rings iff N_c is a subdirect product of simple rings.

Proof: Rad $N=[\text{Rad}(N_c)]$ by Theorem 4 and Property 6, so Rad $N=0$ iff $\text{Rad}(N_c)=0$, otherwise there exists an $n\neq 0$ with $n\circ c=0$ $\forall c\in N_c$.

COROLLARY 7: The polynomial composition-ring $(R[x],+,.,\circ)$ over an infinite integral domain is a subdirect product of simple composition-rings iff R is a subdirect product of fields.

REFERENCES

[1] Jennings, A.S., Substitution groups of formal power series, Canad.J.Math. 6(1954), 325-340.
[2] Kautschitsch, H. and Müller, W.B., Ideale in Kompositionsringen formaler Potenzreihen mit nilpotenten Anfangskoeffizienten, Arch.d.Math. 34(1980), 517-525.
[3] Lausch, H. and Nöbauer, W., Algebra of Polynomials (North-Holland, Amsterdam, 1973).
[4] Pilz, G., Near-rings (North-Holland, Amsterdam, 1977).
[5] Pilz, G., Near-rings, what they are and what they are good for, Contemporary Math., 9(1982), 97-119.

Near-rings and Near-fields, G. Betsch (editor)

D. G. NEAR-RINGS ON THE INFINITE DIHEDRAL GROUP

S.J. MAHMOOD and J.D.P. MELDRUM

Department of Mathematics,
University Studies Centre for Girls,
King Saud University,
P. O. Box 22452,
Riyadh,
Saudi Arabia.

Department of Mathematics,
University of Edinburgh,
King's Buildings,
Mayfield Road,
Edinburgh, EH9 3JZ,
Scotland.

In this paper, we look at the behaviour of all the d. g. near-rings that can be defined on the infinite dihedral group. We are mainly interested in determining which d. g. near-rings are faithful and in characterizing the lower and upper faithful d. g. near-rings for those which are not faithful. The upper and lower defects are also found in all these cases and in three cases they turn out to be not equal.

1. INTRODUCTION.

This paper is concerned with the d. g. near-rings on the infinite dihedral group. These have been determined by J. J. Malone [2]. Using the techniques of J. D. P. Meldrum [5] and [6], we determine the upper and lower faithful d. g. near-rings for all the d. g. near-rings and two sets of generators for each d. g. near-ring. Following [6], we also determine the upper and lower defects and defects in all these cases. Before giving more details, we give a summary of the definitions.

We write maps on the right and hence we use left near-rings. As a general source for definitions and results we refer to J. D. P. Meldrum [7], especially chapter 13. See also G. Pilz [8] for a more complete survey of near-rings. $(R,+,.)$ is a left zero-symmetric near-ring if $(R,+)$ is a group with neutral element 0, $(R,.)$ is a semigroup, $x(y+z) = xy + xz$ for all $x,y,z \in R$ and $0x = x0 = 0$ for all $x \in R$. The standard example of such a near-ring is $M_0(G) := \{f: G \to G; 0_G f = 0_G\}$, and all near-rings are embeddable in $M_0(G)$ for a suitable G. If $s \in R$ satisfies $(x+y)s = xs + ys$ for all $x,y \in R$ then s is called distributive. R is distributively generated (d. g.) if there exists a semigroup $(S,.)$ contained in R such that $(R,+) = Gp\langle S\rangle$. End G, the semigroup of all endomorphisms of G, forms a distributive semigroup in $M_0(G)$ and generates the d. g. near-ring (E(G),End G). We always write d. g. near-rings in the form (R,S) as which semigroup S is taken to be the distributive generating set for R is very important.

Let (R,S), (T,U) be two d. g. near-rings. Then a near-ring homomorphism $\theta: (R,S) \to (T,U)$ such that $S\theta \leq U$ is called a d. g. (near-ring) homomorphism. We say that (R,S) is faithful if there is a d. g. monomorphism $\theta: (R,S) \to (E(G),\text{End } G)$ for some group G. Not all d. g. near-rings are faithful, but to each d. g. near-ring (R,S) are associated two canonical faithful d. g. near-rings. The lower faithful d. g. near-ring for (R,S) is a d. g. near-ring $(\underline{R},\underline{S})$ and a d. g. epimorphism $\theta: (R,S) \to (\underline{R},\underline{S})$ such that $S\theta = \underline{S}$, and for any d. g.

homomorphism ϕ: (R,S) $\to$ (T,U) where (T,U) is faithful, there exists a unique d. g. homomorphism ψ: ($\underline{R}$,$\underline{S}$) $\to$ (T,U) such that $\phi = \theta\psi$. The upper faithful d. g. near-ring for (R,S) is a d. g. near-ring (R^*,S^*) and a d. g. epimorphism θ: (R^*,S^*) $\to$ (R,S) such that $\theta|_S$ is an isomorphism of semigroups, and for any d. g. homomorphism ϕ: (T,U) $\to$ (R,S) where (T,U) is faithful, there exists a unique d. g. homomorphism ψ: (T,U) $\to$ (R^*,S^*) such that $\phi = \psi\theta$. For these results see Meldrum [3] for the prehistory and Mahmood [1] for the details. There is one minor change of notation from the earlier papers, involving the upper faithful d. g. near-ring. To end the definitions we define $\underline{D}$(R,S) := Ker((R,S)$\to$($\underline{R}$,$\underline{S}$)) and D^*(R,S) := Ker((R^*,S^*)$\to$(R,S)), the lower and upper defects for (R,S) respectively. D(R,S) := Ker((R^*,S^*)$\to$($\underline{R}$,$\underline{S}$)) is the defect for (R,S).

We need the idea of presentations from group theory. For each set X, we can define Fr(X), the free group on X. Let R be a set of elements in Fr(X), N the normal subgroup of Fr(X) generated by R. Then G := Fr(X)/N is said to be given by the presentation G = Gp<X;R>. We write <X;R> for Gp<X;R>. Presentations are far from unique: for a given group G, different generating sets X can be chosen, and for each X different sets R of defining relations can be used. If we start with a presentation R = <S;Y> for the d. g. near-ring (R,S) we wish to obtain presentations of $R^* = <S^*;Y^*>$, $\underline{R} = <\underline{S};\underline{Y}>$ for the upper and lower faithful d. g. near-rings. Multiplication in R^* and $\underline{R}$ is defined completely by the multiplication in S^* and $\underline{S}$ (12.7, [7]), which in turn is given by that in S. In this paper the group we start with is D_∞, the infinite dihedral group. J. J. Malone [2] has determined all the d. g. near-rings (R,S) such that (R,+) is an infinite dihedral group. For those that are not faithful, we determine the upper and lower faithful d. g. near-ring for two sets S of generators, and the lower faithful d. g. near-ring in a number of other cases. We cover the lower faithful d. g. near-ring first, in section 2, then the upper faithful d. g. near-ring, in section 3. Finally we consider defects in section 4.

Before we leave this section we quote the results we need from Malone [2]. We write (R,+) as the infinite dihedral group, with what is probably the standard presentation

$$(R,+) = \langle a,b;\ 2b,\ a + b + a + b\rangle. \tag{1.1}$$

From [2] we know that there are six d. g. near-rings defined on (R,+):

(i) R_0, the zero d. g. near-ring on (R,+). This is a distributive near-ring, i. e. all elements are distributive.

(ii) R_1, with multiplication determined by the following table

	a	b
a	b	b
b	b	b

This near-ring is also distributive.

(iii) R_2, with multiplication determined by the following table, again giving a distributive near-ring.

	a	b
a	b	0
b	0	0

(iv) R_3, with multiplication determined by the following table, another distributive near-ring.

	a	b
a	0	0
b	0	b

(v) R_4, with multiplication given by

$$\begin{aligned} xy &= y \quad \text{if } x \notin Gp\langle a\rangle, \\ &= 0 \quad \text{if } x \in Gp\langle a\rangle. \end{aligned}$$

The distributive elements are precisely the elements of finite order.

(vi) R_5, with multiplication given by

$$\begin{aligned} xy &= y \quad \text{if } x \notin Gp\langle 2a,a+b\rangle, \\ &= 0 \quad \text{if } x \in Gp\langle 2a,a+b\rangle. \end{aligned}$$

The distributive elements are again precisely the elements of finite order.

Note that both R_4 and R_5 have left identities, namely any element not in $Gp\langle a\rangle$ for R_4, any element not in $Gp\langle 2a,a+b\rangle$ for R_5. So they are both faithful (13.12, [7]). For the rest of the paper, therefore, we will only be interested in R_0, R_1, R_2 and R_3.

2. THE LOWER FAITHFUL D. G. NEAR-RING.

We first have to establish the method we will use. Essentially we are using a form of the method used in section 3 of [5], based on theorems 2.5 and 2.6 of that paper, and on lemma 4.7 of [3]. See also theorem 13.32 of [7]. We adapt these results as follows.

Let (R,S) be a d. g. near-ring without identity. Without loss of generality, assume that (R,S) is not faithful, and has lower faithful d. g. near-ring $(\underline{R},\underline{S})$. Adjoin an identity to $(\underline{R},\underline{S})$ as described in theorem 13.32 of [7] to obtain (T,U) say, where $U = \{\underline{S}\} \cup \{1\}$. Then (T,+) is the free product of $(\underline{R},+)$ and a free group on the single element 1, say

$$(T,+) = (\underline{R},+) * \langle 1\rangle.$$

In fact the free product, denoted by $*$, can be taken to be in any variety to which (R,+) belongs. Which variety we choose does not affect the argument. From the construction of $(\underline{R},\underline{S})$, it follows that (T,+) is the free (R,S)-module on the element 1 (lemma 13.20 and theorem 13.21 of [7]). Let $N = \text{Ker }\theta$, where $\theta : (R,S) \to (\underline{R},\underline{S})$, is the canonical homomorphism. Then $w(s_1, \ldots, s_n)$, a word in the elements of S, is in N precisely when $1w(s_1, \ldots, s_n) = 0$ in (T,+). Using theorems 2.5 and 2.6 of [5], we find those words $w(s_1, \ldots, s_n)$ such that $w(s_1, \ldots, s_n) \in (T',+)r$, where $r \in X$ for a presentation $(R,+) = <S;X>$, and

$$(T',+) = (R,+) * <1>.$$

We add these words to the relations in the presentation of (R,+). We then repeat the process starting now with the new presentation. Eventually we reach a presentation for $(\underline{R},+)$. This is the method described in theorem 2.6 of [5]. In practice, by judicious choice of the relations to add on in the first stage, the process takes only two steps, not a possibly infinite number.

Familiarity with the details of the structure of the infinite dihedral group is assumed. For those who are not familiar with it, any reasonable text-book or tame group theorist colleague will supply the necessary information. We will start by listing some properties of the four d. g. near-rings (R_i,S), $0 \leq i \leq 3$, which are easily deducible, and are some help in subsequent work, though not all of them are completely necessary for what we do later. We use the presentation given in (1.1).

To show that R_j, for j = 0,1,2,3, is not faithful, we use (R^*,S^*). The definition of R^* is given by theorem 2.2 of [6] or theorem 13.26 of [7], namely for each $s \in S$, define s^* as an endomorphism of $(T^*,+) := (R,+) * <1>$ by $1s^* = s \in R$, $rs^* = rs$, and extending, as we may by the properties of free products. Then the d. g. near-ring generated by S^* in $M_0(T^*)$ is (R^*,S^*).

Lemma 2.1. *Using the presentation* (1.1), *we have that* a + b + a + b *is not a relation in* (R_j^*,S^*) *for* j = 0,1,2,3.

Proof. We just need to find $g \in T^*$ such that $g(a+b+a+b) \neq 0$ in $(R_j^*,+)$.

For R_0, let g = 1 + 1. Then

$$g(a^*+b^*+a^*+b^*) = 2a + 2b + 2a + 2b = 4a.$$

For R_1, let g = 1 + a − 1. Then

$$\begin{aligned} g(a^*+b^*+a^*+b^*) &= a + b - a + b + b - b + a + b - a + b + b - b \\ &= a + b - a - b + a + b - a - b = 4a. \end{aligned}$$

For R_2, let g = 1 + a. Then

$$g(a^*+b^*+a^*+b^*) = a + b + b + 0 + a + b + b + 0 = 2a.$$

For R_3, let g = 1 + b. Then

$$g(a^*+b^*+a^*+b^*) = a + 0 + b + b + a + 0 + b + b = 2a.$$

Corollary 2.2. (R_j,S) *is not faithful for* $j = 0,1,2,3$.

Proof. This follows directly from lemma 2.1, since $a + b + a + b$ is a relation in $(R_j,+)$, but not in $(R_j^*,+)$.

The normal subgroups of D_∞ are well-known. This enables us to determine all the ideals of R_j, for $j = 0,1,2,3$. First we list the normal subgroups of D_∞. This list can be found in [2] or in [4].

Lemma 2.3. *The normal subgroups of* $(R,+)$ *given by* (1.1) *are* R, $\{0\}$ *and*

$N_1 := \langle 2a,a+b \rangle$,

$N_2 := \langle 2a,b \rangle$,

$hA := \langle ha \rangle$, *for* $h \geq 1$.

Corollary 2.4. *The complete set of ideals of* (R_0,S) *is given by* $\{R_0, N_1, N_2, hA; h \geq 0\}$.

Proof. This follows immediately from lemma 2.3 since in R_0, all products are 0.

Theorem 2.5. (i) *The complete set of ideals of* (R_1,S) *is given by* $\{R_1, N_1, N_2, hA;$ *h an even integer*$\}$.

(ii) *The complete set of ideals of* (R_2,S) *is given by* $\{R_2, N_2, A, hA;$ *h an even integer*$\}$.

(iii) *The complete set of ideals of* (R_3,S) *is given by* $\{R_3, N_2, hA; h \geq 0\}$.

Proof. All three near-rings are distributive. So the ideals are normal subgroups H such that $R_jH \leq H$, $HR_j \leq H$ for each R_j. The results are then found by a simple checking process, using the list in lemma 2.3.

Before we start on the determination of lower faithful near-rings, we remark that $R/\langle 2a \rangle$ is an abelian group of order 4.

Theorem 2.6. $R_j/\langle 2a \rangle$ *for* $j = 0,1,2,3$, *is always faithful, being a ring.*

Proof. This follows since a d. g. near-ring with an abelian group structure is a ring (Corollary 9.26 of [7]).

Another remark that needs to be made is that since R_j, $j = 0,1,2,3$, is distributive, the semigroup S of distributive generators can be taken to be any multiplicative semigroup that generates R. For each near-ring, we find the lower faithful d. g. near-ring for a number of different generating semigroups.

Theorem 2.7. *Consider* (R_0,S), *the zero d. g. near-ring on the infinite dihedral group.*

(i) *If* $S \leq$ *the set of elements of order* 2, *then* (R_0,S) *is faithful.*

(ii) *If* $S = \{a,b\}$, *then* $(\underline{R}_0,\underline{S})$ *is* $(R_0/\langle 4a \rangle,\underline{S})$, *the zero d. g. near-ring on the dihedral group of order* 8.

(iii) *If* $S > \{a,b\}$ *and* S *contains only one element of order* 2, *namely* b, *then* $(\underline{R}_0,\underline{S})$ *is the same as in* (ii).

(iv) *If* $S \geq \{a,b,a+b\}$, *then* $(\underline{R}_0,\underline{S})$ *is* $(R_0/\langle 2a \rangle,\underline{S})$, *the zero ring on the group of order* 4, *exponent* 2.

Proof. We use lemma 3.3 of [5]. This states that (R,S) is faithful if and only if the presentation $(R,+) = \langle S,X \rangle$ is integrally closed, i. e. if $w(s_1, \ldots, s_n) \in X$ then

$w(ms_1, \ldots, ms_n)$ is a relation in R for all integers m. If S consists entirely of elements of order 2, this is automatically satisfied, as $w(ms_1, \ldots, ms_n) = 0$ if m is even, and is $w(s_1, \ldots, s_n)$ if m is odd. This proves (i).

To proceed, we note that

$$(1+1)(a+b+a+b) = a + a + b + b + a + a + b + b = 4a.$$

So 4a must be a relation in $(\underline{R_j},\underline{S})$ for j = 1,2,3, as long as $\{a,b\} \leq S$. The second example on p. 55 of **[5]** gives us (ii). In (iii), $S = \{a, n_i a, b\}$ for $i \in I$. The relations are $a + b + a + b$, $n_i a + b + n_i a + b$, $n_i a - a - \ldots - a$ or $n_i a + a + \ldots + a$, with an appropriate number of summands, depending on whether n_i is positive or negative. As for the second example on p. 55 of **[5]**, we get again that the lower faithful d. g. near-ring is the same in case (iii) as in case (ii).

Finally we deal with case (iv). We are now in the position described in lemma 13.7 of **[7]**. So in $(\underline{R_0},\underline{S})$ we must have $\underline{a} + \underline{b} = \underline{b} + \underline{a}$. This is sufficient to prove the result.

We now turn to R_1. From theorem 2.5, we know that the ideals of R_1 are R_1, N_1, N_2 and hA for all even integers h. The following result simplifies some calculations.

Lemma 2.8. $Ann(R_1) = N_1$.

This result is immediate from the multiplication table. Now a bit of notation which will be used consistently in the rest of the section.

Notation. An arbitrary word in (T',+) is of the form

$$w := n_1 1 + r_1 + n_2 1 + \ldots + n_s 1 + r_s$$

where n_1 may be 0, r_s may be 0, $n_i \in Z$, $r_i \in R_1$, for $1 \leq i \leq s$. Write

$$n := n_1 + \ldots + n_s.$$

Theorem 2.9. (R_1,S) *is faithful, where* $S = \{b,a+b\}$.

Proof. A presentation for $(R_1,+)$ is given by $R_1 = \langle b,a+b;\ b+b,\ a+b+a+b\rangle$. Write $x := \underline{b}$, $y := \underline{a+b}$. Consider $w(x+x)$. Then

$$\begin{aligned} w(x+x) &= n_1 b + r_1 b + \ldots + n_s b + r_s b + n_1 b + r_1 b + \ldots + n_s b + r_s b \\ &= 2(n_1+m_1+\ldots+n_s+m_s)b = 0 \end{aligned}$$

where $r_i b = m_i b$, $m_i \in Z$, for $1 \leq i \leq s$, from the multiplication table. Now consider $w(y+y)$. Then

$$\begin{aligned} w(y+y) &= n_1(a+b)+0+\ldots+n_s(a+b)+0+n_1(a+b)+0+\ldots+n_s(a+b)+0 \\ &= 2n(a+b) = 0, \end{aligned}$$

using lemma 2.8. Thus no new relations are introduced and so $(R_1,\{b,a+b\})$ is faithful, as required.

Lemma 2.10. *Let* (R_1,S) *be a d. g. near-ring, where* $2ra + b \in S$. *Then* $4r\underline{a}$ *is a relation in* $(\underline{R}_1,\underline{S})$.

Proof. Let $x := 2ra + b$. Then $x + x$ is a relation in $(R_1,+)$. Consider $(1+b)(x+x)$. We obtain

$$(1+b)(x+x) = 2r\underline{a} + \underline{b} + \underline{b} + 2r\underline{a} + \underline{b} + \underline{b} = 4r\underline{a}.$$

This suffices to prove the result.

We can use this in several cases. First we look at the situation in which S consists of two elements of order 2.

Theorem 2.11. *Let* $S = \{2ra+b,(2r\pm 1)a+b\}$. *Then* $(\underline{R}_1,\underline{S}) = R_1/\langle 4ra\rangle$.

Proof. Lemma 2.10 shows that $4r\underline{a}$ is a relation in $(R_1,+)$. Let $x := 2ra + b$, $y := (2r\pm 1)a + b$. Then $wx = n_1(2r\underline{a}+\underline{b}) + m_1\underline{b} + \ldots + n_s(2r\underline{a}+\underline{b}) + m_s\underline{b}$, where $r_ix = m_i\underline{b}$. Since $4r\underline{a} = 0$, it follows that $2r\underline{a}$ commutes with $\underline{b}$. Hence $wx = n2r\underline{a} + (n+m)\underline{b}$ where $m = m_1 + \ldots + m_s$. Also $wy = n_1((2r\pm 1)\underline{a}+\underline{b}) + 0 + \ldots + n_s((2r\pm 1)\underline{a}+\underline{b}) = n((2r\pm 1)\underline{a}+\underline{b})$. We now have

$$w(x+x) = 2(n2r\underline{a}+(n+m)\underline{b}) = 4nr\underline{a} + 2(n+m)\underline{b} = 0,$$
$$w(y+y) = 2n((2r\pm 1)\underline{a}+\underline{b}) = 0.$$

Thus we have a new relation, namely $4r(x+y)$. Then

$$w(4r(x+y)) = 4r(wx+wy).$$

But in $R_1/\langle 4ra\rangle$ all elements have order dividing $4r$. Hence $w(4r(x+y)) = 0$. Thus $R_1/\langle 4ra\rangle$ is faithful.

Next we consider the case in which $S = \{b,a+b,2ra+b\}$. Denote b by x, a+b by y and 2ra+b by z.

Theorem 2.12. *Let* $S = \{b,a+b,2ra+b\}$. *Then* $(\underline{R}_1,\underline{S}) = R_1/\langle 2ra\rangle$.

Proof. Again lemma 2.10 shows that $4r\underline{a}$ is a relation in $(\underline{R}_1,+)$, and so $2r\underline{a}$ commutes with $\underline{b}$. By the above work, we know that the relations $x + x$, $y + y$ and $z + z$ do not introduce new relations into $(\underline{R}_1,+)$. But we have also a relation in $(R_1,+)$ given by $2r(y+x) + x + z$. Consider $w(2r(y+x)+x+z)$.

$$\begin{aligned} w(2r(y+x)+x+z) &= 2r(wy+wx) + wx + wz \\ &= 2r(n(\underline{a}+\underline{b})+(n+m)\underline{b}) + (n+m)\underline{b} + n2r\underline{a} + (n+m)\underline{b} \end{aligned} \qquad (2.13)$$

where $n = n_1 + \ldots + n_s$, $mb = (r_1+ \ldots +r_s)b = (r_1+ \ldots +r_s)(2ra+b)$, since $2ra \in Ann(R_1)$. We use a similar argument to that used in the proof of theorem 2.11. We divide the consideration of this element into four cases, depending on whether n or m are even or odd.

(i) n even, m even. Then (2.13) is 0 as $4r\underline{a} = 0$.

(ii) n even, m odd. Then (2.13) becomes $2r\underline{b} + \underline{b} + \underline{b} = 0$.

(iii) n odd, m even. Then (2.13) becomes $2r(\underline{a}+\underline{b}+\underline{b}) + \underline{b} + 2r\underline{a} + \underline{b} = 2r\underline{a} + \underline{b} + 2r\underline{a} + \underline{b} = 0$.

(iv) n odd, m odd. Then (2.13) becomes $2r(\underline{a}+\underline{b}) + 2r\underline{a} = 2r\underline{a}$.

Hence $2r\underline{a}$ is a relation in $\underline{R}_1$. Having added this relation we can repeat the process. The new relation is $2r(y+x)$. So $w(2r(y+x)) = 2r(wy+wx)$. But in $R_1/<2ra>$ all elements have order dividing 2r. Hence $w(2(y+x)) = 0$ always. Thus $R_1/<2ra>$ is faithful.

Theorem 2.14. *Let* $S = \{b,a+b,(2r+1)a+b\}$. *Then* $(\underline{R}_1,\underline{S}) = R_1/<2ra>$.

Proof. From the proof of theorem 2.9 we know that b + b and a + b + a + b do not introduce new relations. As for a + b + a + b, we can check that (2r+1)a + b + (2r+1)a + b does not introduce a new relation. So we consider $(2r+1)(y+x) + x + z$, where $x := b$, $y := a + b$ and $z := (2r+1)a + b$. With n and m defined as in theorem 2.12, we obtain, since $z \in \mathrm{Ann}(R_1)$,

$$\begin{aligned} w((2r+1)(y+x)+x+z) &= (2r+1)(wy+wx) + wx + wz \\ &= (2r+1)(n(\underline{a}+\underline{b})+(n+m)\underline{b}) + (n+m)\underline{b} + n((2r+1)\underline{a}+\underline{b}) \end{aligned} \tag{2.15}$$

We consider separately four cases as in theorem 2.12.

(i) n even, m even. Then (2.15) is 0.

(ii) n even, m odd. Then (2.15) becomes $(2r+1)\underline{b} + \underline{b} = 0$.

(iii) n odd, m even. Then (2.15) becomes $(2r+1)(\underline{a}+\underline{b}+\underline{b}) + \underline{b} + (2r+1)\underline{a} + \underline{b} = 0$.

(iv) n odd, m odd. Then (2.15) becomes $(2r+1)(\underline{a}+\underline{b}) + (2r+1)\underline{a} + \underline{b} = \underline{a} + \underline{b} + (2r+1)\underline{a} + \underline{b} = \underline{a} - (2r+1)\underline{a} = -2r\underline{a}$.

Hence $2r\underline{a}$ is a relation in $\underline{R}_1$. Having added this relation we can repeat the process. The new relation is $2r(y+x)$. As in theorem 2.12, this introduces no new relations. Thus $R_1/<2ra>$ is faithful.

There are a number of other results giving the lower faithful d. g. near-ring for (R_1,S) where S consists entirely of elements of order 2. They are somewhat more complicated than the previous cases. So we list them without proof.

Result 2.16. (i) *Let* $S = \{2ra+b,(2r\pm1)a+b,2r'a+b\}$, *with* $r \neq r'$, $r,r' \geq 1$. *Let* $t = \mathrm{H.C.F.}(r,r')$. *If* $(r-r')/t$ *is even, then* $(\underline{R}_1,\underline{S}) = R_1/<4ta>$. *If* $(r-r')/t$ *is odd, then* $(\underline{R}_1,\underline{S}) = R_1/<2ta>$.

(ii) *Let* $S = \{2ra+b,(2r+1)a+b,(2r'+1)a+b\}$, *with* $r \neq r'$, $r,r' \geq 1$. *Let* $t = \mathrm{H.C.F.}(4r,2r-2r')$. *Then* $(\underline{R}_1,\underline{S}) = R_1/<ta>$.

(iii) *Let* $S = \{(2r-1)a+b,2ra+b,(2r'+1)a+b\}$, *with* $r \neq r'$, $r,r' \geq 1$. *Let* $t = \mathrm{H.C.F.}(4r,2(r-r'-1))$. *Then* $(\underline{R}_1,\underline{S}) = R_1/<ta>$.

We now turn to generating sets involving an element of infinite order.

Theorem 2.17. *Let* $S = \{a,ra+b\}$, *where* $r \in Z$. *Then* $(\underline{R}_1,\underline{S}) = R/<4a>$.

Proof. Write $x := a$, $y := ra + b$. Then the relations in a presentation of R_1 are $y + y$, $x + y + x + y$. By considering $(1+1)(x+y+x+y) = 2a + 2(ra+b) + 2a + 2(ra+b) = 4a$, we see that $4\underline{a}$ is a relation in $\underline{R}_1$. We first find wx and wy: $wx = n_1\underline{a} + m_1\underline{b} + \ldots + n_s\underline{a} + m_s\underline{b}$, $wy = n\underline{y}$, if $y \in N_1$, $wy = n_1\underline{y} + m_1\underline{b} + \ldots + n_s\underline{y} + m_s\underline{b}$ if $y \notin N_1$. Now if $y \notin N_1$, then r is even. Since $4\underline{a} = 0$, it follows that $2\underline{a}$ commutes with $\underline{b}$. So $wy = n\underline{y} + m\underline{b}$ if $y \notin N_1$. Consider

$w(y+y)$. If r is odd, then $y \in N_1$, so $w(y+y) = 2ny = 0$ since y has order 2. If r is even, then $w(2y) = 2(n\underline{y}+m\underline{b})$. The possible values of $n\underline{y} + m\underline{b}$ are $r\underline{a}$, $r\underline{a} + \underline{b}$, $\underline{b}$, 0. Since r is even $2(n\underline{y}+m\underline{b})$ is zero in all cases as $4\underline{a} = 0$.

Now look at $w(x+y+x+y)$. If $y \in N_1$, then r is odd, and $w(x+y+x+y) = 2(n_1\underline{a}+m_1\underline{b}+ \dots +n_s\underline{a}+m_s\underline{b}+n(r\underline{a}+\underline{b}))$. Suppose n is even. Then this reduces to $2(n_1\underline{a}+m_1\underline{b}+ \dots +n_s\underline{a}+m_s\underline{b})$ and $n = n_1 + \dots + n_s$. If $m_1 + \dots + m_s$ is odd, this is of the form $2(q\underline{a}+\underline{b})$ which is 0. If $m_1 + \dots + m_s$ is even, this is of the form $2(n_1\underline{a}+\varepsilon_2 n_2\underline{a}+ \dots +\varepsilon_s n_s\underline{a}) = 2(n_1+\varepsilon_2 n_2+ \dots +\varepsilon_s n_s)\underline{a}$ where $\varepsilon_i = \pm 1$ for $2 \leq i \leq s$. But $n_1 + \dots + n_s$ even forces $n_1 + \varepsilon_2 n_2 + \dots + \varepsilon_s n_s$ even. So $w(x+y+x+y)$ reduces to a multiple of $4\underline{a}$, hence is 0. Now suppose that n is odd. Then $w(x+y+x+y) = 2(n_1\underline{a}+m_1\underline{b}+ \dots +n_s\underline{a}+m_s\underline{b}+r\underline{a}+\underline{b})$. Since $n_1 + \dots + n_s + r$ is even (as n and r are odd), we can repeat the above process to show that this is zero. Now let $y \notin N_1$. Then r is even and $w(x+y+x+y) = 2(n_1\underline{a}+m_1\underline{b}+ \dots +n_s\underline{a}+m_s\underline{b}+n\underline{y}+m\underline{b})$. If n is even, this becomes $2(n_1\underline{a}+m_1\underline{b}+ \dots +n_s\underline{a}+(m_s+m)\underline{b})$ and can be treated as above. If n is odd, then this becomes $2(n_1\underline{a}+m_1\underline{b}+ \dots +n_s\underline{a}+m_s\underline{b}+r\underline{a}+\underline{b}+m\underline{b})$. The coefficient of $\underline{b}$ is $m_1 + \dots + m_s + 1 + m = 2m + 1$. So this element is of the form $2(q\underline{a} + \underline{b})$ which is 0. This covers all cases for the relation $x + y + x + y$.

The final relation is $4\underline{a}$, i. e. $4x$. But $w(x+x+x+x) = 4wx = 0$, since in $R_1/\langle 4a\rangle$ all elements have order divisible by 4. Thus no new relations are introduced and we have our result.

Our final look at d. g. near-rings on R_1 is at the case where $S = \{a,b,ra\}$, for $r > 1$.

Theorem 2.18. *Let* $S = \{a,b,ra\}$, *for* $r > 1$. *Then* $(\underline{R}_1,\underline{S})$ *is* $R_1/\langle 2a\rangle$ *if* r *is odd and* $(r-1)/2$ *is also odd, or* r *is even and* $r/2$ *is odd. Otherwise* $(\underline{R}_1,\underline{S})$ *is* $R_1/\langle 4a\rangle$.

Proof. Let $x := a$, $y := b$, $z := ra$. As at the beginning of the proof of theorem 2.17 we see that $4\underline{a}$ is a relation in $\underline{R}_1$. The relations in a presentation for R_1 can be taken as $y + y$, $x + y + x + y$, $rx - z$. We first find what effect x, y and z have on w: $wx = n_1\underline{a}+m_1\underline{b}+ \dots +n_s\underline{a}+m_s\underline{b}$, $wy = (n+m)\underline{b}$, as in the proof of theorem 2.17, and $wz = n_1\underline{z}+m_1\underline{b}+ \dots +n_s\underline{z}+m_s\underline{b}$ if $z \notin N_1$, i. e. r is odd, and $wz = n\underline{z}$ if $z \in N_1$, i. e. r is even. We consider the innocuous relations first. We have $w(y+y) = 2(n+m)\underline{b} = 0$ in all cases. Also $w(x+y+x+y) = 2(n_1\underline{a}+m_1\underline{b}+ \dots +n_s\underline{a}+m_s\underline{b}+(n+m)\underline{b})$. This is of the form $2(q\underline{a}+\underline{b}) = 0$ unless n is even since $m_1 + \dots + m_s = m$. But then an argument similar to that in the proof of theorem 2.17 shows that we get 0.

Now we deal with $rx - z$. We have $w(rx-z) = r(n_1\underline{a}+m_1\underline{b}+ \dots +n_s\underline{a}+m_s\underline{b}) - wz$. If r is even then we have $r(n_1\underline{a}+m_1\underline{b}+ \dots +n_s\underline{a}+m_s\underline{b}) - nr\underline{a}$. If $4|r$, then this is 0 as $R_1/\langle 4a\rangle$ has exponent 4. If $2|r$ but $r/2$ is odd, then take $n_1 = 1$, $m_1 = 1$, $s = 1$. We get $r(\underline{a}+\underline{b}) - r\underline{a} = -r\underline{a}$ as r is even. But $4\underline{a} = 0$, $-r\underline{a} = 0$ implies $t\underline{a} = 0$, where $t = \text{H.C.F.}(4,r) = 2$, by hypothesis. So $R_1/\langle 2a\rangle$ is $(\underline{R}_1,\underline{S})$ as it is a ring (theorem 2.6). Now let r be odd. Then $w(rx-z) = r(n_1\underline{a}+m_1\underline{b}+ \dots +n_s\underline{a}+m_s\underline{b}) - (n_1r\underline{a}+m_1\underline{b}+ \dots +n_sr\underline{a}+m_s\underline{b})$. But $n_1\underline{a}+m_1\underline{b}+ \dots +n_s\underline{a}+m_s\underline{b} = q\underline{a} + \underline{b}$, where $m_1 + \dots + m_s = m$, $q = n_1 + \varepsilon_2 n_2 + \dots +\varepsilon_s n_s$ for some choice of $\varepsilon_i = \pm 1$, $2 \leq i \leq s$. Then $n_1r\underline{a}+m_1\underline{b}+ \dots +n_sr\underline{a}+m_s\underline{b} = qr\underline{a} + m\underline{b}$. So $w(rx-z) = r(q\underline{a}+\underline{b}) - (qr\underline{a}+m\underline{b})$. If m is odd this is $r(q\underline{a}+\underline{b}) + qr\underline{a} + \underline{b} = q\underline{a} + \underline{b} + qr\underline{a} + \underline{b} = q(1-r)\underline{a}$. By taking $n_1 = 1 = m_1$, s = 1, we get $(1-r)\underline{a} = 0$. If $(1-r)/2$ is odd, as before this leads to $R_1/\langle 2a\rangle = (\underline{R_1},\underline{S})$. If $4|(r-1)$, then this is always zero. If m is even $w(rx-z)$ becomes $rq\underline{a} - qr\underline{a} = 0$ for all choices of q.

The final check is whether $4\underline{a}$ introduces new relations in the two cases $4|r$, $4|(r-1)$. But as before, since $R_1/\langle 4a\rangle$ has exponent 4, no new relations are introduced.

This finishes our look at the lower faithful d. g. near-rings on R_1. We turn to R_2 next. Fortunately the work is much easier here. The first result parallels lemma 2.8.

Lemma 2.19. $\mathrm{Ann}(R_2) = N_2$.

Again this is obvious from the multiplication table. The following result is useful.

Lemma 2.20. *Let* (R_2,S) *be a d. g. near-ring. If* r *is odd and* $ra + b \in S$, *then* $2r\underline{a} = 0$ *in* R_2.

Proof. Put $x := ra + b$. Then $x + x$ is a relation. So $(1+a)(x+x) = x + ax + x + ax = ra + b + rb + ra + b + rb = 2ra$ since $b + rb = 0$.

Theorem 2.21. *If* $a + b \in S$, *then* $(\underline{R_2},\underline{S}) = R_2/\langle 2a\rangle$.

Proof. Use lemma 2.20 and theorem 2.6.

Theorem 2.22. *If* $a \in S$, *then* $(\underline{R_2},\underline{S}) = R_2/\langle 2a\rangle$.

Proof. Note that S must contain at least one element of order 2, say $ra + b$. Put $x := a$, $y := ra + b$. Consider $(1+1)(x+y+x+y)$. As before this forces $4\underline{a} = 0$. If r is odd, then lemma 2.20 forces $2r\underline{a} = 0$, which together with $4\underline{a} = 0$ gives $2\underline{a} = 0$. Another appeal to theorem 2.6 gives the result. So assume that r is even. Consider $(1+a)(x+y+x+y) = \underline{x} + \underline{b} + \underline{y} + \underline{x} + \underline{b} + \underline{y}$, since $(1+a)x = \underline{x} + \underline{b}$, $(1+a)y = \underline{y}$. So $(1+a)(x+y+x+y) = 2(\underline{a}+\underline{b}+r\underline{a}+\underline{b}) = 2(1-r)\underline{a} = 0$. Together with $4\underline{a} = 0$ we again get $2\underline{a} = 0$ and the result follows from theorem 2.6.

Finally we look at the case when $S = \{2ra+b,(2r\pm 1)a+b\}$ and $r > 0$. This gives the only non-ring lower faithful d. g. near-ring for R_2.

Theorem 2.23. *Let* $S = \{2ra+b,(2r\pm 1)a+b\}$. *Then* $(\underline{R_2},\underline{S}) = R_2/\langle 2(2r\pm 1)a\rangle$.

Proof. Put $t = 2r\pm 1$. That $2t\underline{a} = 0$ follows from lemma 2.20. We just need to show that no more relations come in. Let $x := 2ra + b$, $y := ta + b$. The relations we do have are $x + x$, $y + y$, $2t(x+y)$. Also $wx = n\underline{x}$, $wy = n_1(t\underline{a}+\underline{b}) + m_1\underline{b} + \ldots + n_s(t\underline{a}+\underline{b}) + m_s\underline{b}$. As $2t\underline{a} = 0$, $t\underline{a}$ commutes with $\underline{b}$. So $wy = n\underline{y} + (n+m)\underline{b}$. Further $w(x+x) = 2n\underline{x} = 0$ as $2x = 0$. Also $w(y+y) = 2(n\underline{y}+(n+m)\underline{b})$. If $n + m$ is even, this is $2n\underline{y}$ which is 0. If $n + m$ is odd, we get either $2\underline{b}$ (m odd, n even) or $2(\underline{y}+\underline{b}) = 2t\underline{a}$ (m even, n odd). In either case $w(y+y) = 0$. Finally $w(2t(x+y)) = 2tw(x+y)$. $R_2/\langle 2ta\rangle$ has exponent 2t. So $w(2t(x+y)) = 0$. This finishes the proof.

This brings us to the final d. g. near-ring to consider, namely R_3. We first note the annihilator of R_3.

Lemma 2.24. $\mathrm{Ann}(R_3) = A$.

This is again obtained directly from the multiplication table.

Lemma 2.25. *If* $ra + b \in S$, *then* $2r\underline{a}$ *is a relation in* R_3.

Proof. Consider $(1+b)(ra+b+ra+b) = ra + b + b + ra + b + b = 2ra$, since $b(ra+b) = b$.

Lemma 2.26. *If* $\{a,ra+b\} \leq S$, *then* $2(r+a)\underline{a}$ *is a relation in* R_3.

Proof. Consider $(1+b)(a+ra+b+a+ra+b) = \underline{a} + 0 + \underline{ra} + \underline{b} + \underline{b} + \underline{a} + 0 + \underline{ra} + \underline{b} + \underline{b} = 2(r+1)\underline{a}$.

This leads to our final result in this section.

Theorem 2.27. *If either* $a \varepsilon S$ *or* $a + b \varepsilon S$ *or* $\{ra+b,(r+1)a+b\} \leq S$, *then* $(R_3,\underline{S}) = R_3/<2a>$.

Proof. If $a \varepsilon S$, then S must contain an element of the form ra + b as well. By lemmas 2.25 and 2.26 it follows that $2r\underline{a}$ and $2(r+1)\underline{a}$ are relations, hence so is $2\underline{a}$, since H.C.F.(r,r+1) = 1. If $a + b \varepsilon S$, then lemma 2.25 gives the result. Finally if $\{ra+b,(r+1)a+b\} \leq S$, then $2r\underline{a}$ and $2(r+1)\underline{a}$ are relations, hence so is $2\underline{a}$ as before. Just apply theorem 2.6 now.

3. THE UPPER FAITHFUL D. G. NEAR-RINGS.

We will now determine the upper faithful d. g. near-rings for R_0, R_1, R_2 and R_3 for two sets of generators in each case, namely S = {a,b} and S = {b,a+b}. This is done as the determination of upper faithful d. g. near-rings is harder than that of lower faithful d. g. near-rings. For $(R_0,\{b,a+b\})$ and $(R_1,\{b,a+b\})$, the upper faithful d. g. near-rings are R_0 and R_1 respectively. So we only deal with the remaining cases. The method we use for finding upper faithful d. g. near-rings is that of theorem 2.2 of **[6]**.

A useful result in calculating upper faithful d. g. near-rings is theorem 13.12 of [7], which says that a free group in any variety is faithful. Hence it follows that if (R,+) lies in a variety, then so does $(R^*,+)$. We quote this as a lemma.

Lemma 3.1. *Let* (R,S) *be a d. g. near-ring. If* (R,+) *lies in a variety, then so does* $(R^*,+)$.

An application of this to our case is as follows. Since $(R_i,+)$ is metabelian, i. e. has abelian derived group, then so is $(R_i^*,+)$. Also R_i is generated by two elements in all the cases we consider. So $(R_i^*,+)$ will be a homomorphic image of the free metabelian group on two generators, a group which is easy to handle. In practice this means that to determine $(R_i^*,+)$ we need only determine the orders of the two generators, x and y say, find out what $[x,y] := -x - y + x + y$, is, and what the conjugates of [x,y] are. We write conjugates in the form $x^y := -y + x + y$.

We establish the following notation for the rest of this section. Let $T := (D_\infty,+) * <1>$, be the free product of an infinite dihedral group given by (1.1) and an infinite cyclic group generated by 1. Since we will generally be identifying D_∞ with $(R_j,+)$ for some j, we write a general element of T as

$$w = n_1 1 + r_1 + \ldots + n_s 1 + r_s, \tag{3.2}$$

where $n_i \varepsilon Z$, $r_i \varepsilon D_\infty$, for $1 \leq i \leq s$ and the first and last elements may be zero, $s \geq 1$. We remind the reader that theorem 2.2 of **[6]** shows that if we define the elements of S as endomorphisms of T by

$$1 . s = s, \quad r_i . s = r_i s$$

then S generates R_j^* in $M_0(T)$. We start with $(R_0,\{a,b\})$. Note that in R_0, we may assume without loss of generality that w = n1 since $r_i s = 0$ in all cases with R_0.

Theorem 3.3. *Let* S = {a,b}. *Then* (R_0^*,S) *is given by*

$$R_0^* = \langle a,b;\ 2b,\ [a,b]^a = [a,b],\ [a,b]^b = [b,a]\rangle.$$

Proof. We first check that 2b is a relation in R_0^*. But n1.(b+b) = 2nb = 0. Hence $(R_0^*,+) = \langle a,b\rangle$, a has infinite order and b has order 2, and by lemma 3.1, R_0^* is metabelian. Consider an arbitrary element of $(R_0^*,+)$. It is of the form $m_1a + b + m_2a + b + \ldots + m_ta + b + m_{t+1}a$, where $m_i \in Z$, $t \geq 1$, and m_1 and m_{t+1} may be zero. Then it is a relation if

$$n1(m_1a + b + \ldots + m_ta + b + m_{t+1}a) = m_1na + nb + \ldots + m_tna + nb + m_{t+1}na = 0,$$

for all n. If n is even, this means $n(m_1 + \ldots + m_{t+1})a = 0$ for all even values of n. As a has infinite order, this forces $m_1 + \ldots + m_{t+1} = 0$. If n is odd, then $m_1na + b + \ldots + m_tna + b + m_{t+1}na = 0$. This means that there must be an even number of b's, i. e. t = 2q and $m_1na - m_2na + \ldots - m_{2q}na + m_{2q+1}na = 0$. As before we must have $m_1 - m_2 + \ldots - m_{2q} + m_{2q+1} = 0$. Conversely it is easy to check that if $m_1a + b + \ldots + m_{2q}a + b + m_{2q+1}a$ satisfies $\Sigma_i m_i = 0$ and $\Sigma_i (-1)^i m_i = 0$, then it annihilates n1 for all n. This is equivalent to $\Sigma_{i=1}^q m_{2i} = 0$, $\Sigma_{i=0}^q m_{2i+1} = 0$. So the set of relations for R_0^* is

$$\{\Sigma_{i=1}^{2q}(m_ia+b) + m_{2q+1}a;\ \Sigma_{i=1}^q m_{2i} = 0,\ \Sigma_{i=0}^q m_{2i+1} = 0\}.$$

We now check what happens to the commutators. All commutators are words of the form $\Sigma_{i=1}^t(m_ia+b) + m_{t+1}a$ in which $\Sigma_{i=1}^{t+1} m_i = 0$. A standard technique from group theory shows that words in commutators are precisely words of this form. So all the relations in R_0^*, other than 2b, are in the derived group. We now consider $[a,b]^a - [a,b]$ = $-2a + b + a + b + a + b - a + b + a$, i. e. $m_1 = -2$, $m_2 = m_3 = 1$, $m_4 = -1$, $m_5 = 1$. This makes it easy to check that this is a relation. Also we look at $[a,b]^b - [b,a] = 0$, simply using 2b = 0. This finally shows that the presentation in the theorem is that of $(R_0^*,+)$.

We now move on to R_1. Here again we have only one case, namely S = {a,b}.

Theorem 3.4. *Let* S = {a,b}. *Then* (R_1^*,S) *is given by*

$R_1^* = \langle a,b;\ 2b,\ c := [a,b],\ d := [a,b]^a,\ d^a = c,\ c^b = -c,\ d^b = -d\rangle$.

Proof. We first check that 2b is a relation in R_1^*. Let $w \in T$ as in (3.2). Consider w.2b. Then w.2b = 2fb for some integer f, as all products lie in $\langle b\rangle$. As 2b = 0, this shows that 2b is a relation in R_1^*. An arbitrary element of R_1^* is $m_1a + b + \ldots + m_ta + b + m_{t+1}a = u$ as in theorem 3.3. Consider wu. First we point out that from the multiplication table, we have $r_ia = r_ib = k_ib$ say. So wa = $n_1a + k_1b + \ldots + n_sa + k_sb$, wb = (n+k)b, where $n = \Sigma_i n_i$ and $k = \Sigma_i k_i$. Hence wu = $m_1(wa) + (n+k)b + \ldots + m_t(wa) + (n+k)b + m_{t+1}(wa)$. If n + k is even this gives

$(m_1+ \ldots +m_{t+1})(wa) = 0$. Take $n_1 = 3$, $n_2 = 1$, $k_1 = 1 = k_2$. Then $wa = 3a + b + a + b = 2a$. Hence we must have $(m_1+ \ldots +m_{t+1})2a = 0$ and so $\Sigma_i\, m_i = 0$. Then $wu = 0$ for all cases in which $n + k$ is even.

If $n + k$ is odd, this gives $m_1(wa) + b + \ldots + m_t(wa) + b + m_{t+1}(wa) = 0$. Take $n_1 = 1$, $k_1 = 2$. Then $wa = a$ and we must have $m_1a + b + \ldots + m_ta + b + m_{t+1} = 0$. As in the proof of theorem 3.3, this gives us $t = 2q$, $m_1 + m_3 + \ldots + m_{2q+1} = 0$, $m_2 + m_4 + \ldots + m_{2q} = 0$. Then $wu = 0$ for all cases in which $n + k$ is odd and $wa \in \langle a\rangle$. By taking $n_1 = 2$, $k_1 = 1$, we have $wa = 2a + b$. So $wu = m_1(2a+b) + b + \ldots + m_{2q}(2a+b) + b + m_{2q+1}(2a+b)$. The coefficients of b add up to $m_1 + \ldots + m_{2q+1} + 2q = 2q$. Hence wu will always lie in $\langle a\rangle$ in this case. We rewrite wu as $m_1(2a+b) + m_2(-2a+b) + \ldots + m_{2q}(-2a+b) + m_{2q+1}(2a+b)$. Note that $m_i(\pm 2a+b) = 0$ if m_i is even and is $\pm 2a + b$ if m_i is odd. So wu reduces to $\Sigma_{i\in J}((-1)^{i-1}2a+b)$, where $i \in J$ if and only if m_i is odd. Since $\Sigma_{i=1}^{2q+1}\, m_i = 0$, and we have left out all m_i such that m_i is even, it follows that $|J|$ is even, say 2f. This can be rewritten as $\Sigma_{j\in J}(\varepsilon_j 2a+b)$, where $\varepsilon_j = \pm 1$. So $wu = \varepsilon_1 2a - \varepsilon_2 2a + \ldots + \varepsilon_{2f-1}2a - \varepsilon_{2f}2a = 2a(\varepsilon_1-\varepsilon_2+ \ldots -\varepsilon_{2f})$. So we need $\varepsilon_1 - \varepsilon_2 + \ldots - \varepsilon_{2f} = 0$. But if $wa \notin \langle a\rangle$, we have $wa = da + b$, say. Then $wu = da(\Sigma_{j\in J}\varepsilon_j(-1)^{j-1}) = 0$. We conclude that the full set of relations is

$$\{\ \Sigma_{i=1}^{2q}\ (m_ia+b) + m_{2q+1}a;\ \Sigma_{i=1}^{q}\ m_{2i} = 0,\ \Sigma_{i=0}^{q}\ m_{2i+1} = 0,\ \Sigma_{j\in J}(-1)^{i+j} = 0\},$$

where J is defined by $i \in J$ if and only if m_i is odd, and j is a dummy variable which enumerates the elements of J.

We now check what happens to the commutators. The relations in this case are a subset of those in $(R_0^*,+)$. Hence $(R_0^*,+)$ is a homomorphic image of $(R_1^*,+)$. So all relations are in the derived group. Let $c := [a,b]$. Consider $d := [a,b]^a$. We show that no power of d is equal to a power of c. If not then $nd = n'c$ say, i. e. we have a relation $nd - n'c$, or $n(-2a-b+a+b+a) - n'(-a-b+a+b) = n(-2a+b+a+b+a) + n'(b-a+b+a) = -a + n(-a+b+a+b) + a + n'(b-a+b+a)$. Without loss of generality we may assume that $n > 0$. We have to consider separately $n' > 0$ and $n' < 0$.

First let $n' > 0$. Then we have a relation $m_1a + b + \ldots + m_{2q}a + b + m_{2q+1}a$ in which $q = n + n'$, $m_1 = -2$, $m_i = (-1)^i$ for $2 \leq i \leq 2n$, $m_i = (-1)^{i+1}$ for $2n < i \leq 2(n+n')+1$, as can easily be checked. So $J = \{2, \ldots, 2q+1\}$, and $(-1)^{i+j} = (-1)^{i+i-1} = -1$. Thus $\Sigma_{i\in J}(-1)^{i+j} = -2q \neq 0$ and this cannot be a relation. Now let $n' < 0$. Then we have a relation $m_1a + b + \ldots + m_{2q}a + b$ in which $q = n - n' - 1$, $m_1 = -2$, $m_i = (-1)^i$ for $2 \leq i \leq 2n-1$, $m_{2n} = 2$, $m_i = (-1)^i$ for $2n+1 \leq i \leq 2q$. But then all m_i for i odd are negative. Hence $\Sigma_{i=1}^{q}\, m_{2i-1} \neq 0$, and this cannot be a relation. So c and d are independent elements.

Consider $c^b = -b - a - b + a + b + b = -b - a + b + a = [b,a] = -c$. Also, since R_1^* is metabelian, it follows that $d^b = c^{a+b} = c^{b+a} = (-c)^a = -(c^a) = -d$. The final consideration is d^a. We wish to show that $d^a - c$ is a relation to finish the proof. But $d^a - c = c^{a+a} - c = -2a - a - b + a + b + 2a - (-a - b + a + b) = -3a - b + a + b + 2a + b - a + b + a$. We get, using the general form for u, $m_1 = -3$, $m_2 = 1$, $m_3 = 2$, $m_4 = -1$, $m_5 = 1$. So $-3+2+1=0$, $1-1=0$, $J = \{1,2,4,5\}$. So $\Sigma_{j \varepsilon J}(-1)^{i+j} = (-1)^{1+1} + (-1)^{2+2} + (-1)^{4+3} + (-1)^{5+4} = 1 + 1 - 1 - 1 = 0$. Hence $d^a - c$ is a relation.

The next near-ring to consider is (R_2,S) with $S = \{a,b\}$ and $\{b,a+b\}$. We consider $(R_2^*,\{a,b\})$ first.

Theorem 3.5. *Let* $S = \{a,b\}$. *Then* (R_2^*,S) *is given by*

$R_2^* = <a,b;\ 2b,\ c := [a,b],\ d := [a,b]^a,\ d^a = c,\ c^b = -c,\ d^b = -d>$.

Proof. We first check that 2b is a relation in R_2^*. Let $w \ \varepsilon\ T$ as in (3.2). Consider $w(b+b)$. From the multiplication table $wb = nb$. Hence $w(b+b) \ \varepsilon\ 2<b> = \{0\}$, for all $w \ \varepsilon\ T$. We now consider wa. We can write $wa = n_1a + k_1b + \ldots + n_sa + k_sb$, where $r_ia = k_ib$, $1 \le i \le s$. Then $wa = m_1(wa) + nb + \ldots + m_t(wa) + nb + m_{t+1}(wa)$. From now on the proof follows a line very similar to that of theorem 3.4 and leads to the same additive structure.

Now we turn to $(R_2^*,\{a+b,b\})$. We, perhaps surprisingly, obtain the same additive structure as before.

Theorem 3.6. *Let* $S = \{a+b,b\}$. *Write* $x := a + b$, $y := b$. *Then* (R_2^*,S) *is given by*

$R_2^* = <x,y;\ 2y,\ c := [x,y],\ d := [x,y]^x,\ d^x = c,\ c^y = -c,\ d^y = -d>$.

Proof. First we show that x has infinite order. We start by calculating wx. We see that $wx = n_1x + k_1b + \ldots + n_sx + k_sb$, where $r_ix = k_ib$, $1 \le i \le s$. So $wx = n_1(a+b) + k_1b + \ldots + n_s(a+b) + k_sb$. Take $n_1 = 1$, $k_1 = 1$ to obtain $wx = a + b + b = a$. Hence $w(fx) = f(wx) = fa$ is never 0 unless $f = 0$. Thus x has infinite order. Next we show that y has order 2. But $wy = nb$ as in theorem 3.5. Hence $2y = 0$. We now see that $wu = m_1(wx) + nb + \ldots + m_t(wx) + nb + m_{t+1}(wx)$. The rest follows as in the proof of the two previous theorems.

The last d. g. near-ring that needs to be considered is (R_3,S). We look at two cases, namely $S = \{a,b\}$ and $S = \{a+b,b\}$, as we have mentioned earlier. To maintain continuity, we consider $S = \{a+b,b\}$ first.

Theorem 3.7. *Let* $S = \{a+b,b\}$. *Write* $x := a + b$, $y := b$. *Then* (R_3^*,S) *is given by*

$R_3^* = <x,y;\ 2y,\ c := [x,y],\ d := [x,y]^x,\ d^x = c,\ c^y = -c,\ d^y = -d>$.

Proof. We start by looking at wx and wy: $wx = n_1x + k_1b + \ldots + n_sx + k_sb$, where $r_ix = k_ib$, $1 \le i \le s$. As in the proof of theorem 3.6, this forces x to have infinite order. Also $wy = n_1b + k_1b + \ldots + n_sb + k_sb \ \varepsilon\ <b>$. As in the proof of theorem 3.4, we get 2y as a relation. The rest of the pattern should be familiar from the last three theorems.

This brings us to the last of the upper faithful d. g. near-rings, namely $(R_3^*,\{a,b\})$.

Theorem 3.8. *Let* $S = \{a,b\}$. *Then*

$R_3^* = <a,b;\ 2b,\ [a,b]^a = [a,b],\ [a,b]^b = [b,a]>$.

Proof. We can see that 2b is a relation as in the proof of the previous theorem. In this case though, we have $wa = na$ from the multiplication table. Again $wb = (n+m)b$, in the same way. So $wu = m_1na + (n+m)b + \ldots + m_tna + (n+m)b + m_{t+1}na$. It now only needs a slight modification of the proof of theorem 3.3 to finish this proof.

4. DEFECTS.

In this last section we find the upper defect, lower defect and defect in the six cases in which the upper faithful d. g. near-ring has been found.

Theorem 4.1. *Let* S = {a,b}. *Then*

(i) $\underline{D}(R_0,S)$ *is a zero ring on an infinite cyclic group,*

(ii) $D^*(R_0,S)$ *is a zero ring on an infinite cyclic group,*

(iii) $D(R_0,S)$ *is the direct sum of* $\underline{D}(R_0,S)$ *and* $D^*(R_0,S)$.

Proof. From theorem 2.7, $\underline{D}(R_0,S)$ is <4a>, i. e. an infinite cyclic group. From theorem 3.3 we see that $D^*(R_0,S)$ = <[a,b] + 2a>, again an infinite cyclic group. Finally we see, again from theorem 3.3, that 4a and [a,b] commute additively. Hence the additive structures of the three defects are as stated. The rest follows since all products are zero in R_0.

Theorem 4.2. *Let* S = {a,b}. *Then*

(i) $\underline{D}(R_1,S)$ *is a zero ring on an infinite cyclic group,*

(ii) $D^*(R_1,S)$ *is a zero ring on the direct sum of two infinite cyclic groups,*

(iii) $D(R_1,S)$ *is a zero ring on the direct sum of three infinite cyclic groups.*

Proof. From theorem 2.17, $(\underline{D}(R_1,S),+)$ = <4a>. Products are zero since 4na.4ma = 16nmb = 0. From theorem 3.4, $(D^*(R_1,S),+)$ = <[a,b] + 2a, $[a,b]^a$ + 2a>. Consider (-a+b+a+b)(-2a+b+a+b+a). Since all products of a and b are equal to b, we see that the product above is 24b = 0. Similarly all other products are zero. Again from theorem 3.4, $(D(R_1,S),+)$ = <[a,b], $[a,b]^a$, 4a>. Since 2a commutes with [a,b], it follows that $(D(R_1,S),+)$ is the direct sum of three infinite cyclic groups. As above any product in $D(R_1,S)$ results in an element of the form 2nb = 0. This finishes the proof.

Theorem 4.3. *Let* S = {a,b}. *Then*

(i) $\underline{D}(R_2,S)$ *is a zero ring on an infinite cyclic group,*

(ii) $D^*(R_2,S)$ *is a zero ring on the direct sum of two infinite cyclic groups,*

(iii) $D(R_2,S)$ *is a zero ring on the direct sum of three infinite cyclic groups.*

Proof. From theorem 2.22, $(\underline{D}(R_2,S),+)$ = <2a> and from theorem 3.5, $(D^*(R_2,S),+)$ = <[a,b] + 2a, $[a,b]^a$ + 2a>. The rest follows as in the proof of theorem 4.2.

Theorem 4.4. *Let* S = {b,a+b}. *Then*

(i) $\underline{D}(R_2,S)$ *is a zero ring on an infinite cyclic group,*

(ii) $D^*(R_2,S)$ *is a zero ring on the direct sum of two infinite cyclic groups,*

(iii) $D(R_2,S)$ *is a zero ring on the direct sum of three infinite cyclic groups.*

Proof. From theorem 2.21, $(\underline{D}(R_2,S),+)$ = <2a>. All products are zero, since 2na.2ma = 4nmb = 0. From theorem 3.6 we can see that, in the notation of that theorem, $(D^*(R_2,S),+)$ = <2x, c+d>, as can be checked by writing these elements in terms of a and b and using the known structure of D_∞. 2x commutes with both c and d as we see from theorem 3.6. In (R_2^*,S) the products are given by $x^2 = y$, $y^2 = xy = yx = 0$. As 2y = 0, it follows that, as before, $D^*(R_2,S)$ is a zero ring. Finally, again from theorem 3.6, and in the notation of that theorem, $(D(R_2,S),+)$ = <2x, c, d> and as above 2x, c, d all commute, and all products are zero.

Theorem 4.5. *Let* $S = \{a,b\}$. *Then*

(i) $\underline{D}(R_3,S)$ *is a zero ring on an infinite cyclic group,*

(ii) $D^*(R_3,S)$ *is a zero ring on an infinite cyclic group,*

(iii) $D(R_3,S)$ *is a zero ring on the direct sum of two infinite cyclic groups.*

Proof. From theorem 2.27, $(\underline{D}(R_3,S),+) = \langle 2a \rangle$. The multiplication table shows that this is a zero ring. From theorem 3.8 we get that $(D^*(R_3,S),+) = \langle [a.b] + 2a \rangle$ and $(D(R_3,S),+) = \langle [a,b], 2a \rangle$. Again the multiplication table shows that all products are zero, since $2b = 0$ and all elements involving b have an even number of b's in them.

Theorem 4.6. *Let* $S = \{b,a+b\}$. *Then*

(i) $\underline{D}(R_3,S)$ *is a zero ring on an infinite cyclic group,*

(ii) $D^*(R_3,S)$ *is a zero ring on the direct sum of two infinite cyclic groups,*

(iii) $D(R_3,S)$ *is a zero ring on the direct sum of three infinite cyclic groups.*

Proof. From theorem 2.27 we see that $\underline{D}(R_3,S)$ is the same as in theorem 4.5. We now consider theorem 3.7. From this we again see that $(D^*(R_3,S),+) = \langle 2x, c+d \rangle$ as in the last theorem, and the remainder of the theorem follows as before.

In [6] all defects were calculated for some d. g. near-rings. In all cases the defects were zero rings, $\underline{D}(R,S) \simeq D^*(R,S)$ and $D(R,S) = \underline{D}(R,S) \oplus D^*(R,S)$. Here we have examples in which $\underline{D}(R,S)$ is not isomorphic to $D^*(R,S)$. In [5], theorem 3.5, we have an example in which $(\underline{D}(R,S),+)$ is not an abelian group, hence $\underline{D}(R,S)$ is not a ring. But the other statement still holds. This statement is most unlikely to be generally true. So a next stage would be to seek counter examples to this statement.

REFERENCES.

[1] S. J. Mahmood. Limits and colimits in categories of d. g. near-rings. Proc. Edinburgh Math. Soc. 23 (1980), 1–7.

[2] J. J. Malone. D. g. near-rings on the infinite dihedral group. Proc Royal Soc. Edinburgh 78A (1977), 67–70.

[3] J. D. P. Meldrum. The representation of d. g. near-rings. J. Austral. Math. Soc. 16 (1973), 467–480.

[4] J. D. P. Meldrum. The endomorphism near-ring of an infinite dihedral group. Proc. Royal Soc. Edinburgh 76A (1977), 311–321.

[5] J. D. P. Meldrum. Presentations of faithful d. g. near-rings. Proc. Edinburgh Math. Soc. 23 (1980), 49–56.

[6] J. D. P. Meldrum. Upper faithful d. g. near-rings. Proc. Edinburgh Math. Soc. 26 (1983), 361–370.

[7] J. D. P. Meldrum. Near-rings and their links with groups. Pitman Research Notes No. 134, London 1985.

[8] G. Pilz. Near-rings. 2nd. Edition. North-Holland. Amsterdam 1983.

Near-rings and Near-fields, G. Betsch (editor)

NEAR-RINGS ASSOCIATED WITH COVERED GROUPS

C.J. MAXSON

Mathematics Department
Texas A&M University
College Station, TX 77843

From the time of Descartes, early in the 17th century, mathematicians have been interested in associating algebraic structures with geometric structures and investigating the interplay. In this paper we continue this line of investigation by associating near-rings to generalized translation spaces and noting how the geometry influences the algebraic structure.

I. GEOMETRY

We start with some geometric background and some definitions. Recall that an affine plane A is a translation plane if its group T of translations operates transitively on the set of points of A. In this case T is an abelian group with congruence partition $\{T_i\}$, i.e., each T_i is a subgroup of T, $|T_i| = |T_j|$ $T = \bigcup_i T_i$, $T_i \cap T_j = (0)$, $i \neq j$ and $T_iT_j = T$, $i \neq j$. On the other hand let $C = \{G_\alpha\}$ be a non-trivial congruence partition (at least two components) consisting of subgroups G_α of a group G. From this congruence partition of G a translation plane C(G) is obtained by taking the elements of G as points, the cosets of the components of C as lines and setting xG_α parallel to yG_β if and only if $\alpha = \beta$. It is well known that every translation plane has the form C(G) for some abelian group G with a suitable partition. More will be said about the equivalence of certain geometries and groups with special properties later. For material on translation planes see [1], [4] or [10].

We now define generalized translation structures. Let $\Sigma = (P, L, \parallel)$ where P is a set of points, L a collection of subsets of P called lines, with an incidence relation defined in $P \times L$ and a parallelism defined in $L \times L$ satisfying the following axioms:

(A1) Every two points a, b in P are incident with at least one line;

(A2) $|L| \geq 2$ and for each $A \in L$, $|A| \geq 2$;

(A3) Parallelism is an equivalence relation;

(A4) For each $x \in P$, $A \in L$ there exists a unique line containing x and parallel to A.

Further let Φ be a one-one map of P into the collineations of the incidence structure Σ such that $\Phi(P)$ is a point transitive group of fixed point free collineations which map lines onto parallel lines. We say $\langle\Sigma,\Phi\rangle$ is a generalized translation structure.

For a generalized translation structure $\langle\Sigma,\Phi\rangle$, let 0 denote the element in P such that $\Phi(0)$ is the identity map in the group $\Phi(P)$. For each a in P there exists a unique map ϕ_a in $\Phi(P)$ such that $\phi_a(0) = a$. Without loss of generality we take $\Phi(a) = \phi_a$.

Note that if every two points determine a unique line then we have a translation structure studied by André [2]. For other work on translation structures see the paper by Biliotti and Herzer [3] and the references given there. If further we require $|A| = |B|$ for each $A, B \in L$ then we have a Sperner translation space ([11]).

We now extend the relationship between translation planes and congruence partitions to generalized translation structures. Let $\langle G,+\rangle$ be a group written additively, but not necessarily abelian, with identity 0. A collection $C = \{H,K,\ldots\}$ of subgroups of G is a cover if

(C1) $\cup C = G$

(C2) $\forall H \in C$, $H \neq \{0\}$, $H \neq G$ and $\forall K \in C$, if $H \neq K$ then $H \not\subseteq K$.

When C is a cover of G the pair (G,C) is called a covered group. The elements of C are called cells of the cover. If further,

(C3) $\forall H, K \in C$, $H = K$ or $H \cap K = \{0\}$.

Then C is called a fibration (or partition) and the pair (G,C) is a fibered group.

Let (G,C) be a covered group. By taking $P(G) = G$, $L(G) = \{x + H \mid H \in C,\ x \in G\}$ and setting $x + H \parallel y + K$ iff $H = K$, it is routine to verify that $\Sigma(G) = \langle P(G), L(G), \parallel\rangle$ satisfies (A1) - (A4). Further, by associating to each $a \in G$ the left translation λ_a, denoted by $\Phi(G)\colon a \to \lambda_a$ we find that we have a generalized translation structure $(\Sigma(G), \Phi(G))$.

Conversely every generalized translation structure arises in this manner (see [12]). From this we see that a generalized translation structure may be considered as a covered group $\langle G,C)$ and we henceforth do so.

Specializing, we note that each translation structure arises from a fibered group, each Sperner space arises from a group with an equal fibration, i.e., $|H| = |K|$ $\forall H, K \in C$, and as noted above each translation plane arises from a group with a congruence fibration. Thus fibrations, equal fibrations and congruence fibrations "tighten" the structure. We now tighten the structure in an alternate fashion namely by requiring that there be a semigroup of operators.

Let $\langle\Sigma,\Phi\rangle$ be a generalized translation structure and recall that an endomorphism of Σ is an incidence preserving function $\sigma\colon P \to P$, that is, for $A \in \mathcal{L}$, $\sigma(A) \subseteq B \in \mathcal{L}$. The collection end Σ of endomorphisms of Σ is a semigroup with identity under the operation of composition. Suppose S is a subsemigroup of end Σ with the following properties:

(01) The identity id and the constant map $w(a) = 0$ (where 0 has been previously identified) for each $a \in \mathcal{P}$ are both in S;

(02) For each $\sigma \in S$ the following diagram commutes

$$\begin{array}{ccc} \mathcal{P} & \longrightarrow & \text{end } \Sigma \\ \downarrow & & \downarrow \\ \text{end } \Sigma & \longrightarrow & \text{end } \Sigma \end{array} \qquad\qquad \begin{array}{ccc} a & \longrightarrow & \phi_{\sigma(a)} \\ \downarrow & & \searrow \\ \phi_a & \longrightarrow & \sigma\phi_a = \phi_{\sigma(a)}\sigma. \end{array}$$

Then we say $T = \langle\Sigma,\Phi,S\rangle$ is a <u>generalized translation space with operators</u>, henceforth denoted by GTSO.

If $\langle G,C\rangle$ is the group with cover giving rise to the GTSO $\mathcal{T} = \langle\Sigma,\Phi,S\rangle$ then property (02) ensures that each $\sigma \in S$ is an endomorphism of G. But then since $\sigma(0) = 0$, each line through 0 is mapped by σ into a line through 0. Hence, S is a semigroup of endomorphisms satisfying

(01)' The identity endomorphism and zero endomorphism are in S;

(02)' For each $\sigma \in S$, for each $G_i \in C$, there exists $G_j \in C$ with $\sigma(G_i) \subseteq G_j$.

Conversely if G is a group with a cover C and a semigroup of endomorphisms satisfying (01)' and (02)' then we obtain a GTSO. Furthermore every GTSO arises in this manner, thus we identify the GTSO $\langle\Sigma,\Phi,S\rangle$ with the associated group with operators and cover $\langle G,C,S\rangle$.

We now show how to associate near-rings with GSTO's, $\langle G,C,S\rangle$. First, we consider the set $E(G,C) = \{f \in \text{End } G \mid f(H) \subseteq H, \forall H \in C\}$. (Note that the semigroup of operators plays no role here.) Under composition of functions $E(G,C)$ is a semigroup with identity, called the semigroup of <u>dilatations</u> of $\langle G,C,S\rangle$. We consider the near-ring distributively generated by $E(G,C)$, denoted by dg $E(G,C)$. For our second associated near-ring we consider $M_S(G,C) = \{ f \in M_0(G) \mid f(H) \subseteq H, \forall H \in C, \ f\sigma = \sigma f, \ \forall\sigma \in S\}$ which is a zero-symmetric near-ring with identity under the operations of function addition and composition. Unfortunately the word "kernel" has been used in the literature for both of these near-rings. In this paper we call dg $E(G,C)$ the <u>kernel of $\langle G,C,S\rangle$</u> and $M_S(G,C)$ the <u>centralizer of $\langle G,C,S\rangle$</u>. In the next section we discuss the centralizer and in Section III the kernel.

II. THE CENTRALIZER OF $\langle G,C,S\rangle$

We begin this section with the result that centralizers of GTSO's are very general, indeed every zero-symmetric near-ring with identity arises in this manner. The next result, whose proof is due to Peter Fuchs extends the corresponding result in [12] where it was established only in the finite case.

Theorem II.1. [12]. Let N be a zero-symmetric near-ring with identity in which $ab = 1$ implies $ba = 1$, $a,b \in N$. Then there exists a GTSO, $T = \langle G,C,S \rangle$ such that $N \cong M_S(G,C)$.

Proof. Let $G = N \times N$. Clearly G is a finitely generated N-group (e.g., $\{(1,0), (0,1)\}$ is a generating set). Hence maximal N-subgroups of $N \times N$ exist and every N-subgroup $H \subsetneq N \times N$ is contained in a maximal one. Let $C = \{H_\alpha\}$ denote the set of all maximal N-subgroups of G. C is a cover of G since $H_\alpha \subseteq H_\beta$ implies $H_\alpha = H_\beta$ and if $(a,b) \in G$, then $(a,b) \in N(a,b) \subseteq H_\alpha$ for some maximal N-subgroup H_α. Therefore $\cup H_\alpha = G$.

For each $a \in N$, let $\rho_a\colon G \to G$ be the function defined by $\rho_a(u,v) = (ua,0)$ and let $t\colon G \to G$ be the function defined by $t(u,v) = (v,0)$. Then ρ_a, t are N-endomorphisms of G. We let S be the semigroup of endomorphisms generated by $\{id\} \cup \{\rho_a \mid a \in N\} \cup \{t\}$. For each $\sigma \in S$ and $H_\alpha \in C$, $\sigma(H_\alpha)$ is an N-subgroup of G so $\sigma(H_\alpha) \subseteq H_\beta$ for some $H_\beta \in C$. Thus S is a semigroup of operators. Proceeding as in [12] one then shows $N \cong M_S(G,C)$.

As with most representation results the above leads to several interesting avenues of investigation. Here we consider only the problem of characterizing structural properties of $M_S(G,C)$ in terms of the geometry $\langle G,C,S \rangle$. (Further representation results are considered in [12].)

Convention: For the remainder of this section, all structures are finite and all near-rings are zero-symmetric with identity.

We first characterize when $M_S(G,C)$ is a near-field. For this we need some further concepts. Let $T = \langle G,C,S \rangle$ be a GTSO. A set $Y = \{y_1,\dots,y_t\}$ is a generating set if $G = \cup Sy_i$ and $Sy_i \not\subseteq Sy_j$ if $i \neq j$. Since G is finite, generating sets exist. For each $y_i \in Y$, define $I_i = \cap \{H_\alpha \mid H_\alpha \in C, y_i \in H_\alpha\}$ and let $I = \{\text{sequences } (x_1,\dots,x_t) \mid x_i \in I_i \text{ and } F(y_i, y_j) \subseteq F(x_j, x_j)\}$ where $F(u,v) = \{(\alpha,\beta) \in S \times S \mid \alpha u = \beta v\}$. Further let p_i be the i-th projection map on I. We say $\langle G,C,S \rangle$ is kernel dependent if, for $\sigma \in S$, $p_i I \cap \operatorname{Ker} \sigma \neq \{0\}$ implies $p_i I \subseteq \operatorname{Ker} \sigma$ (see [11]). Recall also that $x, y \in G^* = G \setminus \{0\}$ are connected if there exist $w_1,\dots,w_{n-1} \in G$ and $\sigma_1,\dots,\sigma_n$, $t_1,\dots,t_n \in S$ such that

$$\begin{aligned} \sigma_1 x &= t_1 w_1 \neq 0 \\ \sigma_2 w_1 &= t_2 w_2 \neq 0 \\ &\vdots \\ \sigma_n w_{n-1} &= t_n y \neq 0 \end{aligned}$$

This relation is an equivalence relation on G^* and the equivalence classes are called the connected components. G is said to be connected if G^* is a connected component.

Theorem II.2. [12]. Let $T = \langle G,C,S\rangle$. Then $M_S(G,C)$ is a near-field iff T is connected and kernel dependent.

To obtain definitive structural results one places some restrictions on the semigroup of operators. One considers the situation in which S is a group of automorphisms (with 0 adjoined). As one might suspect (from the previous work on centralizer near-rings [17]) the orbits of the action and stabilizers of elements in G^* play an important role. For results in the case of fibrations see [14] and for the general results see [5].

Next, one considers the situation in which S is a cyclic semigroup, $S = \langle\alpha\rangle \cup \{0, \mathrm{id}\}$. We write $M_\alpha(G,C)$ for $M_S(G,C)$. When α is not invertible and not nilpotent, $M_\alpha(G,C)$ is not a simple near-ring. Therefore one restricts to the case of a nilpotent operator.

Theorem II.3. [5], [14]. Let α be a nilpotent operator on (G,C).

(II.3.1) There are $|\mathrm{Ker}\ \alpha| - 1$ connected components.
(II.3.2) There is a unique generating set, Y.
(II.3.3) If C is a fibration, $\mathrm{Ker}\ \alpha$ is contained in a unique cell, say H_0.

When G is connected much can be said.

Theorem II.4. [5], [14]. Let α be a nilpotent operator on (G,C). The following are equivalent.

(II.4.1) $M_\alpha(G,C)$ is a near-field.
(II.4.2) $M_\alpha(G,C)$ is a simple near-ring.
(II.4.3) $M_\alpha(G,C)$ is a 2-semisimple near-ring.
(II.4.4) $M_\alpha(G,C) \cong \mathbb{Z}_2$.

Further, if C is a fibration the above are equivalent to

(II.4.5) $M_\alpha(G,C)$ is a local near-ring.
(II.4.6) $Y \cap H_0 = \emptyset$.

If G is not connected and α is a nilpotent operator, necessary and sufficient conditions in terms of the geometry $\langle G,C,S\rangle$ have been determined which characterize when $M_\alpha(G,C)$ is simple. Again for the case in which C is a fibration see [14] and for the general case see [5].

We conclude this section with two suggestions for further research, first a rather specific problem and then a more general program.

Sug 1: Characterize 2-semisimplicity in terms of $\langle G,C,S\rangle$ for $S = \langle\alpha\rangle \cup \{0, \mathrm{id}\}$.

Sug 2: For other classes of operators, (e.g., inverse semigroups, completely 0-simple semigroups) characterize the structure of $M_S(G,C)$ in terms of $\langle G,C,S\rangle$.

III. THE KERNEL OF $\langle G,C,S\rangle$.

We recall from Section 1 that the kernel of $\langle G,C,S\rangle$ is the near-ring dg E distributively generated by the dilatations $E = E(G,C)$. When C is a fibration, the structure of the semigroup E is well-known [3], [6], [7], [10], [15].

Theorem III.1. If $\langle G,C\rangle$ is a finite fibered group then $E^* = E\setminus\{0\}$ is a group of fixed point free automorphisms of G. Further, E^* is a cyclic group.

A now classical result states that in the case $\langle G,C\rangle$ is a translation plane, G is an abelian group. Thus from the above theorem, if $\langle G,C\rangle$ is a finite translation plane, dg E is a finite field [3], [10]. Therefore to generalize to arbitrary fibrations one is lead to determining the structure of fibered groups. Such a study has been carried out in [7] and [8]. For our particular case we specialize one of the structural results (see also [6]).

Theorem III.2. If $\langle G,C\rangle$ is a finite fibered group with $E^* \neq \{id\}$ then G is a finite p-group for some prime p, of exponent p and of nilpotency class at most 2.

Using the above result concerning E and G along with some work of Herzer [6] (recently generalized in [13]) the following rather surprising result has been obtained.

Theorem III.3. [15]. If $\langle G,C\rangle$ is a finite fibered group then dg E is a commutative ring. If, further, $E^* \neq \{id\}$ then dg E is a finite field.

If $E^* = \{id\}$ then clearly dg $E = \mathbf{Z}_n$ where n is the exponent of G. Examples ([7]) show that in this case n need not be a prime. Theorem III.3, together with the classical work of André ([1]), shows that whether or not G is abelian, whenever $E^* \neq \{id\}$ there is a field associated with the geometry $\langle G,C\rangle$ in a natural manner. We note that in the non-abelian case, although the elements of the associated field are sums of automorphisms, they need not themselves be morphisms.

We also mention that in the abelian case the field dg E has geometric applications. The significance of the field dg E in the non-abelian case is still unknown.

Recently ([9], [16]) a study of the near-ring dg E for finite covered groups $\langle G,C\rangle$ has been initiated. In general, very little is known. In fact, in the above studies G is restricted to be a finite elementary abelian p-group and the structure of the ring dg E is investigated. In [9] several basic results, several examples and a complete solution to the case $G = (\mathbb{Z}_p)^3$ are presented. In [16] necessary and sufficient conditions in terms of an associated lattice are given for the kernel dg E to be a field or a simple ring or a semisimple ring. The surface has just been scratched, there is still much to be investigated here.

REFERENCES

[1] André, J., Über nicht-Desarguessche Ebenen mit tansitiver Translationsgruppe, Math. Zeitschr., 60 (1954), 156-186.

[2] __________, Über Parallelstrukturen, II: Translationsstrukturen, Math. Zeitschr., 76 (1961), 155-163.

[3] BILIOTTI, M. and Herzer, A., Zur Geometrie der Translationsstrukturen mit eigentlichen Dilatationen. Abh. Math Sem. Univ. Hamburg 52 (1983), 1-27.

[4] Dembowski, H. P., Finite Geometries, Springer-Verlag, New York-Heidelberg-Berlin, 1968.

[5] Fuchs, P. and Maxson, C. J., Kernels of Covered Groups with Operators (in preparation).

[6] Herzer, A., Endliche nichtkommutative Gruppen mit Partition Π und fixpunktfreiem π-Automorphismus. Arch. Math. 34 (1980), 385-392.

[7] Karzel, H. J. and Maxson, C. J., Fibered groups with non-trivial centers, Result. der Math., 7 (1984), 192-208.

[8] __________, Fibered p-groups. Abh. Math. Sem. Univ. Hamburg (to appear).

[9] Karzel, H., Maxson, C. J. and Pilz, G. F., Kernels of Covered Groups. Result. der Math. (to appear).

[10] Luneburg, H., Translation Planes, Springer-Verlag, New York-Heidelberg-Berlin, 1980.

[11] Maxson, C. J., Near-rings Associated with Sperner Spaces, Jour. of Geom., 20 (1983), 128-145.

[12] __________, Near-rings associated with Generalized Translation Structures. Jour. of Geom. 24 (1985), 175-193.

[13] Maxson, C. J. and Meldrum, J. D. P., D G Near-rings and Rings (submitted).

[14] Maxson, C. J. and Oswald, A., Kernels of Fibered Groups with Operators, (submitted).

[15] Maxson, C. J. and Pilz, G. F., Near-rings determined by fibered groups, Arch. Math. 44 (1985), 311-318.

[16] Maxson, C. J. and Pilz, G. F., Kernels of Covered Groups II, (submitted).

[17] Pilz, G. F., Near-rings, Revised Edition, North-Holland, New York, 1983.

Near-rings and Near-fields, G. Betsch (editor)

KRULL DIMENSION AND TAME NEAR-RINGS

J.D.P. MELDRUM and A.P.J. VAN DER WALT

Department of Mathematics,
University of Edinburgh,
Mayfield Road,
Edinburgh EH9 3JZ,
Scotland.

Department of Mathematics,
University of Stellenbosch,
Stellenbosch 7600,
South Africa.

A tame near-ring R is one which has a faithful module in which all R-modules are R-ideals. Such a near-ring is very ring-like. In this paper we extend some results on rings with Krull dimension to tame near-rings. Krull dimension is a generalization of the minimal condition. The main results are as follows. (i) A tame right noetherian near-ring has a possibly transfinite power of its J_2 radical zero. (ii) The J_2 radical of a tame near-ring does not contain a non-zero idempotent ideal. (iii) The nil radical of a tame near-ring, is residually nilpotent.

1. INTRODUCTION AND DEFINITIONS.

Contrary to what is suggested by their name, general near-rings are in some respects very far from rings. One such is the fact that none of the obvious generalizations of the Jacobson radical to near-rings is necessarily nilpotent even in the finite case. This is one reason for introducing tame near-rings, a reasonably large class of near-rings which behave very like rings. They were first introduced by Scott **[10]** and **[11]** and he showed how close to rings they were in their behaviour. In particular tame near-rings with the descending chain condition have all three J (for Jacobson) radicals nilpotent. In this paper we generalize results about rings with Krull dimension to near-rings. Apart from those results that are essentially lattice theoretic in nature, the proofs of the results about near-rings are not straight generalizations of the proofs from ring theory. This is mainly due to the fact that the ring-like properties lie in a faithful module, whereas the near-ring has the Krull dimension.

A (left) near-ring is a set R with two operations, + and ., such that (R,+) is a not necessarily abelian group, (R,.) is a semigroup and $x(y+z) = xy + xz$, $0x = 0$, for all $x,y,z \in R$. If (G,+) is a group written additively then $M_0(G) = \{f : G \to G, 0_G f = 0_G\}$ is a near-ring under point-wise addition and composition of functions. A homomorphism from R into $M_0(G)$ defines G as an R-module. A submodule H of the R-module G is a subgroup of G such that $HR \leq H$. An R-ideal of G is a normal subgroup K of G such that $(g+k)r - gr \in K$ for all $g \in G$, $k \in K$, $r \in R$. The group (R,+) is an R-module, denoted R_R, under the right regular representation. A right invariant subnear-ring of R is an R-submodule of R_R and a right ideal of R is an R-ideal of R_R. A left invariant subnear-ring of R is a subgroup S of (R,+) such that $RS \leq S$ and a left ideal of R is a normal left invariant subnear-ring. An ideal is a left ideal which is also a right ideal. An R-module G is monogenic if $G = gR$ for some $g \in G$. A monogenic R-module G is of type 0 if it has no non-trivial proper R-ideals, of type 1 if it is of type 0 and $g \in G$ implies $gR = G$ or $gR = \{0\}$, of type 2 if it has no non-trivial proper R-submodules. For $\nu = 0,1,2$, define $J_\nu(R)$ as the intersection of all annihilators of R-modules of type ν. These definitions follow closely those to be found in Meldrum **[8]**, or in Pilz **[9]**, except that he uses right near-rings, and his nomenclature for the substructures of R-modules is different. Most standard results can be found there.

Recall that a tame R-module G is an R-module in which all R-submodules are R-ideals. A near-ring is tame if it has a faithful tame R-module. Two large families of tame near-rings are given by firstly, the subnear-rings of $M_0(G)$ generated by a semigroup S of endomorphisms of G containing Inn G, the inner automorphisms of G, and secondly, the zero-symmetric parts of near-rings of polynomials. We now come to Krull dimension. Defined originally for commutative rings, it has been extended to arbitrary rings where it has been used extensively (Gordon and Robson [2]). In this context it is essentially a condition on the lattice of substructures generalizing the descending chain condition. Let G be an R-module. Denote the Krull dimension of G by kd(G). Then kd(G) = −1 if and only if $G = \{0\}$. If α is an ordinal and $kd(G) \nless \alpha$ then $kd(G) = \alpha$ if there is no infinite descending chain $G = G_0 \geq G_1 \geq G_2 \geq \ldots$ of R-ideals of G such that $kd(G_i/G_{i+1}) \nless \alpha$ for $i = 0,1,2,\ldots$. In G_i/G_{i+1} we define the Krull dimension relative to the R-ideals of G/G_{i+1}. Note that kd(G) = 0 corresponds to the descending chain condition on R-ideals. We say that R has Krull dimension if the R-module R_R has Krull dimension. Any given R-module need not necessarily have Krull dimension.

Finally for an ideal I of the near-ring R we define $I^0 = R$, $I^{\beta+1} = I^\beta . I$, $I^\alpha = \bigcap_{\beta<\alpha} I^\beta$ for limit ordinals α and $I_0 = I$, $I_{\beta+1} = \bigcap_{n=1}^{\infty} (I_\beta)^n$ and $I_\alpha = \bigcap_{\beta<\alpha} I_\beta$ for limit ordinals α.

In the next section we gather together results about Krull dimension and powers of ideals needed later. In the last section we present our results about tame near-rings with Krull dimension.

2. PRELIMINARY RESULTS.

In the applications we need a number of results about modules with Krull dimension from Gordon and Robson [2] and Lenagan [7]. The difficulty in using these results is to unravel those results that arise from the properties of the lattice of substructures from those which use results about rings. Here we list the results we need with indications of proof where necessary.

Lemma 2.1. ([2] Lemma 1.1(i)). *Let* G *be a tame* R-*module with Krull dimension,* H *an* R-*ideal of* G . *Then* kd(G) = sup{kd(G/H), kd(H)} *if either side exists.*

Note that we need tameness since R-ideals of H are not necessarily R-ideals of G. Indeed the result is no longer true if G is not tame.

Example 2.2. The wreath product G of two infinite cyclic groups satisfies the maximal condition on normal subgroups but contains a normal subgroup H which is the direct sum of an infinite number of infinite cyclic groups. Take R to be the integers with 1 acting as the identity map on G. Then G has Krull dimension (since it satisfies the maximal condition on R-ideals, see lemma 2.4) but H does not (see lemma 2.6).

Lemma 2.3. *Let* G *be an* R-*module with Krull dimension. Then every homomorphic image of* G *has Krull dimension less than or equal to the Krull dimension of* G.

See proposition 1.2(i) in [2]. This result follows because the lattice of R-ideals of a homomorphic image of G is a sublattice of the lattice of R-ideals of G, and it does not need tameness of the R-module.

Lemma 2.4. *Every* R-*module with the maximal condition on* R-*ideals has Krull dimension.*

This result can be proved by the same methods as proposition 1.3 of [2].

Before we generalize proposition 1.4 of [2], we need the concept of, and a result about uniform dimension. We say that the R-module G has *finite uniform dimension* if there is a fixed (finite) bound on the number of summands in a direct sum of non-zero summands.

Lemma 2.5. *Let* G *be a tame* R-*module which does not contain an infinite direct sum of non-zero* R-*ideals. Then* G *has finite uniform dimension.*

Proof. We first show that any non-zero R-ideal of G contains an R-ideal of G which is uniform, i.e. an R-ideal with the property that any two non-zero R-ideals of G contained in it intersect non-trivially. Let H be a non-zero R-ideal of G. If it is uniform, we are home. If not there exist K_1, L_1 R-ideals of G contained in H such that $K_1 \cap L_1 = \{0\}$. So $K_1 \oplus L_1 \leq H$. If either K_1 or L_1 are uniform, we are home. If not we obtain $K_1 \oplus K_2 \oplus L_2 \leq H$ where K_2, L_2 are non-zero R-ideals of G contained in L_1. Since G does not contain an infinite direct sum of non-zero R-ideals, this process must stop and it can only stop when we have a non-zero uniform R-ideal.

The second step is to show that there is a finite number of uniform R-ideals of G whose sum is direct and is an essential R-ideal of G, i.e. it has non-trivial intersection with every non-zero R-ideal of G. By above, G has a uniform R-ideal H_1. If H_1 is not essential, then there is a non-zero R-ideal K_1 such that $H_1 \cap K_1 = \{0\}$. Then K_1 contains a uniform non-zero R-ideal H_2 and $H_1 \oplus H_2$ is direct. This process must stop after a finite number of steps as before, for the same reason. This gives us the second step.

The last step is to obtain the fixed bound on the number of non-zero direct summands. The method used is a version of the Steinitz Exchange Theorem used with finite dimensional vector spaces. At first we consider direct sums consisting of uniform R-ideals. Let $H_1 \oplus \ldots \oplus H_n$ be a direct sum of uniform R-ideals which is essential. To be able to use this sum for the exchange process, we need to show that if K is an R-ideal of G which intersects each H_i non-trivially, then K is essential in G. This we now do.

For $i = 1, \ldots, n$, define $L_i = K \cap H_i$, all non-zero by hypothesis. Hence $L = L_1 \oplus \ldots \oplus L_n$ is direct and $L \leq K$. We show that L is essential in G. Let $g \in G - \{0\}$ and let gR denote the R-ideal of G generated by g, since G is tame. Since $H_1 \oplus \ldots \oplus H_n$ is essential, we have $gR \cap (H_1 \oplus \ldots \oplus H_n) \neq \{0\}$. Thus there exists $g_1 = h_1 + \ldots + h_n$ where $h_i \in H_i$, $g_1 \in gR$, and some $h_i \neq 0$, without loss of generality $h_1 \neq 0$. Because H_1 is uniform, and L_1

is a non-zero R-ideal in H_1, L_1 is essential in H_1. Hence $h_1R \cap L_1 = \{0\}$. So we have $h_1r_1 = l_1 \neq 0$. Then $g_1r = h'_1 + h'_2 + \ldots + h'_n$ where $h'_1 = l_1 \neq 0$. If $h'_2, \ldots, h'_n$ are all zero we have $0 \neq h'_1 \varepsilon L \cap gR$ and we are home. If $h'_2, \ldots, h'_n$ are not all zero, then without loss of generality $h'_2 \neq 0$. We now repeat the process to get $h'_2R \cap L_2 = \{0\}$ and there exists $r_2 \varepsilon R$ such that $(h'_1 + h'_2 + \ldots + h'_n)r_2 = l'_1 + l_2 + h''_3 + \ldots + h''_n$. Then $l'_1 + l_2 \neq 0$ and $l'_1 + l_2 \varepsilon L \cap gR$. If $h''_3 + \ldots + h''_n = 0$ we are home. In any case we eventually obtain a non-zero element of $gR \cap L$ after at most n steps. Thus L is essential.

Now suppose that $M_1, \ldots, M_m$ are non-zero R-ideals of G such that $M_1 \oplus \ldots \oplus M_m$ is a direct sum. We show that $m \leq n$, which will give us our result. $M_2 \oplus \ldots \oplus M_m$ is not essential since it intersects M_1 trivially. Hence $(M_2 \oplus \ldots \oplus M_m) \cap H_i = \{0\}$ for some i, without loss of generality i = 1. Hence the sum $H_1 \oplus M_2 \oplus \ldots \oplus M_m$ is direct. Similarly $H_1 \oplus M_3 \oplus \ldots \oplus M_m$ is not essential and after renumbering the H_i's if necessary, the sum $H_1 \oplus H_2 \oplus M_3 \oplus \ldots \oplus M_m$ is direct. Proceeding in this way we obtain that $m \leq n$.

This proof is a fairly straightforward adaptation of the proof of the corresponding result in Chatters and Hajarnavis [1], lemma 1.9. We come now to the generalization of proposition 1.4 of Gordon and Robson [2].

Lemma 2.6. *A tame* R-*module with Krull dimension has finite uniform dimension.*

This can be proved as in [2].

We now look at critical R-modules. We call a non-zero R-module G α-*critical* if $kd(G) = \alpha$ and the Krull dimension of any proper homomorphic image of G is strictly less than α. A critical R-module is an R-module which is α-critical for some ordinal α.

Lemma 2.7. *Any non-zero* R-*module with Krull dimension has a critical* R-*ideal.*

The proof is similar to that of proposition 2.1 of [2].

We now present a technical lemma, generalizing Proposition 1 of Lenagan [7], which we will need later.

Lemma 2.8. *Let* G *be a module with an ascending chain of* R-*ideals* $0 < H_1 < H_2 < \ldots$ *such that* $G = \bigcup_{i=1}^{\infty} H_i$. *If there is an infinite descending chain of* R-*ideals* $G = G_0 > G_1 > \ldots$ *such that for all* i,j

$$G_i \not\leq G_{i+1} + H_j \qquad (1)$$

then G *does not have Krull dimension.*

See Proposition 1 of [7] for the proof.

We now come to the question of the length of well-ordered descending chains of R-ideals in modules with Krull dimension. This generalizes work of Gulliksen [3] and Krause [4].

Lemma 2.9. *Let* G *be an* R-*module with Krull dimension* α. *Then* $\alpha \geq$ limit $\{\beta$; G *has a well-ordered descending chain of* R-*ideals of length at least* $\omega^{\beta}\}$. *Further, if* $\alpha < \omega$, *then any* R-*module* G *with the property that* α = limit $\{\beta$; G *has a well-ordered descending chain of* R-*ideals of length at least* $\omega^{\beta}\}$ *has Krull dimension* α.

Proof. We use transfinite induction on α. Consider the case $\alpha = 0$. Then G has Krull-dimension 0 if and only if G has the descending chain condition on R-ideals. That holds if and only if every well ordered descending chain of R-ideals in G is of finite length, i.e. 0 = limit $\{\beta;$ G has a well-ordered descending chain of R-ideals of length at least $\omega^{\beta}\}$.

Now assume that the first result is true for all ordinals $\gamma < \alpha$. Let G have kd(G) = α and consider a well-ordered descending chain of R-ideals in G of length $\sigma = \omega^{\beta}.n_1 + \ldots + n_r$, where $n_1, \ldots, n_r \in Z$, and the exponents of ω are ordinals. We need to show that $\beta \leq \alpha$. So suppose that $\beta > \alpha$. Let G_m be the $\omega^{\alpha}.m$th term of the well-ordered descending chain of length σ. Since $\sigma = \omega^{\beta}.n_1$, and $\beta > \alpha$, we can find G_m for each $m \geq 1$. Then G_i/G_{i+1} has a well-ordered descending chain of R-ideals of G of length ω^{α} inherited from G. By the induction hypothesis the Krull dimension of G_i/G_{i+1} relative to G is not less than α, for each $i \geq 0$. By definition kd(G) $\neq \alpha$. This finishes the proof, by induction, of the first statement.

We now prove the second statement by induction. The case $\alpha = 0$ has already been done. As α is finite in this case, we have limit $\{\beta;$ G has a well-ordered descending chain of R-ideals of length at least $\omega^{\beta}\}$ = max $\{\beta;$ G has a well-ordered descending chain of R-ideals of length at least $\omega^{\beta}\}$. We assume that the result is true for $\gamma < \alpha$. Suppose G has Krull dimension α. Then kd(G) $\geq \alpha$ implies that G has a well-ordered descending chain of R-ideals of length at least ω^{α}, using the induction hypothesis. Using part of the lemma, we have α = max $\{\beta;$ G has a well-ordered descending chain of R-ideals of length at least $\omega^{\beta}\}$. Now suppose that G has the property that α = max $\{\beta;$ G has a well-ordered descending chain of R-ideals of length at least $\omega^{\beta}\}$. Let $G = G_0 \geq G_1 \geq G_2 \geq \ldots$ be an infinite descending chain of R-ideals. If more than a finite number of the factor modules G_i/G_{i+1} are not of Krull dimension $< \alpha$ (relative to G) then an infinite number of these factors have a well-ordered descending chain of R-ideals of length at least ω^{α}. By lifting these chains to G, we get a well-ordered descending chain of length at least $\omega^{\alpha}.\omega = \omega^{\alpha+1}$. This cannot happen by hypothesis. So at most a finite number of the factor modules G_i/G_{i+1} are not of Krull dimension $< \alpha$ and thus kd(G) = α, since obviously kd(G) $\nless \alpha$ by the induction hypothesis. This finishes the proof of the second part of the lemma by induction.

The final set of preliminary results concerns powers of ideals. We list them without proof because they are easy to prove and the proofs parallel exactly those for rings in Krause and Lenagan [5], after account has been taken of the left-right switch in the definition. Note that with near-rings, powers of ideals are just subsets of the near-ring, not ideals in their own right.

Lemma 2.10. *Let* I *be an ideal of the near-ring* R. *Then the following hold*

(i) $I^{\alpha}.I^{\beta} \leq I^{\alpha+\beta}$ *for all ordinals* α, β.

(ii) $(I^{\alpha})^n \leq I^{\alpha.n}$ *for all ordinals* α, *positive integers* n.

(iii) $I_{\alpha} \leq I^{\omega^{\alpha}}$ *for all ordinals* α.

3. TAME NEAR-RINGS WITH KRULL DIMENSION.

Here we extend some results concerning rings with Krull dimension due to Lenagan [6] and [7] and Krause and Lenagan [5] to tame near-rings with Krull dimension. Many of the proofs are suitable amalgams of proofs for the ring case taken from the papers mentioned above, together with one or two special features. We consider three topics: the transfinite powers of the J_2 radical in tame near-rings with the ascending chain condition on right ideals, the non-existence of idempotent ideals contained in $J_2(R)$, where R is tame with Krull dimension and finally the nil radical in tame near-rings with Krull dimension.

For the first of these three topics we consider the following situation. Let R be a right noetherian tame near-ring, i.e. a tame near-ring with the ascending chain condition on right ideals. It will have Krull dimension by lemma 2.4, say $kd(R) = \alpha$. By lemma 2.9 any well ordered descending chain of right ideals has length strictly less than $\omega^{\alpha+1}$. Let G be a faithful tame R-module.

Lemma 3.1. *With the notation given above, let* $g \in G$. *Then the* R-*module* gR *has a well ordered descending chain of* R-*ideals which reaches* $\{0\}$ *and such that each factor is* 2-*primitive.*

Proof. By proposition 3.4 of Pilz [9], we have $gR \cong_R R/Ann_R(g)$ where the isomorphism is given by $gr \to Ann_R(g) + r$. By the ascending chain condition on right ideals, we can find a right ideal L_1 of R maximal subject to containing $Ann_R(g)$. Then $L_1/Ann_R(g)$ is a maximal right ideal in $R/Ann_R(g)$ and so gL_1 is a maximal R-ideal in gR. Since gR is tame as the submodule of a tame module, we deduce that gR/gL_1 is 2-primitive. Now repeat the process with gL_1 to obtain gL_2 such that gL_1/gL_2 is 2-primitive. Continuing transfinitely if necessary we obtain the result that we want.

Corollary 3.2. *The conclusion of lemma* 3.1 *holds for any monogenic tame* R-*module for a right noetherian near-ring* R.

Lemma 3.3. *With the same hypotheses as lemma* 3.1, *there exists an ordinal* σ *such that* $J_2^{\sigma} \leq Ann_R(gR)$ *for all* $g \in G$.

Proof. Given $g \in G$, we construct the well-ordered descending chain of R-ideals whose existence is guaranteed by lemma 3.1, say

$$gR \geq gR_1 \geq \ldots \geq gR_\lambda \geq \ldots \geq \{0\}.$$

Since each factor is 2-primitive, we have

$$(gR_\lambda)J_2 \leq gR_{\lambda+1}$$

for all ordinals λ. We will prove by transfinite induction that $(gR)J_2^{\sigma} = \{0\}$ for some ordinal σ by showing that $(gR)J_2^{\lambda} \leq gR_\lambda$ for all ordinals λ.

The case $\lambda = 1$ has already been pointed out. So assume that the result is true for all $\mu < \lambda$. If $\lambda = \mu + 1$ for some μ, then $(gR)J_2^{\lambda} = (gR)J_2^{\mu}.J_2 \leq gR_\mu.J_2$ by the induction hypothesis and $gR_\mu.J_2 \leq gR_{\mu+1} = gR_\lambda$ by the remark above. If λ is a limit ordinal, then $gR_\lambda = \bigcap_{\mu<\lambda} gR_\mu$ and $J_2^{\lambda} = \bigcap_{\mu<\lambda} J_2^{\mu}$. Let $gr \in gR$, $x \in J_2^{\lambda}$. Then for all $\mu < \lambda$,

$x \in J_2^{\mu}$ and by the induction hypothesis $(gr)x \in (gR)J_2^{\mu} \leq gR_{\mu}$. Hence $(gr)x \in \bigcap_{\mu<\lambda} gR_{\mu} = gR_{\lambda}$. This is true for all $gr \in gR$, $x \in J_2^{\lambda}$. Hence $(gR)J_2^{\lambda} \leq gR_{\lambda}$, which is sufficient to complete the induction argument, hence the proof of the result.

Corollary 3.4. $J_2^{\omega^{\omega^{\alpha+1}}} \leq \mathrm{Ann}(GR)$.

Proof. By lemma 2.9, any well-ordered descending chain of right ideals in R has length strictly less than $\omega^{\alpha+1}$. Hence $J_2^{\omega^{\omega^{\alpha+1}}} \leq \mathrm{Ann}(gR)$ for all $g \in G$ by the lemma. This proves the corollary.

Theorem 3.5. *Let* R *be a tame right noetherian near-ring. Then* $J_2(R)^{\omega^{\omega^{\alpha+1}}} = \{0\}$ *if* R *has an identity and has square zero otherwise.*

Proof. If R has an identity, then GR = G, and since G is faithful, it follows from corollary 3.4 that $J_2^{\omega^{\omega^{\alpha+1}}} = \{0\}$. Otherwise $RJ_2^{\omega^{\omega^{\alpha+1}}}$ annihilates G and since G is faithful, we have $(J_2^{\omega^{\omega^{\alpha+1}}})^2 = \{0\}$.

Theorem 3.6. *Let* R *be a tame right noetherian near-ring. Then* $(J_2(R))_{\alpha}$ *is nilpotent if* R *has an identity, and* $(J_2(R))_{\alpha+1}$ *has square zero otherwise.*

Proof. The second part follows from lemma 2.10 and theorem 3.5. The first part follows from lemma 2.10 and lemma 2.9 as in the proof of theorem 3.5 once we observe that by lemma 2.9, $J_2^{\sigma} = \{0\}$ for some $\sigma = \omega^{\alpha}.n + \ldots$ for some positive integer n.

We now turn to the non-existence of idempotent ideals within the J_2 radical. There are two results of this type, one where the module has Krull dimension, the other where the near-ring has Krull dimension.

Let R be a tame near-ring with a faithful tame R-module G. Let G have Krull dimension.

Definition 3.7. The *critical socle series* of G is defined by

$S_0(G) = \{0\}$, $S_{\lambda+1}(G)/S_{\lambda}(G)$ is the sum of all critical submodules of $G/S_{\lambda}(G)$ and $S_{\lambda} = \bigcup_{\mu<\lambda} S_{\mu}(G)$ for limit ordinals λ.

By lemma 2.7 it follows that $S_{\lambda+1}(G) > S_{\lambda}(G)$ for all ordinals λ and hence that $G = S_{\alpha}(G)$ for some ordinal α.

Theorem 3.8. *Let* R *be a tame near-ring with a faithful tame* R*-module* H *with Krull dimension. Then* $J_2(R)$ *does not contain a non-zero idempotent ideal.*

Proof. Let $I \leq J_2(R)$, I be an idempotent ideal. We show that I annihilates all β-critical submodules of H with $\beta \leq \mathrm{kd}(H)$. Note that there is no distinction between submodules and R-ideals in H, since H is tame. The cases $\beta = -1,0$ are both trivial since 0-critical modules are just the 2-primitive submodules of H. Now assume that $LI = \{0\}$ for all modules with $\mathrm{kd}(L) < \beta$. Then if M is a β-critical submodule of H and M′ is a non-zero submodule of M, we have $\mathrm{kd}(M/M') < \beta$. Hence $(M/M')I = \{0\}$, i.e. $M' \geq MI$. Thus MI will generate the unique minimal submodule of M if it is not zero. Since $I \leq J_2(R)$, it follows that $\{0\} = MI.I = MI^2$ and hence $\mathrm{Id}\langle I^2\rangle \leq \mathrm{Ann}\, M$, i.e. $MI = \{0\}$ since $I = \mathrm{Id}\langle I^2\rangle$. This proves the result by induction.

Corollary 3.9. *If* R *is a tame near-ring with a faithful tame* R*-module with Krull dimension then, for any ideal* $I \leq J_2(R)$, *either* $I > I^2 > \ldots$ *is a strictly decreasing sequence or* I *is nilpotent.*

Proof. Otherwise $J_2(R)$ would contain an idempotent ideal contrary to the theorem.

We now turn to the case in which R has the Krull dimension.

Theorem 3.10. *Let* R *be a tame near-ring with Krull dimension. Then* $J_2(R)$ *does not contain a non-zero idempotent ideal.*

Proof. Let G be a faithful tame R-module. Let $g \in G$, $I \leq J_2(R)$, I an idempotent ideal. Then by proposition 3.4 of Pilz [9] we have $gR \cong_R R/Ann_R(g)$ and as R has Krull dimension, so does $R/Ann_R(g)$ by lemma 2.3. Also $gR \leq G$, hence gR is a tame R-module. By theorem 3.8 it follows that $gR.I = \{0\}$. Thus $RI \leq Ann_R(g)$ and so $I^2 \leq Ann_R(g)$. This holds for all $g \in G$, and so $I^2 \leq Ann_R(G)$. As G is faithful, it follows that $I^2 = \{0\}$. This proves the result.

Corollary 3.11. *Let* R *be a tame near-ring with Krull dimension. Let* I *be an ideal contained in* $J_2(R)$. *Then either the sequence* $I > I^2 > \ldots$ *is strictly decreasing or* I *is nilpotent.*

Proof. Otherwise $J_2(R)$ would contain an idempotent ideal contrary to the theorem.

We come finally to the nil radical of a tame near-ring with Krull dimension. Unfortunately we cannot obtain the nilpotence of the nil radical as happens for the ring case. But we do obtain something.

Lemma 3.12. *Let* R *be a tame near-ring with Krull dimension and let* G *be a faithful tame* R*-module, with* $g \in G$. *Write* A *for* $Ann_R(g)$. *If* $(C+A)/A$ *is a critical right ideal of* R/A *and* C *is nil then* $C^2 \leq A$.

Proof. Let $\theta : R \to R/A$ where as before $R/A \cong_R gR \leq G$ by proposition 3.4 of Pilz [9]. By lemma 2.3, R/A has Krull dimension, and is tame as $gR \leq G$ and G is tame. Let C be a nil right ideal of R, such that $C\theta$ is critical. We can find such an ideal unless all nil right ideals of R are contained in A, by lemma 2.7. So without loss of generality we may assume that $C\theta \neq \{0\}$. Let $c\theta \in C\theta$, $c\theta \neq 0$. Then $(cC)\theta$ is an R-ideal of $R\theta$ since $R\theta$ is tame. But $c^n = 0$ for some $n > 0$. So there exists a least $m > 1$ such that $(c\theta)^m = 0$. It follows that $(c\theta)^{m-1} \neq 0$ and $(c\theta)^{m-1}$ annihilates $c\theta$. Thus $Ann_C(c\theta) \geq \{A, c^{m-1}\}$. Hence $c\theta . C\theta \cong C\theta/Ann_{C\theta}(c\theta) \cong C + A/Ann_{C+A}(c\theta)$ and $Ann_{C+A}(c\theta) \geq \{A, c^{m-1}\} > A$. Thus $kd(c\theta . C\theta) < kd(C\theta)$ since $C\theta$ is critical. Hence as $C\theta$ is critical, we have $c\theta . C\theta = 0$, i.e. $(C\theta)^2 = \{0\}$ and $C^2 \leq A$. The fact that a non-zero submodule of a critical tame module with Krull dimension has the same Krull dimension follows from lemma 2.1.

Lemma 3.13. *Let* R *be a tame near-ring with Krull dimension. Let* N *be the nil radical of* R. *If* N *is not nilpotent, then there exists a chain of right ideals* $0 = A_0 < A_1 < \ldots$ *in* N *such that each* $A_i\theta$ *is nilpotent but* $\bigcup_{i=1}^{\infty} A_i\theta$ *is not nilpotent, where* $\theta : R \to R/A$ *for* $A = Ann_R(g)$, $g \in G$, *a faithful tame* R*-module.*

Proof. By hypothesis and using lemma 3.12, we deduce that there is a non-zero right ideal $C\theta$ with square $\{0\}$. By Zorn's Lemma we may choose a maximal right ideal contained in N whose square is zero in R/A. Call this right ideal A_1. Assume that we have chosen A_i, then choose A_{i+1} so that $A_{i+1}\theta/A_i\theta$ is a maximal right ideal contained in $N\theta/A_i\theta$ whose square is zero. Obviously each $A_i\theta$ is nilpotent. If $X\theta = \bigcup_{i=1}^{\infty} A_i\theta$ is nilpotent, then $X\theta = A_n\theta$ for some n, which is impossible by hypothesis and lemma 3.12.

Lemma 3.14. *Let* R *be a tame near-ring with Krull dimension and faithful tame module* G. *Let* N *be the nil radical of* R. *For every* $g \in G$, *there exists* $n = n(g)$ *such that* $N^{n(g)} \leq \mathrm{Ann}_R(g)$.

Proof. Assume the result does not hold. Then by lemma 3.13, there is a right ideal X contained in N , where $X\theta$ is not nilpotent and θ is defined as in lemma 3.13. We also have $X = \bigcup_{i=1}^{\infty} A_i$ and $A_i\theta$ is nilpotent for each $i \geq 1$. Since $N\theta$ and $X\theta$ have Krull dimension, using lemmas 2.1 and 2.4, we can apply lemma 2.8 and deduce that for some n, j we have $(X\theta)^n \leq (X\theta)^{n+1} + A_j\theta$. Hence $(X\theta)^n \leq A_j\theta$ since the powers of $X\theta$ form a strictly decreasing sequence (corollary 3.11). Thus $X\theta$ is nilpotent, a contradiction.

Theorem 3.15. *Let* R *be a tame near-ring with Krull dimension,* N *the nil radical of* R. *Then* $N^{\omega} = \{0\}$.

Proof. Let G be a faithful tame R-module. By lemma 3.14, $N^{\omega} \leq \bigcap_{g \in G} \mathrm{Ann}_R(g) = \{0\}$ since G is faithful.

ACKNOWLEDGMENT.

The work on this paper was done while the first named author was visiting the Department of Mathematics of the University of Stellenbosch. Financial assistance of both the Council for Scientific and Industrial Research and the University of Stellenbosch is gratefully acknowledged.

REFERENCES.

[1] Chatters, A.W. and Hajarnavis, C. R. *Rings with chain conditions.* (Pitman Research Notes in Mathematics, 44. London. 1980).

[2] Gordon, R. and Robson, J.C. *Krull dimension.* Mem. Amer. Math. Soc. 133 (1973).

[3] Gulliksen, T.H. *A theory of length for noetherian modules.* J. Pure and Appl. Algebra 3 (1973), 159-170.

[4] Krause, G. *Descending chains of submodules and the Krull dimension of noetherian modules* J. Pure and Appl. Algebra 3 (1973), 385-397.

[5] Krause, G. and Lenagan, T.H. *Transfinite powers of the Jacobson radical.* Comm. Alg. 7 (1979), 1-8.

[6] Lenagan, T.H. *The nil radical of a ring with Krull dimension.* Bull. London Math. Soc. 5 (1973), 307-311.

[7] Lenagan, T.H. *Reduced rank in rings with Krull dimension.* Ring theory (Proc. Antwerp Conf.), 123-131. (Lecture Notes in Pure and Appl. Math. 51, Dekker. New York. 1979).

[8] Meldrum, J.D.P. *Near-rings and their links with groups.* (Pitman Research Notes No. 134, London, 1985)

[9] Pilz, G. *Near-rings.* (North-Holland, Amsterdam, 1977 (2nd Ed. 1983)).

[10] Scott, S.D. *Near-rings and near-ring modules.* (Doctoral Dissertation), Australian National University) 1970.

[11] Scott, S.D. *Tame near-rings and N-groups.* Proc. Edinburgh Math. Soc. 23 (1980), 275-296.

Near-rings and Near-fields, G. Betsch (editor)
© Elsevier Science Publishers B.V. (North-Holland), 1987

SOLUTION OF AN OPEN PROBLEM CONCERNING 2-PRIMITIVE NEAR-RINGS

J H MEYER and ANDRIES P J VAN DER WALT

Department of Mathematics, University of Stellenbosch, Stellenbosch 7600, South Africa*.

ABSTRACT We construct a 2-primitive non-ring with identity which will clarify an open problem posed in 1971.

In this note we solve an open problem about 2-primitive near-rings raised by Betsch [1] in connection with his theorem 4.8, which reads as follows:

Let R *be a right near-ring with identity* 1 *which is not a ring and which is* 2-*primitive on the* R-*module* G. *If* R *contains an idempotent* e *of rank* 1, *then* R *contains a minimal left ideal or we have*

(+) $$e \neq 1 \quad \text{and} \quad \mathrm{Ann}_R(G \setminus eG) = 0.$$

(By the rank of an element $r \in R$ we mean the cardinality of the set of non-zero orbits of $\mathrm{Aut}_R G$ on rG.)

The problem is to find an example to show that the exceptional case (+) can indeed occur. We shall show that matrix near-rings over some near-fields furnish examples of this phenomenon.

Matrix near-rings were introduced in Meldrum and Van der Walt [2]. For the convenience of the reader we provide the pertinent definitions:

For a natural number n, we define the $n \times n$ *matrix near-ring* $\mathbb{M}_n(R)$ over the near-ring R as the subnear-ring of $M(R^n)$ generated by $\{f^r_{ij}: R^n \to R^n \mid 1 \leq i,j \leq n,\ r \in R\}$, where $f^r_{ij} < r_1, r_2, \ldots, r_n > = < t_1, t_2, \ldots, t_n >$, with $t_i = rr_j$ and $t_k = 0$ if $k \neq i$. Here R^n denotes $\bigoplus_1^n (R,+)$, and n-vectors are written with pointed brackets and are thought of as column vectors. The elements of $\mathbb{M}_n(R)$ are called $n \times n$ *matrices* over R and will usually be denoted by upper case letters. In particular, if R happens to be a (right) near-field, then $\mathbb{M}_n(R)$ is a zerosymmetric (right) near-ring with identity $f^1_{11} + f^1_{22} + \ldots + f^1_{nn}$. See [2] for more details.

In matrix ring theory, the matrices with zeroes in all the rows except in a fixed row, play an important role. We will generalize this concept to matrix near-rings: If $1 \leq k \leq n$, we call a matrix of the form $\sum_{j=1}^{m} f^{r_j}_{k\ell_j} U_j$, with

*Financial assistance by the Council for Scientific and Industrial Research is gratefully acknowledged.

$r_j \in R$ and $U_j \in \mathbb{M}_n(R)$ for all $1 \leqslant j \leqslant m$, a *k-th row matrix*. Lemma 1 tells us more about the construction of such matrices:

Lemma 1 *Let* $1 \leqslant k \leqslant n$. *If* U *is any matrix, then* $f^r_{ki}U$ *is a k-th row matrix for all* $r \in R$, $1 \leqslant i \leqslant n$. *Furthermore, there exists an expression for* $f^r_{ki}U$ *which consists entirely of functions of the form* f^s_{kj} $(s \in R)$, *apart from parantheses and operators.*

Proof From the definition it follows that $f^r_{ki}U$ is a k-th row matrix. Now,

$$f^r_{ki}U = f^r_{ki}\Big(\sum_{t=1}^{n}\sum_{j=1}^{m_t}[r_{tj};\, t,\ell_{tj}]V_{tj}\Big)$$
$$= f^r_{ki}\Big(\sum_{j=1}^{m_i}[r_{ij};\, i,\ell_{ij}]V_{ij}\Big)$$
$$= f^r_{kk}\Big(\sum_{j=1}^{m_i}[r_{ij};\, k,\ell_{ij}]V_{ij}\Big)$$

where $[x;\, p,q]$ denotes f^x_{pq}.

Then, for each $1 \leqslant j \leqslant m_i$, we treat $[r_{ij};\, k,\ell_{ij}]V_{ij} = f^{r_{ij}}_{k\ell_{ij}}\ V_{ij}$ in the same way, *et cetera*. □

The next lemma is quite useful:

Lemma 2 *If* L *is a minimal left ideal of* $\mathbb{M}_n(R)$ *and* V *is an invertible element of* $\mathbb{M}_n(R)$, i.e. V^{-1} *exists, then* $LV := \{LV \mid L \in L\}$ *is also a minimal left ideal of* $\mathbb{M}_n(R)$.

Proof It is routine calculation to check that LV is a left ideal. Suppose it is not minimal. Then we have a proper subset S of L such that SV is a left ideal properly contained in LV. But then $(SV)V^{-1} = S$ is a left ideal properly contained in L, a contradiction. □

Now let F be an infinite near-field which is not a field. The example we are going to construct works for any $n \geqslant 2$, but for the sake of convenience, we stick to the case $n = 2$. Thus, let $R := \mathbb{M}_2(F)$. As was pointed out in [4], R is 2-primitive on $G := F^2$. The next proposition is fundamental in this paper:

Proposition 1 $\mathrm{Aut}_R G \cong (F \setminus \{0\}, \cdot)$ *where the elements of* $F \setminus \{0\}$ *act on* G *by multiplication on the right.*

Proof It is easy to check that $\Phi\colon F \setminus \{0\} \to \mathrm{Aut}_R G$ defined by $\langle \alpha,\beta \rangle \Phi(\lambda) = \langle \alpha\lambda,\beta\lambda \rangle$ is a monomorphism. To see that it is an epimorphism, suppose Δ is an R-automorphism of G, and that $\langle 1,0 \rangle \Delta = \langle \alpha,\beta \rangle$. Then $\alpha \neq 0$, since, if $\langle 1,0 \rangle \Delta = \langle 0,\beta \rangle$, then $\langle 1,0 \rangle \Delta = (f^1_{11} \langle 1,0 \rangle)\Delta = f^1_{11}(\langle 1,0 \rangle\Delta) = f^1_{11} \langle 0,\beta \rangle = \langle 0,0 \rangle$ which implies $\beta = 0$. But only the zero-element $\langle 0,0 \rangle$ can be mapped onto $\langle 0,0 \rangle$ by an R-automorphism. Thus, if $\langle \gamma,\delta \rangle$ is any element of G, $\langle \gamma,\delta \rangle\Delta = (f^\gamma_{11} \langle 1,0 \rangle + f^\delta_{21} \langle 1,0 \rangle)\Delta =$ $(f^\gamma_{11} \langle 1,0 \rangle)\Delta + (f^\delta_{21} \langle 1,0 \rangle)\Delta = f^\gamma_{11}(\langle 1,0 \rangle\Delta) + f^\delta_{21}(\langle 1,0 \rangle\Delta) =$

$f_{11}^{\gamma} < \alpha,\beta > + f_{21}^{\delta} < \alpha,\beta > = < \gamma\alpha,\delta\alpha > = < \gamma,\delta > \Phi(\alpha)$. Thus, $\Phi(\alpha) = \Delta$, with $\alpha \in F \setminus \{0\}$. □

From this proposition it follows that a full set of orbit representatives of $Aut_R G$ on G is given by $\{< 1,\lambda > | \lambda \in F\} \cup \{< 0,1 >, < 0,0 >\}$. Now consider the matrix f_{11}^{1}. It certainly is an idempotent in R and, since $f_{11}^{1} G = \{< \lambda,0 > | \lambda \in F\}$, f_{11}^{1} is of rank 1. Also, $f_{11}^{1} \neq I$, the identity matrix.

At this point we choose a specific F, namely the infinite Dickson near-field $(F,+,\circ)$ arising from $Q(x)$, the field of rational functions over the rationals, by defining multiplication as follows:

$$\frac{g(x)}{h(x)} \circ \frac{p(x)}{q(x)} = \begin{cases} 0 & \text{if } p(x) = 0 \\ \dfrac{g(x+d)}{h(x+d)} \cdot \dfrac{p(x)}{q(x)} & \text{if } p(x) \neq 0 \end{cases}$$

where $d = d(\frac{p(x)}{q(x)}) :=$ degree $(p(x))$ - degree $(q(x))$. See [3], example 8.29, or [5] for more details. We need the following:

Lemma 3 *Let* F *be the infinite near-field as defined above and suppose* $U \in \mathbb{M}_2(F)$ *is a first row matrix. Then* $U< 1,0 > = < \frac{R(x)}{S(x)},0 >$ *for some polynomials* $R(x)$ *and* $S(x)$ *in* $Q[x]$. *Furthermore, there is an infinite sequence of functions* $\{\eta_i\}_{i\in\mathbb{N}} \subseteq F$ *such that*

$$U< 1,\eta_i > = < \frac{R(x)}{S(x)} + \frac{P_i(x)}{Q_i(x)} \circ \eta_i, 0 >$$

for polynomials $P_i(x)$ *and* $Q_i(x)$ *where* $\{d(\frac{P_i(x)}{Q_i(x)})\}$ *is bounded from above, i.e.* $d(\frac{P_i(x)}{Q_i(x)}) \leqslant k$ (say) *for all* $i \in \mathbb{N}$. *Also, we may choose the* η_i *in such a way that* $\{d(\eta_i)\}$ *is not bounded from below.*

Proof The infinite sequence of functions we have in mind here is the set $\{\frac{1}{x^i}\}_{i=m}^{\infty}$, for $m \in \mathbb{N}$ large enough. Note that $d(\frac{1}{x^i}) = -i$, so that $\{d(\frac{1}{x^i})\}_{i=m}^{\infty}$ is not bounded from below. We use the following induction argument to justify our claim: Since U is a first row matrix, it can be built up from matrices of the form f_{11}^{λ} and $f_{12}^{\gamma} (\lambda,\gamma \in F)$, by lemma 1.

If $U = f_{11}^{\frac{r(x)}{s(x)}}$, then $U< 1,0 > = < \frac{r(x)}{s(x)},0 >$ and $U< 1,\frac{1}{x^i} > =$

$< \frac{r(x)}{s(x)} + 0 \circ \frac{1}{x^i},0 >$ for all $i \geqslant 1$. If $U = f_{12}^{\frac{r(x)}{s(x)}}$, then $U< 1,0 > = < 0,0 >$ and $U< 1,\frac{1}{x^i} > = < 0 + \frac{r(x)}{s(x)} \circ \frac{1}{x^i},0 >$ for all $i \geqslant 1$. Note that $d(\frac{r(x)}{s(x)})$ is fixed for all $i \geqslant 1$. Now suppose V and W are first row matrices such that

$$V< 1,0 > = < \frac{r_1(x)}{s_1(x)},0 >,$$

$$V< 1,\frac{1}{x^i} > = < \frac{r_1(x)}{s_1(x)} + \frac{p_i(x)}{q_i(x)} \circ \frac{1}{x^i},0 > \quad \text{for all} \quad i \geqslant m_1$$

with $d(\frac{p_i(x)}{q_i(x)}) \leqslant k_1$ for all $i \geqslant 1$, and that $W< 1,0 > = < \frac{r_2(x)}{s_2(x)},0 >$,

$$W< 1,\frac{1}{x^i} > = < \frac{r_2(x)}{s_2(x)} + \frac{t_i(x)}{u_i(x)} \circ \frac{1}{x^i},0 > \quad \text{for all} \quad i \geqslant m_2$$

with $d(\frac{t_i(x)}{u_i(x)}) \leqslant k_2$ for all $i \geqslant 1$.

Then $(V + W) < 1,0 > = < \frac{r_1(x)s_2(x) + s_1(x)r_2(x)}{s_1(x)s_2(x)},0 >$

and $(V + W) < 1,\frac{1}{x^i} > = < \frac{r_1(x)s_2(x) + s_1(x)r_2(x)}{s_1(x)s_2(x)} + \frac{p_i(x)u_i(x) + q_i(x)t_i(x)}{q_i(x)u_i(x)} \circ \frac{1}{x^i},0>$

for all $i \geqslant \max\{m_1,m_2\}$ and with

$d(\frac{p_i(x)\ u_i(x) + q_i(x)\ t_i(x)}{q_i(x)\ u_i(x)}) \leqslant \max\{k_1,k_2\}$ for all $i \geqslant 1$.

Furthermore, $f_{11}^{\frac{g(x)}{h(x)}}\ V< 1,0 > = < \frac{g(x + \tilde{d})r_1(x)}{h(x + \tilde{d})s_1(x)},0 >$ where $\tilde{d} = d(\frac{r_1(x)}{s_1(x)})$, and

$$f_{11}^{\frac{g(x)}{h(x)}}\ V< 1,\frac{1}{x^i} > = < \frac{g(x)}{h(x)} \circ [\frac{r_1(x)}{s_1(x)} + \frac{p_i(x)}{q_i(x)} \circ \frac{1}{x^i}],0 >$$

$$= < \frac{g(x)}{h(x)} \circ [\frac{r_1(x)q_i(x - i)\ x^i + s_1(x)p_i(x - i)}{s_1(x)q_i(x - i)x^i}],0 >$$

$$= < \frac{g(x + \tilde{d})r_1(x)}{h(x + \tilde{d})s_1(x)} + \frac{g(x + \tilde{d} + i)p_i(x)}{h(x + \tilde{d} + i)q_i(x)} \circ \frac{1}{x^i},0 >$$

for all $i \geqslant m_3$ (say) where $m_3 \geqslant m_1$.

Since $d(\frac{p_i(x)}{q_i(x)}) \leqslant k_1$ and $d(\frac{g(x + \tilde{d} + i)}{h(x + \tilde{d} + i)})$ is independent of i, we must have

a k_3 (say), such that $d(\frac{g(x + \tilde{d} + i)p_i(x)}{h(x + \tilde{d} + i)q_i(x)}) \leqslant k_3$ for all $i \geqslant 1$.

Actually, we also have to consider the case $f_{12}^{\frac{g(x)}{h(x)}}\ V(V \neq I)$, but since we can

assume V to be a first row matrix, by lemma 1, $f_{12}^{\frac{g(x)}{h(x)}}\ V = 0$. If $V = I$, then

$f_{12}^{\frac{g(x)}{h(x)}}\ V = f_{12}^{\frac{g(x)}{h(x)}}$ and we have considered this case already at the beginning of our induction process. Taking lemma 1 into consideration, the result follows. □

We will use lemma 3 to prove the next proposition, from which it will follow that $Ann_R(G \setminus f^1_{11}G) = 0$, showing that the exceptional case (+) can occur.

Proposition 2 *Let* F *be the infinite near-field as in lemma 3. Suppose* $\{B_1, B_2, \ldots, B_m\}$ *is a finite set of orbits of* $Aut_R G$ *on* G. *Then*

$$Ann_R(G \setminus \bigcup_{i=1}^{m} B_i) = 0.$$

Proof We may assume that the B_i are non-zero. Consequently, an orbit B may have representatives: $< 1,0 >$, $< 0,1 >$ or $< 1,\lambda >$ for a non-zero $\lambda \in F$. Now suppose

$$Ann_R(G \setminus \bigcup_{i=1}^{m} B_i) \neq 0.$$

Then there is a matrix $U \in M_2(F)$ such that $U B_j \neq 0$ for a certain $1 \leqslant j \leqslant m$ and $U B_\tau = 0$ for infinitely many orbits B_τ. We have to consider three cases:

(i) B_j *is generated by* $< 1,0 >$.

Suppose $U< 1,0 > = < \alpha,\beta >$ with (say) $\alpha \neq 0$, then $f^{\alpha^{-1}}_{11}$ $U< 1,0 > = < 1,0 >$ and also,

$$f^{\alpha^{-1}}_{11} U< 1,\lambda > = < 0,0 > \text{ for almost}$$

all $\lambda \in F$, i.e. $f^{\alpha^{-1}}_{11}$ $U< 1,\lambda > = < 0,0 >$ for all $\lambda \in F$, except maybe a finite number of $\lambda \in F$. Furthermore, $f^{\alpha^{-1}}_{11}$ U is a first row matrix, according to lemma 1. So, without loss of generality, we may suppose U is a first row matrix such that $U< 1,0 > = < 1,0 >$ and $U< 1,\lambda > = < 0,0 >$ for almost all $\lambda \in F$. By lemma 3 there is an infinite sequence of functions $\{\eta_i\}_{i\in\mathbb{N}}$ such that $U< 1,\eta_i > = < 1 + \frac{P_i(x)}{Q_i(x)} \circ \eta_i, 0 >$ for polynomials $P_i(x)$ and $Q_i(x)$ in $Q[x]$ and such that $d(\frac{P_i(x)}{Q_i(x)}) \leqslant k$ (say) for all $i \geqslant 1$. But since $< 1,\eta_i >$ belong to different orbits for different η_i, we must have $U< 1,\eta_i > = < 1 + \frac{P_i(x)}{Q_i(x)} \circ \eta_i, 0 > = < 0,0 >$ for almost all η_i. This implies that

$$\frac{P_i(x)}{Q_i(x)} = -\eta_i^{-1} \quad \text{(inverse with respect to } \circ)$$

for infinitely many i. This is clearly a contradiction, since $d(\eta_i^{-1}) = -d(\eta_i)$ can be made arbitrarily large, while $d(\frac{P_i(x)}{Q_i(x)}) \leqslant k$.

So B_j cannot be generated by $< 1,0 >$.

(ii) B_j *is generated by* $< 0,1 >$. Then, in the same way as (i), we can find a matrix $V \in M_2(F)$ such that $V< 0,1 > = < 1,0 >$ and $V< 1,\lambda > = < 0,0 >$ for almost all $\lambda \in F$. But then, if $U = V(f_{12}^1 + f_{21}^1 + f_{22}^1)$, we have $U< 1,0 > = < 1,0 >$ and $U< 1,\lambda > = < 0,0 >$ for almost all $\lambda \in F$, which is impossible by (i).

(iii) B_j *is generated by* $< 1,\lambda >$ *for a non-zero* $\lambda \in F$. As before, we can find a matrix $V \in M_2(F)$ such that $V< 1,\lambda > = < 1,0 >$ and $V< 1,\zeta > = < 0,0 >$ for almost all $\zeta \in F$. But then, if $U = V(f_{11}^1 + f_{21}^\lambda + f_{22}^1)$, we have $U< 1,0 > = < 1,0 >$ and $U< 1,\zeta > = < 0,0 >$ for almost all $\zeta \in F$, a contradiction by (i).

Thus, we must have

$$\mathrm{Ann}_R(G \setminus \bigcup_{i=1}^m B_i) = 0. \qquad \square$$

By a result in [1], §5 (pp 89 - 90), there is a mapping $m \in M_A(F^2)$, where $A := \mathrm{Aut}_R G$, such that

$$m< \lambda,0 > = < \lambda,0 > \quad \text{for all } \lambda \in F$$
$$\text{and } m< \lambda,\beta > = < 0,0 > \quad \text{for all } \lambda,\beta \in F,\ \beta \neq 0.$$

Hence, by proposition 2, our example also shows that there are functions in $M_A(F^2)$ which are not matrices.

In the next result we shall see that, although there are minimal R-subgroups in this near-ring, for example $\{f_{11}^\lambda + f_{21}^\delta \mid \lambda,\delta \in F\}$, there are no minimal left ideals in R. This implies that not only does the exceptional case (+) occur in our example, but it occurs exclusively.

Proposition 3 *Let* R *be the same near-ring as before. Then* R *does not contain a minimal left ideal.*

Proof Suppose L is a minimal left ideal of R and that U is a non-zero matrix in L. According to theorem 4.3 [1], any non-zero element of L has rank 1 and consequently, rank U = 1. We shall produce a contradiction by finding an element X in a minimal left ideal such that rank X $\geqslant$ 2. Since U is non-zero, it must be non-zero on an infinite number of orbits of $\mathrm{Aut}_R G$ on G, according to proposition 2. Thus, we can choose $\alpha,\beta \in F \setminus \{0\}$, $\alpha \neq \beta$ such that

$$U< 1,\alpha > = < \gamma,\delta > \neq < 0,0 >$$
$$\text{and } U< 1,\beta > = < \gamma\circ\lambda,\delta\circ\lambda >, \quad \text{with } \lambda \in F \setminus \{0\}.$$

Let T be the matrix $f_{11}^1 + f_{22}^\alpha$. Then

$$UT< 1,1 > = U< 1,\alpha > = < \gamma,\delta >$$
$$\text{and } UT< 1,\alpha^{-1}\circ\beta > = U< 1,\beta > = < \gamma\circ\lambda,\delta\circ\lambda >.$$

Now $UT \in LT$, which is a minimal left ideal according to lemma 2, because $T^{-1} = f_{11}^1 + f_{22}^{\alpha^{-1}}$ exists. Let us define the matrix W as follows:

$$W = \begin{cases} f_{11}^{\gamma^{-1}} + f_{21}^{x_0\gamma^{-1}} & \text{if } \delta = 0 \\ f_{12}^{\delta^{-1}} + f_{22}^{x_0\delta^{-1}} & \text{if } \delta \neq 0 \end{cases}$$

Then, if $\xi := \alpha^{-1} \circ \beta$, we have $\xi \neq 1$ while

$$WUT\langle 1,1 \rangle = W\langle \gamma,\delta \rangle = \langle 1,x \rangle$$

$$\text{and } WUT\langle 1,\xi \rangle = W\langle \gamma\circ\lambda,\delta\circ\lambda \rangle = \langle \lambda,x\circ\lambda \rangle.$$

Moreover, $WUT \in LT$. Now consider the matrix

$$V = f_{11}^{1} + f_{12}^{1} + f_{21}^{\xi} + f_{22}^{1}.$$

The inverse of V is

$$V^{-1} = f_{11}^{(1-\xi)^{-1}}(f_{11}^{1} + f_{12}^{-1}) + f_{22}^{-\xi(1-\xi)^{-1}}(f_{11}^{1} + f_{12}^{-1}) + f_{22}^{1}$$

which can be easily verified by showing that $VV^{-1} = V^{-1}V = I$. According to lemma 2, LTV is a minimal left ideal of R and furthermore

$$WUTV\langle 0,1 \rangle = WUT\langle 1,1 \rangle = \langle 1,x \rangle$$

$$\text{and } WUTV\langle 1,0 \rangle = WUT\langle 1,\xi \rangle = \langle \lambda,x\circ\lambda \rangle.$$

Let $X = Y(I + WUTV) - Y$, where $Y = f_{11}^{1} + f_{22}^{x}$. Then $X \in LTV$, and

$$\begin{aligned} X\langle 0,1 \rangle &= Y(\langle 0,1 \rangle + \langle 1,x \rangle) - Y\langle 0,1 \rangle \\ &= Y\langle 1,1 + x \rangle - \langle 0,x \rangle \\ &= \langle 1,(x + 1)^2 \rangle - \langle 0,x \rangle \\ &= \langle 1,(x + 1)^2 - x \rangle. \end{aligned}$$

Also

$$\begin{aligned} X\langle 1,0 \rangle &= Y(\langle 1,0 \rangle + \langle \lambda,x\circ\lambda \rangle) - Y\langle 1,0 \rangle \\ &= Y\langle 1 + \lambda,x\circ\lambda \rangle - \langle 1,0 \rangle \\ &= \langle 1 + \lambda,x\circ x\circ\lambda \rangle - \langle 1,0 \rangle \\ &= \langle \lambda,((x + 1)x)\circ\lambda \rangle. \end{aligned}$$

Clearly, $\langle 1,(x + 1)^2 - x \rangle$ and $\langle 1,(x + 1)x \rangle$ belong to different orbits of $\mathrm{Aut}_R G$ on XG which implies rank $X \geqslant 2$. Consequently, R does not have any minimal left ideals. □

REFERENCES

[1] Betsch, G., Some structure theorems on 2-primitive near-rings, Colloquia Mathematica Societatis Jánas Bolyai, 6. Rings, modules and radicals. Keszthely (Hungary), 1971.

[2] Meldrum, J.D.P. and Van der Walt, A.P.J., Matrix near-rings. To appear in Archiv der Mathematik.

[3] Pilz, G., Near-rings (North Holland, 1977).

[4] Van der Walt, A.P.J., Primitivity in matrix near-rings. To appear in Quaestiones Mathematicae.

[5] Zemmer, J.L., Near-fields, planar and non-planar, The Math. student, 32 (1964).

Near-rings and Near-fields, G. Betsch (editor)
© Elsevier Science Publishers B.V. (North-Holland), 1987

ARE THE JACOBSON-RADICALS OF NEAR-RINGS M-RADICALS ?

Rainer MLITZ

Institut für Angewandte und Numerische Mathematik, TU Wien
Wiedner Hauptstraße 8-1o
A-1o4o Wien

Let $\mathcal{U}$ be a universal class of right near-rings (i.e. a class of near-rings satisfying the law $(x+y)z = xz+yz$ which is closed under taking homomorphic images and ideals). It is well known that a Kurosh-Amitsur-radical on $\mathcal{U}$ is a map $N \to \rho N \triangleleft N$ ($\triangleleft$ means "ideal of") defined on $\mathcal{U}$ and satisfying :

(ρ1) $(\rho N+I)/I \subset \rho(N/I)$ for all $I \triangleleft N$,

(ρ2) $\rho(N/\rho N) = 0$,

(ρ3) $\rho(\rho N) = \rho N$,

(ρ4) $\rho I = I$, $I \triangleleft N$ imply $I \subseteq \rho N$.

(ρ1) and (ρ2) define the Hoehnke-radicals and a Hoehnke-radical is a Kurosh-Amitsur-radical if and only if it satisfies

(ρ5) $\rho N = 0 \Leftrightarrow (0 \neq I \triangleleft N \Rightarrow \rho I \neq I)$

(see [3] and e.g. [6]). The Jacobson-radicals J_2 and J_3 are Kurosh-Amitsur-radicals (at least for zero-symmetric near-rings), J_0 and J_1 are not because of the lack of (ρ3) - see [4].

Recently, the concept of Kurosh and Amitsur has been generalised to that of M-*radicals* (which include considerably many interesting factorisations of various structures) - see [5]. For our near-ring case, M will be a binary relation on the universal class $\mathcal{U}$ satisfying the following conditions:

(M1) $A \mathrel{M} N \Rightarrow A \neq 0$ is a subnearring of N ,

(M2) $N \mathrel{M} N$ for all $N \neq 0$ in $\mathcal{U}$,

(M3) $A \mathrel{M} N \Rightarrow A \mathrel{M} \varphi N$ or $A = 0$ for every homomorphism φ defined on N ,

(M4) $A \mathrel{M} N$, $I \triangleleft N$ and $A \subseteq I$ imply $A \mathrel{M} I$.

An M-radical is then a Hoehnke-radical satisfying

(ρ5M) $\rho N = 0 \Leftrightarrow (A \mathrel{M} N \Rightarrow \rho A \neq A)$.

Notice that (ρ1),(ρ2) and (ρ5M) always imply

(ρ4M) $\rho A = A$, $A \mathrel{M} N \Rightarrow A \subseteq \rho N$.

Direct application of the above yields

Proposition 1 : *If* M *is any relation satisfying* (M1) *to* (M4) *and containing the relation of being a nonzero ideal, the following assertions are equivalent:*

(1) ρ *is a Kurosh-Amitsur-radical which is* M*-strong (i.e. fulfils* (ρ4M)*) ;*

(2) ρ *is both an* M*-radical and a Kurosh-Amitsur-radical ;*

(3) ρ *is an* M*-radical fulfilling* (ρ3) .

Notice that Proposition 1 is valid for Ω-groups .

A recent result by Anderson, Kaarli and Wiegandt ([1]) then yields

Corollary 1 : *For zero-symmetric near-rings,* J_2 *and* J_3 *are* M*-radicals for every relation* M *satisfying* (M1) *to* (M4) *and situated between the relations of being a nonzero ideal and that of being a nonzero right-invariant subgroup of* N *(i.e. a subgroup* A *of* N *satisfying* $AN \subseteq A$ *).*

By this corollary, Proposition 2.7 in [5] yields

Corollary 2 : *For zero-symmetric near-rings and every relation* M *with* (M1) *to* (M4) *situated between that of being a nonzero ideal and that of being a right-invariant subgroup* ($\neq 0$)*, the radical classes* R_2 *and* R_3 *of* J_2 *and* J_3 *are closed under* M*-extensions :*

$A\ M\ N$, $A \in R$, $N/\langle A\rangle \in R \Rightarrow N \in R$

(where $\langle A\rangle$ *denotes the ideal of* N *generated by* A *) .*

As exhibited in [5], for associative and alternative rings, the M-radicals with respect to the relation of being a nonzero subring are exactly the strong Kurosh-Amitsur-radicals ; thus, for every relation M with (M1) to (M4) situated between that of being a nonzero subring and that of being a nonzero ideal, the M-radicals fulfil (ρ3). This means, that the sum $R_\rho N$ of all radical M-subobjects of N coincides with ρN . In general, we only know that $R_\rho N$ is contained in ρN ; thus, in [5], besides the general case, "*approximable* radicals" (ρN is generated as an ideal by a radical M-subobject) and "*attainable* radicals" ($\rho N = R_\rho N$) were studied . All the natural examples turned out to be attainable. In the following a near-ring example of an approximable, but not attainable M-radical is given .

Example : Following [5], Proposition 4.3 , a class S of near-rings is the semisimple class of an M-radical if and only if $N \in S$ is equivalent to the property that every M-subobject of N has a nonzero homomorphic image in S . From there it is easily seen that for any relation M with (M1) to (M4) which satisfies $N_c\ M\ N$ for all non zero-symmetric near-rings N (where N_c denotes the constant part of N), the class S of all zero-symmetric near-rings is the semisimple class of an M-radical on the class of all near-rings . Natural examples of such relations M are those of being a nonzero subnearring resp. a nonzero (invariant) N-subgroup . The corresponding radical is given by $\rho N = \langle N_c\rangle$ and is therefore approximable . Suppose now that ρ is attainable; then (by Proposition 1) ρ is a Kurosh-Amitsur-radical whose semisimple class is homomorphically closed and contains all near-rings with zero-multiplication . By a result of Betsch and Kaarli ([2], Theorem 3.3) S has to be the class of all near-rings, a contradiction . It follows that ρ is not attainable .

The above example shows that (ρ1),(ρ2) and (ρ5M) do not imply (ρ3); consequently, there may be some hope to fit J_0 and J_1 into a well-described (see [5]) radical theory by the use of a suitable relation M . As we already know, the M-strongness (i.e. (ρ4M)) is a necessary condition for a radical to be an M-radical . It is straightforward to see that (ρ4M) is equivalent to the M-*regularity* of the corresponding semisimple class S , i.e. to :

$N \in S$, A M N $\Rightarrow$ A has a nonzero homomorphic image in S .

Proposition 2 : *The semisimple classes of* J_0 *and* J_1 *are* M-*regular for every relation* M *satisfying* (M1) *to* (M4) *and contained in the relation of being a nonzero left ideal . Consequently, the radical classes of* J_0 *and* J_1 *are* M-*strong for these relations* M .

Proof : Let L be a nonzero left ideal of a J_1-semisimple near-ring; to show the left-ideal-regularity of S_1 (corresponding to J_1), it suffices to show that there is some nontrivial L-group of type 1. To this aim take an arbitrary N-group G of type 1 . LG = 0 implies $L \subsetneq Ann_N G$. Since the intersection of the N-annihilators of all N-groups of type 1 is $J_1(N) = 0$, there must be some G satisfying $LG \neq 0$. In such a group G, there is some g with $Lg \neq 0$, hence Lg = G (since Lg is a normal N-subgroup in the simple N-group G); moreover, this shows that G is a strictly cyclic L-group. Thus, if $\{H_i / i \in I\}$ is a chain of normal L-subgroups of G with $H_i \neq G$ for all $i \in I$, we have $LH_i = 0$ for all $i \in I$, and thus LH = 0 for the union H of the H_i . By Zorn's Lemma, there is a maximal normal L-subgroup K in G satisfying LK = 0 . G/K is then a nontrivial L-group of type 1 . For J_0 the proof is essentially the same (in this case, the $H_i \neq G$ do not contain L-generators of G, implying that K does not contain L-generators of G) .

This proposition gives some hope for J_0 and J_1 to be M-radicals for "good" relations M . However, at least for J_1, this hope is destroyed by the following

Proposition 3 : *Let* ρ *be a Hoehnke-radical on a universal class* U *of near-rings . If* U *contains at least one near-ring* N *satisfying* $\rho N \neq 0$ *and* $\rho(\rho N) = 0$, *then* ρ *is not an* M-*radical on* U *for any relation* M *satisfying* (M1) *to* (M4) .

Proof : Suppose that ρ is an M-radical ; then by (ρ5M), $\rho N \neq 0$ implies the existence of some A M N with $\rho A = A$. Because of (M3) and (ρ1), A is contained in ρN . Now, (M4) implies A M ρN , yielding $\rho(\rho N) \neq 0$ in contradiction to our assumption .

Corollary 3 : *For most of the universal classes* U *of near-rings (including that of all zero-symmetric near-rings with commutative addition and all larger universal classes),* J_1 *is not an* M-*radical for any relation* M *satisfying* (M1) *to* (M4) .

Proof : Following Kaarli ([4]), it suffices that, for some finite group G with a proper nonzero subgroup H, the class $\mathcal{U}$ contains the near-ring of all mappings from G into G which preserve 0 and H (which then fulfils the requirements of Proposition 3 above) .

For J_0 the situation is more complicated: until now there is only one example (given by Kaarli [4]) showing that J_0 is not idempotent . Unfortunately, this example does not fulfil the requirements of Proposition 3 . Nevertheless, it might be used to obtain a partial answer for J_0 . To construct his example, Kaarli starts with a cyclic group A of order 4 (generator : a_0) and the group B of all {0,1}-sequences with a finite number of nonzero entries . On the direct sum G of these groups he considers the (zero-symmetric) near-ring of transformations S generated by a transformation s_0 satisfying $2s_0(a+b) \notin B$ for $a = a_0$ (for more details see [4]) . B is then an S-subgroup of G, $S/(B:G)_S \neq 0$ is 1-primitive, $(B:G)_S^2 = 0$, implying $J_0(S) = (B:G)_S$.

Further, Kaarli considers the (zero-symmetric) near-ring T of all transformations t on G which satisfy :

$$tg = \begin{cases} 0 & \text{for } g \notin B \text{ or } g = 0 \\ \alpha(t)a_0 + b(t) \in 2A+B & \text{for all } g \in B \setminus C \\ \beta(t)a_0 \in 2A & \text{for all } g \in C \setminus \{0\} \end{cases} ,$$

where C denotes the subgroup of B consisting of all elements with an even number of nonzero entries . T is shown to be 0-primitive .

The near-ring N defined on the direct sum of the additive groups S^+ and T^+ by the following multiplication has S as its J_0-radical :

$$(s+t)(s'+t') = ss' + k(s',t)s_0 + tt'$$

with

$$k(s',t) = \begin{cases} 0 & \text{for } s'G \notin B \text{ or } s' = 0 \\ \beta(t)s_0 & \text{for } s'G \in B \text{ , } s' \neq 0 . \end{cases}$$

Now, if L is any J_0-radical left ideal in N, by Proposition 2 it is contained in $J_0(N) = S$ and hence in $J_0(S) = (B:G)_S$. If we take some $m \neq 0$ in $(B:G)_S$ and some $t \in T$ with $\beta(t) = 2$, we obtain

$$(s+t)(m+t') - (s+t)t' = sm + 2s_0 ;$$

sm belongs to the ideal $(B:G)_S$ of S, but $2s_0(a+b) \notin B$ for $a = a_0$. Thus, $(s+t)(m+t') - (s+t)t'$ is not in $(B:G)_S$ and $(B:G)_S$ contains no nonzero left ideal of N . It follows that N is a near-ring without J_0-radical left-ideals which is not J_0-semisimple . Using this example, we obtain :

Proposition 4 : *For most of the universal classes $\mathcal{U}$ of near-rings (including that of all zero-symmetric near-rings with commutative addition and all larger universal classes),* J_0 *is not an* M*-radical for any relation* M *satisfying* (M1) *to* (M4) *and contained in the relation of being a nonzero left ideal* .

The question whether J_0 is an M-radical for some natural relation M remains thus open. In connection with Proposition 3 it would help knowing the solution of the following

Problem : Is there a (zero-symmetric) near-ring N (with commutative addition) satisfying $J_0(N) \neq 0$ and $J_0(J_0(N)) = 0$?

REFERENCES

1 Anderson, T. , Kaarli, K. and Wiegandt, R. , On left strong radicals of near-rings, Manuscript .

2 Betsch, G. and Kaarli, K. , Supernilpotent radicals and hereditariness of semisimple classes of near-rings, Colloquia Math.Soc.Janos Bolyai 38 (Radical Theory, Eger 1982), 47-58, North-Holland Publ.Comp., Amsterdam/ Oxford/New York , 1985 .

3 Hoehnke, H.-J. , Radikale in allgemeinen Algebren, Math.Nachr. 32 (1966), 347-383 .

4 Kaarli, K. , On Jacobson type radicals of near-rings , Acta Math. Hung., to appear .

5 Mârki, L., Mlitz, R. and Wiegandt, R. , A general Kurosh-Amitsur radical theory , Manuscript .

6 Mlitz, R. , Radicals and semisimple classes of Ω-groups, Proc. Edinburgh Math.Soc. 23 (198o), 37-41 .

SUMMARY

An example of a non idempotent M-*radical of near-rings is given . The behaviour of the Jacobson-radicals of near-rings with respect to the concept of* M-*radical is studied . It turns out that* J_2 *and* J_3 *are* M-*radicals for several natural relations* M, *that on most of the universal classes of near-rings* J_0 *is not not an* M-*radical for relations* M *contained in that of being a nonzero left ideal and that* J_1 *is in general not an* M-*radical for any relation* M .

Near-rings and Near-fields, G. Betsch (editor)

ON MEDIAL NEAR-RINGS

Silvia Pellegrini Manara #

ABSTRACT

We study the near-rings N in which xyzt = xzyt for all x,y,z,t ∈ N and we call them medial. We prove that all such near-rings have a subdirect structure very similar to that of the strongly I.F.P. near-rings ([6]). In the particular case of medial regular near-rings, we observe that they generalize the ß-near-rings of Ligh [5] and the semi-rings of Subrahamanyam [13] .

Moreover we obtain a necessary and sufficient condition for a near-ring to be medial, regular and subdirectly irreducible. Afterwards we have studied the set of nilpotents elements of a medial near-ring. We find conditions for the set Q(N) of the nilpotent elements of N to be an ideal. If N is a medial ring, Q(N) is an ideal and it coincides with the radicals of N. In this case N/Q(N) is a subdirect sum of integral domains.

1. INTRODUCTION

The identity "xyzt=xzyt" has been developed in the theory of quasi-groups: an impulse to their investigation goes back to the '40s with Toyoda's theorem (see Toyoda, On axioms of linear functions, Proc. Imp. Acad. Tokyo, 1941, 17). We study the near-rings N in which xyzt = xzyt for all x,y,z,t ∈ N and we call them medial. For near-rings, the mediality is a condition of "partial commutativity" and its study fits into the set of studies of the near-rings with properties of partial commutativity. We see that the medial near-rings generalize the ß-near-rings of Ligh [5] and the semi-rings of Subrahamanyam [13] . The necessary and sufficient condition for a near-ring to be medial, regular and subdirectly irreducible puts in evidence the connection between this structure and the weakly commutative near-rings. In the study of the set of nilpotent elements, the mediality works in a similar way as the I.F.P. We find conditions for the set Q(N) of nilpotent elements of N to be an ideal and show the connection to the various prime ideals of N. If N is a medial ring, Q(N) is an ideal, and it coincides with the radicals of N. In this case the mediality presents itself again as a condition of weak commutativity: in fact

Address: Silvia Pellegrini Manara, Facoltà di Ingegneria dell'Università di Brescia, Viale Europa 39. 25060 BRESCIA.

Work carried out on behalf of M.P.I.

N/Q(N) is a subdirect sum of integral domains.

2. PRELIMINARIES

N will indicate a left near-ring; for the definitions and the fundamental notations we refer to [10] without explicite mention. An element $e \in N$ is idempotent if $e^2 = e$; if $x \in N$ we indicate $A_d(N) = \{ y \in N / xy = 0 \}$ and we call a near-ring $N = N_o + N_c$ "mixed" if $N_o \neq \{0\}$ and $N_c \neq \{0\}$. A near-ring N if I.F.P. if $xy=0$ implies $xzy = 0$ for all $x,y,z \in N$, strongly I.F.P. if every homomorphic image is I.F.P. and regular if $\forall x \in N \ \exists x' \in N / x = xx'x$. In a regular near-ring xx' and $x'x$ are idempotent elements [4] .

We call a near-ring N medial if $xyzt = xzyt$ for all x,y,z,t in N.

3. SUBDIRECLY IRREDUCIBLE MEDIAL NEAR-RINGS

Proposition 1. *A weakly commutative § near-ring is medial.*

In fact if N is weakly commutative $xyzt = zxyt = xzyt$ and N is medial.

Proposition 2. *Let N be medial:*

1. If $e \neq 0$ is an idempotent element of N, $A_d(e)$ is an ideal of N.

2. If N has a left identity, it is weakly commutative.

3. For all $x, y \in N$ and k integer, $(xy)^k = x^k y^k$.

1. We know that $A_d(e)$ is a right ideal of N. For $n \in N$ and $y \in A_d(N)$ we have: $eny = eeny = eney = 0$ and $A_d(e)$ is a left ideal.

2. If u is a left identity of N, $xyz = uxyz = uyxz = yxz$, for all $x,y,z \in N$ and N is weakly commutative.

3. Trivial.

Proposition 3. *If N is a mixed medial near-ring, N_o is an ideal of N, $N_c N_o = \{0\}$ and $N_o N_c = N_c$.*

If N is medial, $\forall x \in N, \forall n_o \in N_o$, $0nn_o = 00nn_o = 0n0n_o = 0n_o = 0$ and N_o is an ideal of N. It follows that $N_c N_o = \{0\}$ and $N_o N_c = N_c$.

Proposition 4. *The homomorphic images of a medial near-ring are medial.*

The proof is trivial.

The constant and the zero-near-rings are medial. In the following we will exclude these trivial cases.

Theorem 1. *If N is a non trivial medial subdirectly irreducible near-ring, N satisfies one of the following properties:*

§ a near-ring N is weakly commutative if $xyz = yxz$ for all $x,y,z \in N$, [10] .

1. *N is zero-symmetric, simple and each non-zero idempotent is a left identity;*
2. *N is zero-symmetric and the intersection of the non-zero ideals of N is non-zero and without non-zero idempotents;*
3. *N is mixed and when the intersection I of the non-zero ideals of N contains an idempotent, then* $I = N_o$.

Let N be medial, subdirectly irreducible and non trivial. If N is simple and zero-symmetric, N satisfies 1.: in fact, if it has an idempotent e, $A_d(e) = \{0\}$ because $A_d(e)$ is an ideal. Therefore $e(en - n) = 0, \forall n \in N$, implies e left identity. Let N be zero-symmetric but non simple and let P be the intersection of the non zero ideals of N. Now $P \neq \{0\}$ and we suppose that P contains an idempotent e. If there is an element $n \in N$ such that $en \neq n$, we have $A_d(e) \neq \{0\}$, hence $n - en \in A_d(e)$. Therefore P is contained in $A_d(e)$ and thus $e \in A_d(e)$, that is $e^2 = e$, a contradiction. Otherwise e is left identity contained in P and P = N because N is zero-symmetric, but this is excluded. Therefore P doesn't contain idempotents and N satisfies 2. Suppose N is mixed. We know that N_o is an ideal of N and so N is non simple. If P is the intersection of the non zero ideals of N, we have $P \subseteq N_o$. If P has an idempotent $e \neq 0$, it cannot be $en \neq n$ for $n \in N$, because otherwise $e \in A_d(e)$ and this is excluded. Therefore, if P has an idempotent $e \neq 0$, this element is left identity, $P = N_o$ and N satisfies 3.

Corollary 1. *A non trivial, medial near-ring is isomorphic to a subdirect sum of near-rings that satisfy one of the properties of Th. 1.*

It follows immediatly from Prop. 4 and Th. 1. In fact each near-ring is isomorphic to the subdirect sum of subdirectly irreducible near-rings (see [2]).

4. SUBDIRECTLY IRREDUCIBLE MEDIAL REGULAR NEAR-RINGS

A zero-near-ring is medial but not regular, a constant near-ring is medial and regular; in the following, an MR-near-ring is a medial regular near-ring, and it is non trivial if it is not constant.

In general a regular near-ring is not strongly I.F.P., this happens if N is medial: in fact:

Proposition 5. *An MR-near-ring is strongly I.F.P.*

Let N be an MR-near-ring and I an ideal of N: if $xy \in I$ then for all $n \in N$, $xny = xx'xny = xx'nxy \in I$.

Proposition 6. *Let N be a non trivial MR-near-ring; then*

1. *N has no nilpotent elements.*
2. *$A_d(x)$ is an ideal of N, for all $x \in N$.*
3. *If $x = xx'x \in N$, then $A_d(xx') = \{0\}$ [$A_d(x'x) = \{0\}$] iff xx' [$x'x$] is a left identity of N.*

1. If n is nilpotent, $n^k = 0$ for some integer k. Now $n = nn'n$ implies $n'n = (n'n)^k = n'^k n^k = 0$ (Prop. 2.3) and hence $n = 0$.

2. For $n \in N$, and $y \in A_d(x)$, $xny = xx'xny = xx'nxy = 0$.

3. In fact $xx'n = xx'xx'n$ [$x'xn = x'xx'xn$] and hence $xx'(n - xx'n) = 0$ [$x'x(n - x'xn) = 0$]. If $A_d(xx') = \{0\}$ [$A_d(x'x) = \{0\}$], xx' [$x'x$] is a left identity. The viceversa is trivial.

Proposition 7. *Let N be a non trivial subdirectly irreducible MR-near-ring:*
1. *If z is a non zero idempotent of N, $A_d(z) \neq \{0\}$ iff z is constant and in this case $A_d(z) = N_o$.*
2. *If x is not constant, and $x = xx'x$, then xx' and $x'x$ are not constant and non-zero idempotents and $A_d(x) = \{0\} = A_d(xx') = A_d(x'x)$.*
3. *N has a left identity.*

1. If z is constant, $A_d(z) \neq \{0\}$. Viceversa, let $A_d(z) \neq \{0\}$ and $A = \cap A_d(h)$ with $A_d(h) \neq \{0\}$. Since N is subdirectly irreducible, there is some $w \in A$, $w \neq 0$. Then also $w'w \in A$ ($ww'w = 0$), hence A is an ideal. If there is an element $0 \neq y \in N$, such that $w'wy = 0$, we get $A_d(w'w) \neq \{0\}$ and so $w'ww'w = 0$, but this is excluded (Prop. 6.1). Therefore for all $y \in N$ ($y \neq 0$) $w'wy \neq 0$ and $A_d(w'w) = \{0\}$. From Prop. 6.3, $w'w$ is a left identity. So for all idempotent z of N such that $A_d(z) \neq \{0\}$ and for all $y \in N$ we have $zy = z(w'w)y = (zw'w)y = 0y$. In particular if $y = z$, $zz = z = 0z$, and z is constant. Finally, if z is constant, $A_d(z) \supseteq N_o$; if $y \in A_d(z)$, $zy = 0 = 0y$ and $y \in N_o$.

2. Let $x = x_o + x_c$ be a not constant element of N. If xx' is constant, we have $x = xx'(x_o + x_c) = xx'x_o + x_c$ with $xx'x_o = 0$ because of $N_c N_o = \{0\}$. Therefore x must be constant, a contradiction. If $xx' = 0$, we have $0x = x$ and x is again constant. In a similar way we prove that $x'x$ is not constant. Finally from the above point 1., $A_d(x) = \{0\} = A_d(xx') = A_d(x'x)$.

3. This follows from the Th. 1., keeping in mind that now N is regular and cannot be of type 2 (see above point 2. and Prop. 6.2)

Theorem 2. *A non-trivial subdirectly irreducible MR-near-ring with a right distributive element h is a commutative field.*

Obviously h is zero-symmetric and so $A_d(h) = \{0\}$. Let $A_s(h) = \{y \in N \ / \ yh=0\}$. Under our hypotheses, $A_s(h)$ is an ideal of N. We suppose $A_s(h) \neq \{0\}$ and we let $A = \cap A_d(x)$ $(A_d(x) \neq \{0\})$. Since N is subdirectly irreducible, $A_s(h) \cap A = L \neq \{0\}$. If $w \in L$ we have $wh = 0$ and therefore $A_d(w) \neq \{0\}$ and w constant, but $w \in A_d(x) = N_o$ and this is excluded. Consequently $A_s(h) = \{0\}$. Moreover N is zero-symmetric: in fact for all $a \in N$, $(0a)h = 0 = 0a0h = 0$ and $A_s(h) = \{0\}$ implies $0a = 0$. If N is zero-symmetric it is integral: if $xy = 0$ and $x \neq 0$, we get $x'xy = 0$ (where $x = xx'x$) and $x'x$ is a zero-symmetric idempotent element, therefore $y = 0$. Also N has identity: in fact $(n - nx'x)h = nh - nx'xh = 0$ because xx' is left identity, but $A_s(h) = \{0\}$ and so $n = nx'x$ for all $n \in N$. In our case N is medial and so for all $x, y \in N$, $xy = $ $= xx'yxx' = xx'yxxx' = yx$. So N is a commutative field.

Theorem 3. *A non trivial MR-near-ring is isomorphic to a subdirect sum of subdirectly irreducible near-rings that are fields or near-rings with left identity in which each non constant element $x = xx'x$ is such that xx' and $x'x$ are left identities.*

It follows from Th. 2, Prop. 7, keeping in mind that each near-ring is isomorphic to the subdirect sum of subdirectly irreducible near-rings ([2]).

If a near-ring in which each elemnt is power of itself ([1], [7]) is called MP-near-ring, we observe that an MP-near-ring is regular and therefore Th. 3 applies. The MP-near-rings generalize the ß-near-rings of Ligh [5] because a weakly commutative near-ring is medial and the ß-near-rings consist of idempotent elements. In this way the results of [5] are corollaries to Th. 3 (see [7]).

We give now a characterization of subdirectly irreducible MR-near-rings (see [8]).

Theorem 4. *A near-ring N is a non trivial subdirectly irreducible MR-near-ring iff:*

1. *it is a zero-symmetric, weakly commutative, N-simple and integral near-ring;*
2. *it is a weakly commutative near-ring, with N_o, regular, minimal ideal, N_o-simple, integral; with N_c, unique N-subgroup of N. Moreover the right annihilator of each non-constant element is zero.*

Let N be a non trivial subdirectly irreducible MR-near-ring. It satisfies the hypotheses of Th. 1, but not the property 2 because N is regular. In fact, if I is the intersection of the non zero ideals of N, for $i \in I$, there is some $i' \in I$

such that ii' and i'i are non zero idempotent in I and this is excluded. Therefore, if N is zero-symmetric it is simple and all its non-zero idempotents are left identities; N is a weakly commutative near-ring (Prop. 2.2) and for all $x \in N$, $A_d(x) = \{0\}$ (Prop. 6.2). Hence N is integral. Moreover N is N-simple: let R be an N-subgroup of N; we have $RN \subseteq R$. If $x \in R$, there is some $x' \in N$ such that $xx'x = x$ with xx' being a left identity of N, belonging to R. Then $N \subseteq R$. Viceversa, if N is a weakly commutative near-ring it is medial (Prop. 1.) and obviously it is subdirectly irreducible because it is simple. Moreover it is regular: in fact for all $a \in N$ $aN = N$ because now N is N-simple and hence there is some $y \in N$ such that $ay = a$ and for all $z \in N$, $ayz = az$ hence $yz = z$, because N is integral. Now $\forall a \in N, \exists a' \in N / aa' = y$ and $aa'a = ya = a$ implies that N is regular.

Otherwise, from the Th. 1. N is mixed and N_0 is the minimal ideal of N. We know that $N = N_0 + N_c$. Let R be a proper non-zero N-subgroup of N and $R \cap N_c \neq R$; then there are $a \in R \setminus N_c$ and a' such that aa' is a non zero and non constant idempotent of N (Prop. 7.2). Therefore $A_d(aa') = \{0\}$, aa' is a left identity and N is weakly commutative (Prop. 2.2). If now $RN \subseteq R$, $aa' \in R$ and $aa'N = N \subseteq R$ which is excluded. Therefore $R \subseteq N_c$ and so $R = N_c$. The unique proper N-subgroup is N_c. Let us prove that N_0 is regular. If $x \in N_0$, $xN = N$ because the unique N-subgroup of N is N_0 and it cannot be $xN = \{0\}$ because x is non constant. Moreover $xN_0 = N_0$, in fact: for all $y \in N_0$ there is $h = h_0 + h_c$ such that $x(h_0 + h_c) = y$. Such y exists because $xN = N$, but now $x(h_0 + h_c) = y$ implies $xh_0 + h_c = y$ and so $y_c = 0$. Therefore N_0 is N_0-simple, it is an integral zero-symmetric weakly commutative near-ring. From the above point 1, N_0 is regular. Finally, $A_d(x) = \{0\}$, for all not constant elements x. Viceversa, if N is a weakly commutative medical near-ring with a minimal ideal, it is subdirectly irreducible. We prove that N is regular. If $n = n_0 + n_c$ is a typical element of N and $0 \neq n_0' \in N_0$, then $nn_0' \neq 0$ because $A_d(x) \neq \{0\}$ iff x is a constant element. Therefore $(x_0 + x_c)N = N$ because the unique N-subgroup of N is N_c, and so there is $y \in N$ such that $ny = n_0 + n_c$ and for all $z \in N$ we have $nyz = nz$ that is $yz = z$. Otherwise there is $y' \in N$ such that $ny' = y$ and therefore $ny'n = n$. If n is a zero-symmetric element there is some n' such that $n = nn'n$, so N_0 is regular; if n is constant the assertion is trivial.

Corollary 2. *The ideals of a non trivial subdirectly irreducible MR-near-ring N are of the form $N_0 + H$, with H being an ideal of N_c.*

If N has ideals, it is of type 2 of Th. 4. Now if I is one of its ideals, it is of the form $I=N_0+I_c$ with I_c an ideal of N_c. If $I = I_0 + I_c$, $I_0 = I \cap N_0$, $I_c = I \cap N_c$. Now N_0 is minimal, so $I_0 = N_0$ and I_c cannot be an ideal but I^+ is normal in N and then I_c^+ is normal in N_c.

Viceversa if I_c is an ideal of N_c, $I = N_0 + I_c$ is an ideal of N. In fact for $i \in I$ and $n \in N$, $n + i - n = n_0 + n_c + i_0 + i_c - n_c - n_0 = n_0 + (n_o + i_0 - n_o) + ((n_c + i_c - n_c) - n_0 - (n_c + i_c - n_c)) + (n_c + i_c - n_c) = \bar{n}_0 + \bar{i}_c$ with $\bar{i}_c \in I_c$ because I_c^+ is normal in N_c^+. Therefore I^+ is a normal subgroup of N^+. Moreover $NI = N(N_0 + I_c) = NN_0 + I_c \subseteq N_0 + I_c = I$ and for all n, n'$\in$ N and $i \in I$: $(n + i)n' - nn' = (n + i)(n'_0 + n'_c) - n(n'_0 + n'_c) = (n + i)n'_0 - n'_c - nn'_o \in N_0 \subseteq I$.

Corollary 3. *A non trivial subdirectly irreducible MR-near-ring is a right duo-near-ring.* §

If N is of type 1 of Th. 4, the assertion is trivial. Let N be of type 2 of the Th. 4 and I a right ideal of N. If $I \cap N_0 = \{0\}$ we have for n, n'$\in$ N and for $i \in I$, $(n + i)n' - nn' = (n + i)(n'_0 + n'_c) - n(n'_0 + n'_c) = (n + i)n'_0 - nn'_0 \in N_0 \cap I = \{0\}$. For n = 0, $in'_0 = 0$ and (Prop. 7.1), $i \in N_c$, that is $I \subseteq N_c$; but then I is a left ideal, too, because $NI \subseteq I$ and this is excluded (N_0 is minimal). Therefore $J = I \cap N_0 \neq \{0\}$, J is right ideal of N_0 and so $J = N_0$. The right ideals of N are the sum of N_0 and of a normal subgroup of N_c, and these are also left ideals.

5. NIL RADICAL OF MEDIAL NEAR-RINGS

We will denote by $N^{\cdot}$ the multiplicative semigroup of N; if K is a subset of N and n is an element of N, we will write $P(K) = \bigcup_{k \in K} A_d(k)$ and $L(K) = N \setminus (K \cup P(K))$. We will denote the prime radicals of type ν (ν = 0, 1, 2) with $P_0(N)$, $P_1(N)$, $P_2(N)$, see for instance [9], [12].

Proposition 8. *If K is a maximal subsemigroup of $N^{\cdot}$, N is the disjoint union of K, P(K) and L(K).*

First of all, $P(K) \cap K = \emptyset$ because K is maximal: in fact if $x \in P(K) \cap K$, there is $y \in K$ with $xy = 0 \in K$ and this is excluded. By definition $L(K) \cap (K \cup P(K)) = \emptyset$ and therefore proposition holds.

Theorem 5. *If N is medial and K is a maximal subsemigroup of $N^{\cdot}$, then for each $y \in L(K)$ there is some $x \in K$ such that xy is nilpotent iff N is zero-symmetric.*

Let's suppose that for $y \in L(K)$ there is $x \in K$ such that xy is nilpotent. From

§ A right duo-near-ring (see [5]) is a near-ring in which all right ideals are ideals.

Prop. 8 we will prove that L(K), K and P(K) don't have constant elements.

$L(K) \subseteq N_o$:

Let y be a constant element of L(K); from the hypotheses there is some $x \in K$ with xy nilpotent, but for the constant element y we get $xy = y$ for all $x \in N$, a contradiction. Let $y = y_o + y_c$ and xy a nilpotent element, with $x \in K$. Then there is an integer k, such that $(x(y_o + y_c))^k = (xy_o + y_c)^k = 0$. We obtain $(xy_o + y_c)^k = \beta y_o + y_c = 0$ with ß element of N. Now $\beta y_o \in N_o$ because N_o is an ideal of N, $y_c \in N_c$ but $N_o \cap N_c = \{0\}$ and therefore $y_c = 0$, so L(K) has only zero-symmetric elements.

$K \cap N_c = \phi$:

Let k be a constant element of K; from the Prop. 3, $A_d(k) \supseteq N_o$ and $A_d(k) \cap L(K) = \phi$, but this is absurd because of the condition $L(K) \subseteq N_o$ above.

$P(K) \cap N_c = \phi$:

In fact, the elements of P(K) are right annihilators of some element of K, and if a constant element y belongs to P(K), we have $xy = y$, for all $x \in N$.

Viceversa, let N be zero-symmetric, $y \in L(K)$ and M the subsemigroup of $N^{\cdot}$ generated by $K \cup \{y\}$. Since $y \notin K$, M contains K properly. Therefore $0 \in M$. Now in K there are $x_1, x_2, \ldots, x_s$ such that $x_1 y^{i_1} x_2 y^{i_2} \ldots x_s y^{i_s} = 0$.
Put $x_1 x_2 \ldots x_s = x$; obviously $x \in K$. Applying the mediality k - 1 times we obtain $x_1 y^{i_1} x_2 y^{i_2} \ldots x_s y^{i_s} = x_1 x_2 \ldots x_s y^{i_1+i_2+\ldots i_s} = 0$, that is $xy^k = 0$ where $k = i_1 + i_2 + \ldots + i_s$. If $xy^k = 0$, also $x^{k-1} xy^k = x^k y^k = (xy)^k = 0$ (Prop. 2.3). Therefore xy is nilpotent and and $xy \neq 0$ because of $y \notin P(K)$.

6. ZERO-SYMMETRIC MEDIAL NEAR-RINGS

In the following our near-rings will be zero-symmetric.

If N has subsemigroups in $N^{\cdot}$, we define $Q(N) = N \setminus \bigcup_{\alpha \in \Lambda} K_\alpha$, where $\{K_\alpha\}_{\alpha \in \Lambda}$ is the family of all maximal subsemigroups of $N^{\cdot}$; if N has no multiplicative subsemigroups in $N^{\cdot}$, we put $Q(N) = N$.

Proposition 9. *The set Q(N) is the set of all nilpotent elements of N.*

If N has no subsemigroups in $N \setminus \{0\}$, each element is nilpotent. We have $Q(N) = N$. If N has subsemigroups in $N^{\cdot}$, let $\{K_\alpha\}_{\alpha \in \Lambda}$ be the family of maximal subsemigroups in $N^{\cdot}$. Since each K_α doesn't contain nilpotent elements, Q(N) contains the nilpotent elements of N. Hence if $x \in Q(N)$, x is nilpotent because otherwise it would generate a subsemigroup of $N^{\cdot}$ contained in a maximal subsemigroup K_α , a contradiction.

We recall that:

$$(\alpha) \qquad P_2(N) \supseteq P_1(N) \supseteq P_0(N)$$

(see [12] Def. 3.1, 3.2, 3.3, 3.4)

In [11] it is proven that $P_0(N)$ has only nilpotent elements. Therefore:

$$(\beta) \qquad P_0(N) \subseteq Q(N)$$

Theorem 6. *If N is a medial near-rings then*

$$(\gamma) \qquad P_2(N) \supseteq P_1(N) \supseteq Q(N) \supseteq P_0(N)$$

We have to prove that each prime ideal of type 1 contains the nilpotent elements of N: from (α) and (β) it will follow (γ). Let P be a prime ideal of type 1 and a_0 a nilpotent element with order of nilpotency k. If $a_0 \notin P$, there is $t_0 \in N$ such that $a_1 = a_0 t_0 a_0 \notin P$; otherwise $a_0 \in P$ since P is prime of type 1. In this way we are able to construct two sequences of elements $\{a_n\}$ and $\{t_n\}$ such that $a_n = a_{n-1} t_{n-1} a_{n-1}$ for $n = 1, 2, \ldots$ and $a_n \notin P$. Moreover each $a_n \neq 0$, because $a_n = a_{n-1} t_{n-1} a_{n-1} = 0$ implies $a_{n-1} \in P$ and this is excluded. The element a_k "contains" the element a_0 2^k times; applying the mediality k-1 times, we have $a_k = 0$, a contradiction. Therefore $a \in P$.

Corollary 4. *If N is a medial near-ring in which each prime ideal of type 0 is a prime ideal of type 1, then Q(N) is an ideal.*

Corollary 5. *If N is medial ring, Q(N) is an ideal.*

7. MEDIAL RINGS

Theorem 7. *If N is a medial ring then $N \setminus K$ is an ideal of N, for each maximal subsemigroup K of $N^{\bullet}$.*

If y_1, y_2 are elements of $N \setminus K$, there are two elements m_1, $m_2 \in K$, such that $m_1 y_1$ and $m_2 y_2$ are zero or nilpotent. Therefore there are two integer $n_1 \geq 1$ and $n_2 \geq 1$, such that $(m_1 y_1)^{n_1} = (m_2 y_2)^{n_2} = 0$. If $(m_1 y_1)^{n_1} = 0$, applying n times the mediality, we obtain $(m_1 m_2 y_1)^{n_1+1} = 0$. If $(m_2 y_2)^{n_2} = 0$, applying n_2 times the mediality we obtain $(m_1 m_2 y_2)^{n_2} = 0$. Let now $n \geq n_1 + n_2 + 1$. One of the terms in the expansion of $(m_1 m_2 y_1 - m_1 m_2 y_2)^n$ is of the form $(m_1 m_2 y_1)^{i_1} (m_1 m_2 y_2)^{i_2} (m_1 m_2 y_1)^{i_3} \ldots (m_1 m_2 y_2)^{i_k}$ where $i_1 + i_2 + \ldots + i_k = n$. Applying the mediality, we have: $(m_1 m_2 y_1)^{i_1} (m_1 m_2 y_2)^{i_2} \ldots (m_1 m_2 y_2)^{i_k} = (m_1 m_2 y_1)^{\bar{i}_1} (m_1 m_2 y_2)^{\bar{i}_2}$. Now $\bar{i}_1 \geq n_1 + 1$ or $\bar{i}_2 \geq n_2$ and this term is zero. Therefore the element $m_1 m_2 (y_1 - y_2)$ is nilpotent and $y_1 - y_2 \in N \setminus K$. Moreover, for all $y \in N \setminus K$ and $n \in N$, it is easy to verify that ny and yn belong to $N \setminus K$.

Theorem 8. *In a medial ring N, an ideal P is minimal prime of type 2 iff* $P = N \setminus K$ *for some maximal subsemigroup K in* $N^{\cdot}$.

Let $P = N \setminus K$. If $xy \in N \setminus K$, x or y belong to $N \setminus K$ because otherwise $xy \in K$ and this is excluded. Therefore $N \setminus K$ is an ideal of type 2. Moreover it is minimal of type 2. We suppose that H is a prime ideal of type 2 contained in $N \setminus K$. Let $y \in N \setminus K$: we prove that $y \in H$. If $y \in N \setminus K$, from Th. 5 there is some $x \in K$ such that xy is zero or nilpotent. If $xy = 0$, then $xy \in H$ and therefore $x \in H$ since H is prime of type 2; but now $x \notin H$ because $x \in K$ and then $y \in H$. If xy is nilpotent, $(xy)^s \in H$ for some integer s. Since H is prime of type 2, $xy \in H$ or $(xy)^{s-1} \in H$. After a finite number of steps we obtain in any case $xy \in H$ and then $x \in H$ or $y \in H$ and since $x \notin H$ it is $y \in H$. Therefore $H = N \setminus K$.

Viceversa let P a minimal ideal prime of type 2. We put $K = N \setminus P$. If $x, y \in K$, also $xy \in K$ because otherwise $xy \in P$ and since P is prime of type 2, $x \in P$ or $y \in P$. Hence K is a subsemigroup of $N^{\cdot}$, and therefore contained in a maximal subsemigroup K' of $N^{\cdot}$. Now $N \setminus K'$ is a minimal ideal prime of type 2 and therefore $N \setminus K' \subseteq P$. So $P = N \setminus K'$ and $P = N \setminus K'$.

Proposition 10. *If N is a medial ring,* $N \setminus K$ *is a minimal prime ideal of type 1, for a maximal subsemigroup K of* $N^{\cdot}$

From Th. 7., $N \setminus K$ is prime of type 2 and therefore prime of type 1. Let us suppose that a prime ideal H of type is contained in $N \setminus K$. Let $y \in N \setminus K$, therefore there is $x \in K$ such that xy is zero or nilpotent. In any case, for some integer s, $(xy)^s = 0$. Applying the medial property s times, we obtain $(xny)^{s+1} = 0$ for all $n \in N$. From Th. 6, H contains the nilpotent elements of N, therefore $xny \in H$, for all $n \in N$. As H is prime of type 1, $xNy \in H$ implies $x \in H$ or $y \in H$, since $x \notin H$, $y \in H$ and $N \setminus K$ is a minimal prime ideal of type 1.

Now we are able to prove:

Theorem 9. *In a medial ring,* $Q(N) = P_2(N) = P_1(N)$.

From the definition, $Q(N) = N \setminus \bigcup_{\alpha \in \Lambda} K_\alpha = \bigcap_{\alpha \in \Lambda} (N \setminus K_\alpha)$ and therefore from the Th. 8, Q(N) is the intersection of the prime ideals of type 2. But $P_2(N) \subseteq \bigcap_{\alpha \in \Lambda} (N \setminus K_\alpha)$ and therefore $Q(N) = P_2(N) = P_1(N)$.

Theorem 10. *A medial ring N without nilpotent elements is a commutative ring.*

We observe that in a medial ring N without non zero nilpotent elements, for all $x, y \in N$, $yx^2y = (yx)^2$. In fact for all $x, y \in N$ $xyxy = xxyy$ because N is medial and therefore $xyxy - xxyy = 0$, that is $x(yx - xy)y = 0$. Since N has no non-zero nilpotent elements, $x(yx - xy)y = 0$ implies $yx(yx - xy) = 0$. It's easy to prove that for all $x, y \in N$, $(xy - yx)^2 = 0$ and since N is without nilpotent elements, $xy = yx$ and N is a commutative ring.

Corollary 6. *If N is medial, N/Q(N) is a subdirect sum of integral domains.*

Trivial.

REFERENCES

[1] H.E. Bell, Near-rings in which each element is power of itself, Bull. Aus. Math. Soc. 2 (1970), 363-368.
[2] G. Grätzer, Universal Algebra, Van Nonstrand 1968.
[3] H.E. Heatherly, Near-rings without nilpotent elements, Publ. Math. Debrecen, 20 (1973), 201-205.
[4] H.E. Heatherly, Regular near-rings, J. Ind. Math. Soc. 38 (1974), 345-354.
[5] S. Ligh, The structure of a special class of near-rings, J. Austr. Math. Soc. 13 (1972), 141-146.
[6] S. Ligh, A special class of near-rings, J. Austral. Math. Soc. 18 (1974), 464-467.
[7] S. Pellegrini Manara, Sui quasi-anelli mediali in cui ogni elemento è potenza di se stesso, Riv. Mat. Univ. Parma, (4) 11 (1985)
[8] S. Pellegrini Manara, Sui quasi-anelli mediali regolari, (to appear)
[9] S. Pellegrini Manara, Sul radicale nil di quasi-anelli mediali, (to appear)
[10] G. Pilz, Near-rings, North-Holland, Amsterdam, N.Y. (1977).
[11] D. Ramakotaiah, Radicals for near-rings, Math. Z. 97, 45-46.
[12] D. Ramakotaiah, G.K. Rao, I.F.P. Near-rings, J. Austral. Math. Soc. (27), (1969), 365-370.
[13] N.V. Subrahamanyam, Boolean semirings, Math. Ann. 148 (1962), 395-401.

SUMMARY

I quasi-gruppi mediali sono stati a lungo studiati; noi introduciamo nei quasi-anelli la 'xyzt = xzyt' per ogni x, y, z, t del quasi-anello e chiamiamo Mediali, in analogia a quanto visto nei quasi-gruppi, i quasi-anelli con questa proprietà. Nei quasi-anelli la Medialità si presenta come una condizione di "commutatività parziale" e il suo studio si inserisce nella serie di studi di quasi-anelli con proprietà di commutatività parziale. In generale dimostriamo che tali quasi-anelli hanno una struttura sottodiretta molto simile a quella dei quasi-anelli fortemente I.F.P. Se, in particolare sono anche regolari, sono isomorfi ad una somma sottodiretta di quasi-anelli sottodirettamente irriducibili che sono campi o quasi-anelli con identità sinistra in cui ogni elemento non costante x = xx'x è tale che x'x ed xx' sono identità sinistre. In questo modo vediamo che i quasi-anelli mediali generalizzano i β-quasi-anelli di Ligh ed i semianelli di Subrahamanyam. Otteniamo anche una condizione necessaria e sufficiente affinchè un quasi-anello sia mediale regolare e sottodirettamente irriducibile. Questo risultato mette in evidenza le analogie fra questa struttura ed i quasi-anelli debolmente commutativi. Infine abbiamo studiato l'insieme degli elementi nilpotenti di un quasi-anello mediale. Anche qui la medialità agisce in modo simile alla I.F.P. Inoltre troviamo legami tra l'insieme degli elementi nilpotenti e i vari radicali primi e condizioni affinchè tale insieme sia un ideale.

Near-rings and Near-fields, G. Betsch (editor)

NEAR-RINGS AND NON-LINEAR DYNAMICAL SYSTEMS

Günter F. PILZ

Institut für Mathematik, Johannes-Kepler-Universität Linz
A-4040 Linz, Austria

ABSTRACT

In this paper we study large classes of non-linear systems admitting a transfer function which completely describes their input-output behaviour. Many concepts which are widely believed to be particular to linear systems turn over to this class of "separable" systems. These are systems $\Sigma = (Q,A,B,F,G)$ in which the map F,G can be separated as $F(q,a) = \alpha(q) + \beta(a)$, $G(q,a) = \gamma(q) + \delta(a)$, where Q,A,B are groups, α,γ maps and β,δ homomorphisms. If we study separable systems with identical input- and output groups then it turns out that the parallel- (the series-) connections of systems correspond to the addition (the composition, respectively) of their transfer functions. Hence these systems form a near-ring (the "non-linear analogue of a ring") w.r.t. to parallel- and series connections. Many system-theoretic questions can be studied in the framework of near-rings. In particular, we study feedbacks, reachability questions, invertibility with delay L and close with a discussion of the realization problem of separable systems.

Definition 1: A (discrete, dynamical, time-invariant) *system* (or *automaton*) Σ is a quintuple $\Sigma = (Q,A,B,F,G)$, where Q is a set (of *states*), A a set (of *inputs*), B a set (of *outputs*), F a function $Q \times A \to Q$ (the *state transition function*) and G a function $Q \times A \to B$ (the *output function*).

The description of Σ in Definition 1 is usually called the "local description" of Σ. In order to obtain the "global" description, we do not consider a single input, but a series of input signals a_i, which enter the system "at time $i \in \mathbb{Z}$". Hence we'll consider input sequences $(a_i)_{i \in \mathbb{Z}}$. It is generally assumed that the inputs don't come in "since eternity"; so we assume that there exists an index $k \in \mathbb{Z}$ such that $a_i = 0$ for all $i < k$. Sequences of this type are usually called "formal Laurent series":

Definition 2: For any set X containing 0, let L(X) be the set of all sequences $(x_i)_{i \in \mathbb{Z}}$, for which there is some $k \in \mathbb{Z}$ such that $x_i = 0$ for all $i < k$. The elements of L(X) are called (*formal*) *Laurent series* of elements of X.

If X is a field then L(X) is just the quotient field (see e.g. [11]) of the ring X[[z]] of *formal power series* $(x_0, x_1, \ldots)$ (with $x_i \in X$). If we define

$z := (0,1,0,0,\ldots)$ then using the usual addition and multiplication of power series, we can write $X[[z]] = \{\sum_{i\geq 0} x_i z^i \mid x_i \in X\}$, and in the same spirit we write $L(X) = \{\sum_{i\geq k} x_i z^i \mid k \in \mathbb{Z},\ x_i \in X\}$.

Hence z is not a "symbol" or "indeterminate", as frequently believed. The representation of series (x_i) by sums $\Sigma x_i z^i$ is usually called the *z-transform* or *discrete Laplace transform*. It was not realized from the beginning that this transform is (at least algebraically) merely a trivial identity. (Actually, there is also no need whatsoever to write $\Sigma a_i z^{-i}$ instead of $\Sigma a_i z^i$.)

In this context, the interpretation is as follows. At a certain "time" $k \in \mathbb{Z}$ (hence k can be negative), the system is in state q_k when the first input a_k arrives. The system produces an output $b_k = G(q_k, a_k)$ and changes its state q_k into $q_{k+1} = F(q_k, a_k)$. Then a_{k+1} arrives, and so on.

Let us remark that for continuous systems, the sequences $(a_k, a_{k+1}, \ldots)$ - which can be considered as functions from $\mathbb{Z}$ into A - have to be replaced by a consistent and causal family of input functions from some time set T into A.

Σ, as in Definition 1, is called *linear* if Q,A,B are vector spaces over some field K and F,G are linear maps on the product space $Q \times A$. In this case, F and G can be decomposed into linear functions $\alpha: Q \to Q$, $\beta: A \to Q$, $\gamma: Q \to B$ and $\delta: A \to B$ such that $F(q,a) = \alpha(q) + \beta(a)$, $G(q,a) = \gamma(q) + \delta(a)$ hold for all $(q,a) \in Q \times A$. If the vector spaces in question are finite dimensional, $\alpha, \beta, \gamma, \delta$ are usually represented by matrices.

It is not true, however, that these decompositions are only possible for linear systems. We are going to introduce "separable" systems now. They are much more general than linear ones, allow highly non-linear transition and output functions, but we can do with these systems most things we can do with linear ones.

Definition 3: Σ (as in Definition 1) is called *separable* if Q,A,B are groups (written additively, but not necessarily abelian) and if there are maps $\alpha: Q \to Q$, $\gamma: Q \to B$, and homomorphisms $\beta: A \to Q$, $\delta: A \to B$ such that $F(q,a) = \alpha(q) + \beta(a)$, $G(q,a) = \gamma(q) + \delta(a)$ hold for all $q \in Q$, $a \in A$. We then denote Σ by $(Q,A,B,\alpha,\beta,\gamma,\delta)$ or simply by $(\alpha,\beta,\gamma,\delta)$.

Clearly, each linear system is separable. If $Q = A = B = (\mathbb{R},+)$, $F(q,a) = q^2 + \sin q + 3a$, $G(q,a) = e^q - 1 + \pi a$ determine a non-linear separable system.

Separable systems fit into the classes of non-linear systems described by D.L. Casti ([3], p. 19).

The system $M(G)$ of all maps from an (additive) group G into itself is a "near-ring" w.r.t. pointwise addition and composition. A *near-ring* N is a generalized ring in which we do not assume that addition is commutative and in which just one distributive law (for this paper: $(a+b)c = ac+bc$) is assumed to hold. We say that $n \in N$ is *zero-symmetric* if $n0 = 0$ holds; N is *zero-symmetric* if every $n \in N$ has this property. $M_0(G) := \{f \in M(G) \mid f(0) = 0\}$ is an example of a zero-symmetric near-ring. See [14] for the theory of near-rings.

For discrete systems (or automata) in which the state set Q is a group, there is a natural associated near-ring $N(\Sigma)$; it is the subnearring of $M(Q)$ generated by all functions f_a $(a \in A)$ with $f_a(q) = F(q,a)$ and by the identity function id (because we don't want the system to change if no inputs arrive). For separable systems, $N(\Sigma)_0$ is generated by id and α if Q is abelian. This construction of a "syntactic near-ring" is the obvious extention of the monoid of an automaton (see e.g. [11]) in the presence of a group structure on Q. $N(\Sigma)$ is typically not a ring, not even in the case of linear systems or automata (see [6]).

In this paper, however, we show that separable systems itself form a near-ring by means of series/parallel connections of these systems. For linear systems we do get a ring, see [16]. In fact, our treatment follows partly the ideas in Sain's book [16]. In order to perform this construction, we have to represent a system Σ by its global input-output map f_Σ. This is always possible for separable systems (see below).

Even if A is "only" a group, we may write $(a_k, a_{k+1}, \dots)$ (with $k \in \mathbb{Z}$) as $\sum_{i \geq k} a_i z^i$. It is one of the fundamental ideas of algebraic systems theory that the shift-to-the-left operator ("unit delay") $(a_k, a_{k+1}, \dots) \to (a_k, a_{k+1}, \dots)$ (with a_k now at the (k-1)st place, etc.) just corresponds to the multiplication of $\Sigma a_i z^i$ by z^{-1}. Hence we denote the shift operator by $\bar{z}$. Note that $L(A)$ (see Definition 2) is an additive group, too, under component-wise addition; in fact, $L(A)$ is a subgroup of $A^{\mathbb{Z}}$.

The map $\alpha: Q \to Q$ can be extended to a map (also denoted by α) from $L(Q)$ into $L(Q)$ by $\alpha(\sum_{i \geq k} q_i z^i) := \sum_{i \geq k} \alpha(q_i) z^i$. Also, we can extend β to a map from $L(A)$ into $L(Q)$, and so on.

Proposition 1: For a separable system $\Sigma = (Q,A,B,\alpha,\beta,\gamma,\delta)$, the map $-\alpha+\bar{z}: L(Q) \to L(Q)$ is always bijective. Also, $-\alpha+\bar{z}$ is zero-symmetric iff α is zero-symmetric.

Proof: We have $(-\alpha+\bar{z})(\sum_{i\geq k} q_i z^i) = \sum_{i\geq k} -\alpha(q_i)z^i + \sum_{i\geq k} q_i z^{i-1} = q_k z^{k-1} + \sum_{i\geq k} (-\alpha(q_i)+q_{i+1})z^i$. Now $(-\alpha+\bar{z})(\sum_{i\geq k} q_i z^i) = (-\alpha+\bar{z})(\sum_{i\geq k'} q_i' z^i)$ implies that $q_i z^i$ and $q_i' z^i$ both have the first non-zero coefficient at the same index n (say), and $q_n = q_n'$, $-\alpha(q_n)+q_{n+1} = -\alpha(q_n')+q_{n+1}'$, $-\alpha(q_{n+1})+q_{n+2} = \alpha(q_{n+1}') + q_{n+2}', \ldots$ implies $q_i = q_i'$ for all i. Hence $-\alpha+\bar{z}$ is injective. If $\sum_{i\geq k} q_i z^i \in L(Q)$ then it is easy to see that $(-\alpha+\bar{z})(\sum_{i\geq k} q_i' z^i) = \sum_{i\geq k} q_i z^i$ if the q_i' are given inductively as $q_k' := 0$ and $q_{i+1}' := \alpha(q_i') + q_i$ for $i > k$. This shows that $-\alpha+\bar{z}$ is bijective. If $\alpha(0) = 0$, it is clear from the formula for $-\alpha+\bar{z}$ that $-\alpha+\bar{z}$ maps the zero element $\Sigma 0z^i$ of L(Q) into itself. Conversely, if $(-\alpha+\bar{z})(\Sigma 0z^i) = \Sigma 0z^i$ then we get $\alpha(0) = 0$. □

Now we are in a position to give a formula which directly relates the output sequence of a system to its input sequence. Observe that all calculations have nothing to do with linearity. All we need is separability.

Theorem 1: In a separable system $\Sigma = (Q,A,B,\alpha,\beta,\gamma,\delta)$, the following formulas hold, if the first non-zero input arrives at time $k \in \mathbb{Z}$ in which the system is in state 0:

(i) $\sum_{i\geq k} q_i z^i = (-\alpha+\bar{z})^{-1}\beta(\sum_{i\geq k} a_i z^i)$

(ii) $\sum_{i\geq k} b_i z^i = [\gamma(-\alpha+\bar{z})^{-1}\beta+\delta](\sum_{i\geq k} a_i z^i)$

Proof: (i) Since $q_{i+1} = \alpha(q_i) + \beta(a_i)$ we get $\sum_{i\geq k} q_{i+1} z^i = \alpha(\sum_{i\geq k} q_i z^i) + \beta(\sum_{i\geq k} a_i z^i)$. But $\sum_{i\geq k+1} q_i z^i = \sum_{i\geq k} q_{i+1} z^{i+1} = z \sum_{i\geq k} q_{i+1} z^i$, hence $\sum_{i\geq k+1} q_i z^i = z(\alpha(\sum_{i\geq k} q_i z^i) + \beta(\sum_{i\geq k} a_i z^i))$, whence $-\alpha(\sum_{i\geq k} q_i z^i) + \bar{z}(\sum_{i\geq k+1} q_i z^i) = \beta(\sum_{i\geq k} a_i z^i)$. If $q_k = 0$ then we can write $(-\alpha+\bar{z})(\sum_{i\geq k} q_i z^i) = \beta(\sum_{i\geq k} a_i z^i)$. By Proposition 1, $-\alpha+\bar{z}$ is bijective, hence the result.

(ii) follows from (i), since $b_i = \gamma(q_i) + \delta(a_i)$, hence $\sum_{i\geq k} b_i z^i = \gamma(\sum_{i\geq k} q_i z^i) + \delta(\Sigma a_i z^i)$. □

Definition 4: In a separable system $\Sigma = (\alpha,\beta,\gamma,\delta)$, the function $f_\Sigma := \gamma(-\alpha+\bar{z})^{-1}\beta+\delta$ from L(A) into L(B) is called the *transfer function* of Σ.

Observe that f_Σ is zero-symmetric if this applies to α and γ (by Proposition 1 and the fact that each group homomorphism (hence β and δ) is automatically zero-symmetric).

f_Σ completely characterizes the input-output behaviour of Σ if Σ starts in state 0. If Σ starts in a state $q \neq 0$ which is reachable from 0 by means of an input sequence $a_1 \dots a_r$, we simply start at time k-r, and then in state 0. Since it does not make much sense to start from non-reachable states, f_Σ "characterizes" Σ itself. We call Σ *zero-symmetric* if this applies to f_Σ.

Now we connect two systems $\Sigma_i = (Q_i, A_i, B_i, F_i, G_i)$ $(i = 1,2)$. The *series connection* $\Sigma_1 \mathbin{+\!\!\!+} \Sigma_2$ requires $B_1 = A_2$. Then $\Sigma_1 \mathbin{+\!\!\!+} \Sigma_2 := (Q_1 \times Q_2, A_1, B_2, F, G)$ with $F((q_1,q_2),a) = (F_1(q_1,a), F_2(q_2, G_1(q_1,a))$ and $G((q_1,q_2),a) = G_2(q_2, G_1(q_1,a))$. The *parallel connection* $\Sigma_1 /\!/ \Sigma_2$ works with $A_1 = A_2 =: A$, $B_1 = B_2 =: B$ and gives $\Sigma_1 /\!/ \Sigma_2 := (Q_1 \times Q_2, A, B, F', G')$ with $F'((q_1,q_2),a) = (F_1(q_1,a), F_2(q_2,a))$ and $G'((q_1,q_2),a) = G_1(q_1,a) + G_2(q_2,a)$.

If $\Sigma_i = (Q_i, A_i, B_i, \alpha_i, \beta_i, \gamma_i, \delta_i)$ are separable then in $\Sigma_1 \mathbin{+\!\!\!+} \Sigma_2$ we get $F((q_1,q_2),a) = (\alpha_1(q_1) + \beta_1(a), \alpha_2(q_2) + \beta_2(\gamma_1(q_1) + \delta_1(a))) =$ $= (\alpha_1(q_1), \alpha_2(q_2) + \beta_2\gamma_1(q_1)) + (\beta_1(a), \beta_2\delta_1(a))$. Note that $a \to (\beta_1(a), \beta_2\delta_1(a))$ is a homomorphism. Similarly, $G((q_1,q_2),a) = \gamma_2(q_2) + \delta_2(\gamma_1(q_1) + \delta_1(a)) =$ $= (\gamma_2(q_2) + \delta_2\gamma_1(q_1)) + \delta_2\delta_1(a)$. Hence $\Sigma_1 \mathbin{+\!\!\!+} \Sigma_2$ is again separable. In $\Sigma_1 /\!/ \Sigma_2$ we get $F'((q_1,q_2),a) = (\alpha_1(q_1) + \beta_1(a), \alpha_2(q_2) + \beta_2(a)) = (\alpha_1(q_1), \alpha_2(q_2)) + (\beta_1(a), \beta_2(a))$, and $G'((q_1,q_2),a) = \gamma_1(q_1) + \delta_1(a) + \gamma_2(q_2) + \delta_2(a)$. If B is abelian, the latter can be written as $(\gamma_1(q_1) + \gamma_2(q_2)) + (\delta_1 + \delta_2)(a)$, and $\delta_1 + \delta_2$ is again a homomorphism. In summary, we have shown

Proposition 2: If Σ_1 and Σ_2 are separable systems, the same applies to $\Sigma_1 \mathbin{+\!\!\!+} \Sigma_2$ and (if the output groups are abelian) to $\Sigma_1 /\!/ \Sigma_2$. If Σ_1 and Σ_2 are zero-symmetric then this also applies to $\Sigma_1 \mathbin{+\!\!\!+} \Sigma_2$ and $\Sigma_1 /\!/ \Sigma_2$.

Theorem 2: If Σ_1, Σ_2 are separable then

(i) $f_{\Sigma_1 /\!/ \Sigma_2} = f_{\Sigma_1} + f_{\Sigma_2}$ (if the output groups are abelian)

(ii) $f_{\Sigma_1 \mathbin{+\!\!\!+} \Sigma_2} = f_{\Sigma_2} \circ f_{\Sigma_1}$

Proof: These formulas are already motivated by the definitions of $\Sigma_1 /\!/ \Sigma_2$ and $\Sigma_1 \mathbin{+\!\!\!+} \Sigma_2$. The equations $b_{i+1} = \gamma(q_i) + \delta(a_i)$ etc. only relate the input at time i to the output at the same time i. But inputs at time i also have influence on later outputs (we'll investigate this in this paper as well), and this influence is described by f_Σ. So we give direct proofs. Note that the proof of Proposition 1 shows that $(-\alpha+\bar{z})^{-1}(q_k,q_{k+1},\ldots) = (0,\alpha(0)+q_k,\alpha(\alpha 0+q_k)+q_{k+1},\ldots)$, where the first 0 appears at time k.

(i) Let $\Sigma_1 /\!/ \Sigma_2 = (\alpha,\beta,\gamma,\delta)$, where $\alpha,\beta,\gamma,\delta$ is given as in the lines preceding Proposition 2.

$$f_{\Sigma_1/\!/\Sigma_2}(a_k,a_{k+1},\ldots) = (\gamma(-\alpha+\bar{z})^{-1}\ \beta+\delta)(a_k,a_{k+1},\ldots) =$$
$$= \gamma(-\alpha+\bar{z})^{-1}((\beta_1(a_k),\beta_2(a_k),(\beta_1(a_{k+1}),\beta_2(a_{k+1})),\ldots) + (\delta_1(a_k),\delta_1(a_{k+1}),\ldots) +$$
$$+ (\delta_2(a_k),\delta_2(a_{k+1}),\ldots) = \gamma((0,0),(\alpha_1(0),\alpha_2(0)) + (\beta_1(a_k),\beta_2(a_k),\ldots) +$$
$$+ (\delta_1(a_k),\delta_1(a_{k+1}),\ldots) + (\delta_2(a_k),\delta_2(a_{k+1}),\ldots) =$$
$$= (\gamma_1(0)+\gamma_2(0),\gamma_1(\alpha_1(0)+\beta_1(a_k))+\gamma_2(\alpha_2(0)+\beta_2(a_k)),\ldots) +$$
$$+ (\delta_1(a_k),\delta_1(a_{k+1}),\ldots) + (\delta_2(a_k),\delta_2(a_{k+1}),\ldots) =$$
$$= (\gamma_1(0)+\delta_1(a_k),\gamma_1(\alpha_1(0)+\beta_1(a_k))+\delta_1(a_{k+1}),\ldots) +$$
$$+ (\gamma_2(0)+\delta_2(a_k),\gamma_2(\alpha_2(0)+\beta_2(a_k))+\delta_2(a_{k+1}),\ldots) =$$
$$= (\gamma_1(-\alpha_1+\bar{z})^{-1}\ \beta_1+\delta_1)(a_k,a_{k+1},\ldots) + (\gamma_2(-\alpha_2+\bar{z})^{-1}\ \beta_2+\delta_2)(a_k,a_{k+1},\ldots) =$$
$$= (f_{\Sigma_1}+f_{\Sigma_2})(a_k,a_{k+1},\ldots).$$

(ii) Let $\Sigma_1 \# \Sigma_2$ be described by the maps $\bar{\alpha},\bar{\beta},\bar{\gamma},\bar{\delta}$, as indicated before Proposition 2. Then $f_{\Sigma_1\#\Sigma_2}(a_k,a_{k+1},\ldots) = (\bar{\gamma}(-\bar{\alpha}+\bar{z})^{-1}\ \bar{\beta}+\bar{\delta})(a_k,a_{k+1},\ldots)=$

$$= (\bar{\gamma}(-\bar{\alpha}+\bar{z})^{-1}((\beta_1(a_k),\beta_2\delta_1(a_k),(\beta_1(a_{k+1}),\beta_2\delta_1(a_{k+1}),\ldots) +$$
$$+ (\delta_2\delta_1(a_k),\delta_2\delta_1(a_{k+1}),\ldots,) =$$
$$= (\bar{\gamma}((0,0),(\alpha_1(0),\alpha_2(0)+\beta_2\gamma_1(0)) + (\beta_1(a_k),\beta_2\delta_1(a_k),\ldots) +$$
$$+ (\delta_2\delta_1(a_k),\delta_2\delta_1(a_{k+1}),\ldots) =$$
$$= (\gamma_2(0)+\delta_2\gamma_1(0),\gamma_2(\alpha_2(0)+\beta_2\gamma_1(0)+\beta_2\delta_1(a_k))+\delta_2\gamma_1(\alpha_1(0)+\beta_1(a_k)),\ldots) +$$
$$+ (\delta_2\delta_1(a_k),\delta_2\delta_1(a_{k+1}),\ldots) =$$
$$= (\gamma_2(0)+\delta_2\gamma_1(0)+\delta_2\delta_1(a_k),\gamma_2(\alpha_2(0)+\beta_2\gamma_1(0)+\beta_2\delta_1(a_k)) +$$
$$+ \delta_2\gamma_1(\alpha_1(0)+\beta_1(a_k))+\delta_2\delta_1(a_{k+1}),\ldots) =$$
$$= (\gamma_2(0)+\delta_2(\gamma_1(0)+\delta_1(a_k)),\gamma_2(\alpha_2(0)+\beta_2(\gamma_1(0)+\delta_1(a_k)))+\delta_2(\gamma_1(\alpha_1(0)+$$
$$+ \beta_1(a_k))+\delta_1(a_{k+1})),\ldots) =$$
$$= (\gamma_2(0),\gamma_2(\alpha_2(0)+\beta_2(\gamma_1(0)+\delta_1(a_k)),\ldots) +$$
$$+ (\delta_2(\gamma_1(0)+\delta_1(a_k)),\delta_2(\gamma_1(\alpha_1(0)+\beta_1(a_k))+\delta_1(a_{k+1})),\ldots) =$$
$$= (\gamma_2(-\alpha_2+\bar{z})^{-1}(\beta_2(\gamma_1(0)+\delta_1(a_k)),\beta_2(\gamma_1(\alpha_1(0)+\beta_1(a_k))+\delta_1(a_{k+1}),\ldots) +$$
$$+ (\delta_2(\gamma_1(0)+\delta_1(a_k)),\delta_2(\gamma_1(\alpha_1(0)+\beta_1(a_k))+\delta_1(a_{k+1})),\ldots) =$$

$= f_{\Sigma_2}(\gamma_1(0) + \delta_1(a_k), \gamma_1(\alpha_1(0) + \beta_1(a_k)) + \delta_1(a_{k+1}), \ldots) =$

$= f_{\Sigma_2}(\gamma_1(0, \alpha_1(0) + \beta_1(a_k), \ldots) + (\delta_1(a_k), \delta_1(a_{k+1}), \ldots)) =$

$= f_{\Sigma_2}(\gamma_1(-\alpha_1 + \bar{z})^{-1}(\beta_1(a_k), \beta_1(a_{k+1}), \ldots) + (\delta_1(a_k), \delta_1(a_{k+1}), \ldots)) =$

$= f_{\Sigma_2}((\gamma_1(-\alpha_1 + \bar{z})^{-1}\ \beta_1 + \delta_1)(a_k, a_{k+1}, \ldots)) = f_{\Sigma_2}(f_{\Sigma_1}(a_k, a_{k+1}, \ldots)) =$

$= (f_{\Sigma_2} \circ f_{\Sigma_1})(a_k, a_{k+1}, \ldots)$, which was to be shown. □

Now let us assume that we have a family $(\Sigma_i)_{i \in I}$ of separable systems at hand which all have the same abelian group A as input and output group. Under these assumptions we get the following Corollary to Theorem 2.

Corollary: The set S of transfer functions of those systems which can be constructed from the Σ_i $(i \in I)$ by means of series and parallel connections coincides with the seminear-ring $<f_{\Sigma_i} \mid i \in I>$ generated by $\{f_{\Sigma_i} \mid i \in I\}$ in M(L(A)). If A is of finite exponent then S is the near-ring generated by the f_{Σ_i}'s.

This holds because if e is the exponent of A then $eA = 0$, whence ef_{Σ} is the zero map, and so $(e-1)f_{\Sigma} = -f_{\Sigma}$ can be obtained by (e-1) parallel connections of Σ.

Note that this applies in particular to finite groups A. Of course, S = the near-ring generated by $\{f_{\Sigma_i} \mid i \in I\}$ if $(\Sigma_i)_{i \in I}$ contains with each $\Sigma_i = (\alpha, \beta, \gamma, \delta)$ also the system $-\Sigma_i = (\alpha, \beta, -\gamma, -\delta)$ which yields - as can be easily seen - the transfer function $f_{-\Sigma} = -f_{\Sigma}$. Also, this can be achieved by "external" devices which transfer elements of A into their negatives.

Hence it makes sense to impose the following

Convention for the rest of this paper: $(\Sigma_i)_{i \in I}$ is a family of separable systems which have the same abelian input- and output group A and which contains with each Σ_i also $-\Sigma_i$.

Hence $<f_{\Sigma_i} \mid i \in I>$ is a subnear-ring of M(L(A)). If each f_{Σ_i} is zero-symmetric, so is $<f_{\Sigma_i} \mid i \in I>$ (as a subnear-ring of $M_0(L(A))$). If each f_{Σ_i} is an endomorphism on L(A) then $<f_{\Sigma_i} \mid i \in I>$ is even a ring.

If we identify a separable system Σ with its transfer function f_{Σ} (as described earlier in this paper), we can restate our results as follows.

Theorem 3: Let S(A) be the set of all separable systems with abelian input- and output groups A. Then (S(A),// ,++) is a near-ring.

By standard means of universal algebra (see e.g. [4]), $\langle f_{\Sigma_i} \mid i \in I\rangle =$ $= \{t(f_{\Sigma_{i_1}},\dots,f_{\Sigma_{i_n}}) \mid t \in T_n\}$, where T_n is the set of all terms (polynomials) in the variety of all near-rings.

In order to learn more about $\langle f_{\Sigma_i} \mid i \in I\rangle$, we try to find a more explicite formula for the action of f_Σ. This can be achieved by the following

Definition 5: If $\Sigma = (\alpha,\beta,\gamma,\delta)$, the *n-th Markov symbol* (with $n \geq -1$) $[\alpha,\beta;n]$ is the map from L(A) into Q defined inductively by

$$[\alpha,\beta;-1]\left(\sum_{i\geq k} a_i z^i\right) := 0, [\alpha,\beta;0]\left(\sum_{i\geq k} a_i z^i\right) := \alpha(0) + \beta(a_k)$$

$$[\alpha,\beta;n+1]\left(\sum_{i\geq k} a_i z^i\right) := \alpha\left([\alpha,\beta;n]\left(\sum_{i\geq k} a_i z^i\right)\right) + \beta(a_{k+n+1})$$

$$=: \alpha[\alpha,\beta;n] \sum_{i\geq k} a_i z^i \quad + \beta(a_{k+n+1})$$

Note that $[\alpha,\beta;n] \sum_{i\geq k} a_i z^i$ only depends on $a_k,\dots,a_{k+n}$; hence we can also write $[\alpha,\beta;n]\ (a_k,\dots,a_{k+n})$. Obviously we get

Proposition 3: If α is zero-symmetric then each $[\alpha,\beta;n]$ is zero-symmetric and $[\alpha,\beta;n]\left(\sum_{i\geq k} a_i z^i\right) = \alpha(\alpha(\dots\alpha(\beta(a_k) + \beta(a_{k+1})\dots)) + \beta(a_{k+n-1})) + \beta(a_{k+n})$.

Since $[\alpha,\beta;0]\ (a_k) = \alpha(0) + \beta(a_k)$ is the state q_{k+1} (if $q_k = 0$), we get by the definition of $[\alpha,\beta;n]$ the following

Proposition 4: If a separable system $\Sigma = (\alpha,\beta,\gamma,\delta)$ starts at time k in the state $q_k = 0$ and if the input sequence $(a_k,a_{k+1},\dots)$ arrives then Σ assumes new states $q_{k+1},q_{k+2},\dots$ such that, for $n \in \mathbb{N}$

$$q_{k+n} = [\alpha,\beta;n+1]\ (a_k,\dots,a_{k+n-1})$$

From this, we get the desired formula for f_Σ:

Proposition 5: If $f_\Sigma\left(\sum_{i\geq k} a_i z^i\right) = \sum_{i\geq k} b_i z^i$ then

$$b_k = \gamma(0) + \delta(a_k) = \gamma[\alpha,\beta;-1](-) + \delta(a_k)$$

$$b_{k+n} = \gamma[\alpha,\beta;n-1]\ (a_k,\dots,a_{k+n-1}) + \delta(a_{k+n}) \qquad (n \in \mathbb{N})$$

Note that Theorem 2 could also be proved by induction using this formula.

Now we investigate, which functions $f \in M(L(A))$ can arise as $f = f_\Sigma$ for some separable system $\Sigma = (\alpha,\beta,\gamma,\delta)$. The determination of $\alpha,\beta,\gamma,\delta$ from f is then known as "realization problem". It is useful to establish the following

<u>Notation:</u> $L_k(A) := \{\sum_{i \geq k} a_i z^i \mid a_i \in A\}$.

Of course, $L(A) = \bigcup_{k \in \mathbb{Z}} L_k(A)$. Furthermore, let, for $B,C \subseteq X$, $(B:C)$ be the set of all functions f from X to X such that $f(C) \subseteq B$. This coincides with the "noetherian quotients" as defined in 1.41 of [14]. Since $f_\Sigma(L_k(A)) \subseteq L_k(A)$ for each separable system Σ, we get

<u>Proposition 6:</u> Let Σ_i $(i \in I)$ be separable systems.

(i) Each $f_{\Sigma_i} \in \bigcap_{k \in \mathbb{Z}} (L_k(A) : L_k(A)) =: N(L(A))$.

(ii) $N(L(A))$ is a near-ring.

(iii) $\langle f_{\Sigma_i} \mid i \in I\rangle$ is a subnear-ring of $N(L(A))$.

(iv) If each Σ_i is zero-symmetric then $\langle f_{\Sigma_i} \mid i \in I\rangle$ is a subnear-ring of the zero-symmetric part $N_0(L(A))$ of $N(L(A))$.

If Σ is zero-symmetric then it makes no difference if we consider $f(\sum_{i \geq k} a_i z^i)$ or $f(\sum_{i \in \mathbb{Z}} a_i z^i)$ with $a_i = 0$ for $i < k$. Hence in this case f_Σ can be regarded as a member of $\{f \in M(A^{\mathbb{Z}}) \mid f(i) = 0$ for all $i < k$, where k is some integer (depending on f)$\}$.

<u>Notation:</u> If $(A,+)$ is an abelian group and $(Q,+)$ any group, let $N(A,Q)$ be the set of all f_Σ, where $\Sigma = (\alpha,\beta,\gamma,\delta)$ is any separable system on (A,A,Q). Let $\bar{N}(A,Q)$ be the subnear-ring $\langle N(A,Q)\rangle$ generated by $N(A,Q)$ in the near-ring $N(A)$ of all separable systems on (A,A,Q) for some Q. If Σ is as above, we denote f_Σ also by $f_{\alpha,\beta,\gamma,\delta}$.

Recall that we get $N(A,Q) \subseteq \bar{N}(A,Q) \leq N(A) \leq N(L(A)) \leq M(L(A))$. All members of this chain (except the first one) are always near-rings.

By Proposition 5, we get $f_{\alpha,\beta,\gamma,\delta}(a_k,a_{k+1},\ldots) =$

$$= (\gamma(0) + \delta(a_k),\ \gamma(\alpha(0) + \beta(a_k)) + \delta(a_{k+1}),\ \gamma(\alpha(\alpha(0) + \beta(a_k)) + \beta(a_{k+1})) + \delta(a_{k+2}),\ldots)$$

Now if $f \in M(L(A))$ should coincide with this f_Σ, we can write $f = (f^0, f^1, \ldots)$ such that, for all $\Sigma a_i z^i \in L(A)$, $f(\Sigma a_i z^i) = (f^0(\Sigma a_i z^i), f^1(\Sigma a_i z^i), \ldots)$. This imposes quite a number of restrictions on the functions f^n $(n \in \mathbb{N}_0)$ and hence on f itself. By comparison of components, we see that

(0) $f^0(\sum_{i \geq k} a_i z^i) = \gamma(0) + \delta(a_k)$. Hence f^0 is an affine function depending only on a_k, and we can solve for δ, since $\delta = f^0_0$ = the zero-symmetric part of f^0. Also, $\gamma(0) = f^0_c$ is already fixed at this stage.

(1) $f^1(\sum_{i \geq k} a_i z^i) = \gamma(\alpha(0) + \beta(a_k)) + \delta(a_{k+1}) = \gamma(\alpha(0) + \beta(a_k)) + f^0_0(a_{k+1})$. Hence there must be a function $g^1 \in M(A)$ such that $f'(\sum_{i \geq k} a_i z^i) = g^1(a_k) + f^0_0(a_{k+1})$. We can always solve if $\alpha(0) \notin \operatorname{Im} \beta$, and also in case $\alpha(0) \in \operatorname{Im} \beta$ if we have $g^1(a_k) = f^0_c(a_k) = \gamma(0)$ for each a_k such that $\beta(-a_k) = -\beta(a_k) = \alpha(0)$.

$\vdots \quad \vdots \quad \vdots \quad \vdots \quad \vdots$

(n) The further equations determine α, β, γ more and more, and show that each f^n is a function just depending on the first n+1 coefficients $a_k, \ldots, a_{k+n}$ in $\sum_{i \geq k} a_i z^i$. In fact, there must be functions $g^n : A \to A$ such that
$f^n(\sum_{i \geq k} a_i z^i) = g^n(a_k, \ldots, a_{k+n-1}) + f^0_0(a_{k+n})$.

"In practice" it turns out that α, β, γ (and δ, anyhow) are already determined after comparatively few steps. This is the realization procedure of calculating $\alpha, \beta, \gamma, \delta$ out of a given $f = f_{\alpha,\beta,\gamma,\delta}$. We illustrate this in the case $Q = \mathbb{Z}_2$. Recall that in this case $\alpha: \mathbb{Z}_2 \to \mathbb{Z}_2$, $\beta: A \to \mathbb{Z}_2$, $\gamma: \mathbb{Z} \to A$, $\delta: A \to A$, and that $\gamma(0)$ and δ are fixed in step (0) as $\delta = f^0_0$, $\gamma(0) = f^0_c$. In step (1) we get $\gamma(\alpha(0) + \beta(a_n)) = g^1(a_k)$. This fixes (case (i)) γ completely unless (case (ii)) $\alpha(0) = 0$, $\beta = 0$ (= zero map) (in which case we must have $g^1(a_k) = f^0_c$ for all a_k).

Note that in case (i) α and β may remain "hidden" if e.g. γ is constant (this is only possible if each g^n is constant!). This case will appear in the Example below. In case (ii), $\alpha(0)$, β and δ are fixed, and we get that $\gamma(0) = f^0_c = g^n$, so g^n must be constant; then $\gamma(1)$ can (and need) not be determined. We see:

<u>Remark:</u> The equation $f = f_{\alpha,\beta,\gamma,\delta}$ is solvable iff $f \in N(A)$. In this case, $\gamma(0)$ and δ are always uniquely determined, while this might not be the case for α, β, and the other values of $\gamma(q)$ (Q is thought to be known and fixed). Cf. also the lines after Theorem 7.

How big is N(A,Q)? Although L(A), hence M(L(A)), and also N(L(A)) in Proposition 6 are always infinite (unless $A = \{0\}$), we get

Proposition 7: N(A,Q) is finite if A and Q are finite.

This follows from the fact that there are only finitely many functions $\alpha,\beta,\gamma,\delta$ which determine separable systems on (A,A,Q).

We now determine $N(\mathbb{Z}_2, \mathbb{Z}_2)$ in detail; this will also show how one can proceed in more general cases.

Example 1: In order to describe $N := N(\mathbb{Z}_2, \mathbb{Z}_2)$, we need the following four functions of $M(L(\mathbb{Z}_2))$.

$z := \bar{z}^{-1}$ ("unit shift to the right")

$\underline{1} : \sum_{i\geq k} a_i z^i \to \sum_{i\geq k} 1z^i$

$S : \sum_{i\geq k} a_i z^i \to \sum_{i\geq 1} (a_k + a_{k+1} + \ldots + a_{k+i-1}) z^{k+i}$ ("summation operator")

$i : \sum_{i\geq k} a_i z^i \to \sum_{i\geq 0} i\ z^{k+i}$

Now β and δ can each be the zero map 0 or the identity map id, while α and γ can be 0,1 (= constant map with values 1), id, or 1 + id. It turns out that γ plays a crucial role.

If $\gamma = 0$ then each $f_{\alpha,\beta,\gamma,\delta} = \delta$ (in particular, $f_{0,0,0,0} = 0$, $f_{0,0,0,id} = id \in N$)

If $\gamma = 1$ then $f_{\alpha,\beta,1,\delta} = \underline{1} + \delta$ (in particular, $\underline{1} = f_{0,0,1,0} \in N$)

If $\gamma = id$ then it is easy to see that

$f_{0,\beta,id,\delta} = z \circ \beta + \delta$ (in particular, $z = f_{0,id,id,0} \in N$)

$f_{1,\beta,id,\delta} = z \circ (\underline{1} + \beta) + \delta$

$f_{id,\beta,id,\delta} = S \circ \beta + \delta$ (in particular, $S = f_{id,id,id,0} \in N$)

$f_{1+id,\beta,id,\delta} = i + S \circ \beta + \delta$ (in particular, $i = f_{1+id,0,id,0} \in N$)

If $\gamma = 1 + id$, $f_{\alpha,\beta,1+id,\delta} = \underline{1} + f_{\alpha,\beta,id,\delta}$.

This, and the fact that $\beta,\delta \in \{0,id\}$ shows that N is generated by $0, id, \underline{1}, z, S, i$. Since $id + id = 0$, $S \circ \underline{1} = i$, we can even say that

$$N(\mathbb{Z}_2, \mathbb{Z}_2) \subset \langle id, \underline{1}, z, S \rangle.$$

Note that $N(\mathbb{Z}_2,\mathbb{Z}_2)$ is neither closed w.r.t. addition (since e.g. $z+S \notin N(\mathbb{Z}_2,\mathbb{Z}_2)$), nor w.r.t. composition ($z \circ z \notin N(\mathbb{Z}_2,\mathbb{Z}_2)$).

We now give a list of the functions in $N(\mathbb{Z}_2, \mathbb{Z}_2)$, along with their values in the first three components (which shows that all of them are different).

	0	i	$z \circ \underline{1}$	z	S	$z \circ (\underline{1}+id)$	$i+S$
$a_k \to$	0	0	0	0	0	0	0
$a_{k+1} \to$	0	1	1	a_k	a_k	a_k+1	a_k+1
$a_{k+2} \to$	0	0	1	a_{k+1}	a_k+a_{k+1}	$a_{k+1}+1$	a_k+a_{k+1}

	$z \circ \underline{1}+\underline{1}$	$\underline{1}+i$	$\underline{1}$	$z \circ (\underline{1}+id)+1$	$\underline{1}+i+S$	$\underline{1}+z$	$\underline{1}+S$
$a_k \to$	0	0	1	1	1	1	1
$a_{k+1} \to$	0	0	1	a_k	a_k	a_k+1	a_k+1
$a_{k+1} \to$	0	1	1	a_{k+1}	$a_k+a_{k+1}+1$	$a_{k+1}+1$	$a_k+a_{k+1}+1$

	id	$z \circ \underline{1}+id$	$i+id$	$z+id$	$S+id$	$z \circ (\underline{1}+id)+id$	$i+S+id$
$a_k \to$	a_k	a_k	a_k	a_k	a_k	a_k	a_k
$a_{k+1} \to$	a_{k+1}	a_k+1	$a_{k+1}+1$	a_k+a_{k+1}	a_k+a_{k+1}	$a_k+a_{k+1}+1$	$a_k+a_{k+1}+1$
$a_{k+2} \to$	a_{k+1}	$a_{k+1}+1$	a_{k+2}	$a_{k+1}+a_{k+2}$	$a_k+a_{k+1}+a_{k+2}$	$a_{k+2}+a_{k+1}+1$	$a_k+a_{k+1}+a_{k+2}$

	$\underline{1}+z\circ\underline{1}+id$	$\underline{1}+i+id$	$\underline{1}+id$	$\underline{1}+z\circ(\underline{1}+id)+id$	$\underline{1}+i+S+id$	$\underline{1}+z+id$	$\underline{1}+S+id$
$a_k \to$	a_k+1	a_k+1	a_k+1	a_k+1	a_k+1	a_k+1	a_k+1
$a_{k+1} \to$	a_k	a_{k+1}	$a_{k+1}+1$	a_k+a_{k+1}	a_k+a_{k+1}	$a_k+a_{k+1}+1$	$a_k+a_{k+1}+1$
$a_{k+2} \to$	a_{k+1}	$a_{k+2}+1$	$a_{k+2}+1$	$a_{k+1}+a_{k+2}$	$a_k+a_{k+1}+a_{k+2}+1$	$a_{k+1}+a_{k+2}+1$	$a_k+a_{k+1}+a_{k+2}+1$

Hence we get the following

Corollary: $N(\mathbb{Z}_2, \mathbb{Z}_2)$ has 28 elements and $\bar{N}(\mathbb{Z}_2, \mathbb{Z}_2)$ can be generated by four elements. Hence every separable system constructible by series/parallel connections of separable systems on $(\mathbb{Z}_2, \mathbb{Z}_2, \mathbb{Z}_2)$ can be obtained by means of series/parallel connections from $\Sigma_{id} = (0,0,0,id)$, $\Sigma_1 = (0,0,1,0)$, $\Sigma_z = (0,id,id,0)$, $\Sigma_S = (id,id,id,0)$, which are systems on $(\mathbb{Z}_2, \mathbb{Z}_2, \mathbb{Z}_2)$.

In order to describe a more general situation, we need the following

Definition 6: For $n \in \mathbb{N}_0$ and $A = B = Q = K =$ a field, let
$$M^n : s = \sum_{i \geq k} a_i z^i \to (0,[\alpha,\beta;0]^n(s),[\alpha,\beta;1]^n(s),[\alpha,\beta;2]^n(s),\ldots),$$
(where the exponent n denotes the one w.r.t. the multiplication in F) be the n-th *Markov map* on K.

Theorem 4: Let K be a finite (Galois) field and $A = B = Q = K$. Then

(i) $N(K,K)$ always contains id,z,$\underline{1}$,S,i of the preceding Example.

(ii) If $N(K,K)$ is contained in the near-ring generated by functions f_i ($i \in$ some index set I) then $\bar{N}(K,K) = \langle f_i \mid i \in I\rangle$.

(iii) $N(K,K)$ is always finite, $\bar{N}(K,K)$ always infinite.

(iv) $\bar{N}(K,K) = \langle h_k, M^0, M^1, \ldots, M^{p-1}\rangle$, where $p = \text{char } K$ and $h_k: x \to kx$ (for all $x \in K$), k being a primitive element (i.e. a generator of the cyclic group $(K^*,.)$).

(v) The near-ring $\bar{N}(K,K)$ fulfills the ascending chain condition (ACC) on subnear-rings.

Proof: (i) is clear since $id = f_{0,0,0,id}$, $z = f_{0,id,id,0}$, $\underline{1} = f_{0,0,1,0}$, $S = f_{id,id,id,0}$, $i = S \circ \underline{1}$.

(ii) If $N(K,K) \subset \langle f_i \mid i \in I\rangle \leq \bar{N}(K,K)$ then $\langle N(K,K)\rangle = \bar{N}(K,K) = \langle f_i \mid i \in I\rangle$.

(iii) follows from Proposition 3.

(iv) $\bar{N}(K,K)$ is generated by $N(K,K)$, i.e. the set of all functions $\gamma(-\alpha+\bar{z})^{-1}\beta+\delta$, with $\alpha,\gamma \in M(K)$, $\beta,\delta \in \text{Hom}(K,K)$ (which is K-isomorphic to K). If we take $\gamma = 0$, we get all functions in $\text{Hom}(K,K)$, and hence all $\gamma(-\alpha+\bar{z})^{-1}\beta$ are also in $\bar{N}(K,K)$. Now a finite field is known to be "polynomially complete" (see [10] or [11]), hence there are $k_0,k_1,\ldots,k_{p-1} \in K$ such that $\gamma(x) = k_0 + k_1x + k_2x^2 + \ldots + k_{p-1}x^{p-1}$ for all $x \in K$. Hence we can generate all $\gamma(-\alpha+\bar{z})^{-1}\beta$ additively from $kx^n(-\alpha+\bar{z})^{-1}\beta$, which in turn can be obtained as the composition of $x^n(-\alpha+\bar{z})^{-1}\beta$ with the homomorphism $h_k: x \to kx$. Now $x^n(-\alpha+\bar{z})^{-1}\beta$ sends $s = \sum_{i \geq k} a_i z^i$ into $(0,[\alpha,\beta;0]^n(s), [\alpha,\beta;1]^n(s),\ldots)$, which means that $x^n(-\alpha+\bar{z})^{-1}\beta = M^n$. Finally, all elements in $\text{Hom}(K,K)$ can be obtained from a single map $h_k: x \to kx$, in which k is a primitive element in K.

(v) follows from the fact that $\bar{N}(K,K)$ is finitely generated. □

Remarks: (i) A similar, but more complicated result as for finite fields (Theorem 3) holds for finite vector spaces. It does not seem to be too well-known that each map (we need it for α,γ) from a finite vector space Q (over a field F) into itself is of the form
$q = (q_1,\ldots,q_n) \to (p_1(q_1,\ldots,q_n),\ldots,p_n(q_1,\ldots,q_n))$, where the p_i are polynomial functions over K and $n = \dim_F Q$. This is basically due to the polynomial completeness of finite fields (see [10]).

(ii) If A is a field, N(A) can be considered as near-ring of mappings on a field, namely on the quotient field of the integral domain A[[x]] of formal power series. Theorem 3 (iv) indicates the appearence of addition, multiplication, as well as composition. The suitable frame for these considerations is the one of "composition rings", whose study was initiated in [12] and [1].

(iii) If N(A) is a ring and generated by f_{Σ_i} $(i \in I)$ then every element on N(A) (= every system on (A,A,Q) for some Q) can be obtained by a parallel connection of (one or several) series connections of the Σ_i. This is not the case any more if N(A) is a "proper" near-ring (then one has to take parallel connections of series of parallel of ... connections).

(iv) If $N = N(A)$, or $N = \bar{N}(A,Q)$, can be generated by functions which are either constant or homomorphisms (this was the case for $\bar{N}(\mathbb{Z}_2, \mathbb{Z}_2)$ in our Example!) then N is an abstract affine near-ring (see [14], ch. 9c)). In this case, N_o is a ring and N_c is a module over N_o. Affine near-rings are the "most simple" ones among the proper near-rings.

(v) Several other near-ring theoretic concepts have meaningful interpretations in the context of systems theory. L(A) is, in a natural way, an $N = N(A)$ or $N = N(A,Q)$ - group (= (near-) module). A subgroup L of L(A) is an N-subgroup if $f_\Sigma(\Sigma a_i z^i) \in L$ whenever $f_\Sigma \in N$ and $\Sigma a_i z^i \in L$. Note that L(A) is never a monogenic N(A)- or N(A,Q)-group (there is no $\Sigma a_i z^i$ such that each $\Sigma b_i z^i \in L(A)$ is of the form $\Sigma b_i z^i = f_\Sigma(\Sigma a_i z^i)$ for some $f_\Sigma \in N(A)$ or $\in N(A,Q)$). Ideals I on N are basically subsets of N which are closed w.r.t. parallel connections, and w.r.t. parallel connections with arbitrary systems in N. Elements in N(A,Q) can be nilpotent: take e.g. $A = B = Q = \mathbb{Z}_4$, $\delta(a) = 2a$, $f = f_{0,0,0,\delta}$. Then $f \circ f = 0$, which means that the series composition of the corresponding system Σ with itself sends every input sequence into the zero sequence (0,0,...). The appearance of nilpotent (or nil) N-subgroups indicates that the radical of N(A) of N(A,Q) is non-zero ([14], e.g. 5.45).

We now turn to separable systems with a (separable) feedback.

Let $\Sigma \in S(A)$ have a *separable feedback function* $\Phi: Q \times A \to A$, $(q,a) \to \phi(q) + \psi(a)$, ψ a homomorphism. If we encode the state $q \in Q$ and the output $a \in A$ to $\Phi(q,a)$ and add it to the input, we get a new system Σ^Φ. The following result is easy to see.

<u>Proposition 8:</u> If $\Sigma \in \mathbf{S}(A)$ has maps $\alpha,\beta,\gamma,\delta$ and if Φ is a separable feedback, $\Phi(q,a) = \phi(q) + \psi(a)$, then Σ^{Φ} is again in $\mathbf{S}(A)$ and has maps $\alpha + \beta\phi$, $\beta\psi$, $\gamma + \delta\phi$, $\delta\psi$. If Σ and Φ are zero-symmetric (i.e. $\phi(0) = 0$) then this also applies to Σ^{Φ}.

Hence the class of separable systems is closed w.r.t parallel/series connections and separable feedbacks.

We now turn to questions of reachability. Let A^* be the free monoid over A; so A^* consists of all finite strings of elements of A (including the empty sequence Λ). If $\bar{a} = a_1a_2\ldots a_n \in A^*$, we write $n = |\bar{a}|$. If Σ is any system (separable or not) such that the state set is a group $(Q,+)$, we let $N_i(\Sigma)$ be the subgroup of $M(Q)$ generated by $\{f_{\bar{a}} \mid \bar{a} \in A^*, |\bar{a}| \leq i\}$. Then $N_0(\Sigma) \leq N_1(\Sigma) \leq \ldots$ and $N_\infty(\Sigma) := \bigcup_{i \in \mathbb{N}_0} N_i(\Sigma)$ are groups, too. If $q \in Q$ then $N_i(\Sigma)q =: \mathrm{Reach}_i(q)$ consists of all states which can be reached from q by "(multiple) input strings of length $\leq i$". Clearly, each $\mathrm{Reach}_i(q)$ is a subgroup of Q and $\mathrm{Reach}_0(q) \subseteq \mathrm{Reach}_1(q) \subseteq \ldots$ holds. $\mathrm{Reach}_\infty(q) := \bigcup_{i \in \mathbb{N}_0} \mathrm{Reach}_i(q) = N_\infty(\Sigma)q$ is called the *subgroup reachable from* q. If $(Q,+)$ fulfills the ascending chain condition (ACC) for subgroups (which is equivalent to say that $(Q,+)$ is finitely generated) then there is some $n \in \mathbb{N}_0$ such that $\mathrm{Reach}_n(q) = \mathrm{Reach}_\infty(q)$.

One might be interested how far one can get in precisely i steps. Let $\mathrm{Rch}_i(q) := \{f_{\bar{a}}q \mid |\bar{a}| = i\}$. Then it is not true for separable systems that $\mathrm{Rch}_i(q) \leq \mathrm{Rch}_{i+1}(q)$ (as it is for linear ones):

<u>Example 2:</u> Let $Q = A = (\mathbb{Z}_{12},+)$, $F(q,a) = (q+1)^2 + 4a$. Then
$\mathrm{Rch}_1(0) = \{F(0,a) \mid a \in \mathbb{Z}_{12}\} = \{1 + 4a \mid a \in \mathbb{Z}_{12}\} = \{1,5,9\}$ and
$\mathrm{Rch}_1(1) = \mathrm{Rch}_1(5) = \mathrm{Rch}_1(9) = \{0,4,8\}$. Hence $\mathrm{Rch}_1(0) \not\subseteq \mathrm{Rch}_2(0)$, and we have even $\mathrm{Rch}_1(0) \cap \mathrm{Rch}_2(0) = \emptyset$!

The trouble came from the fact that the system in Example 2 was not zero-symmetric:

<u>Proposition 9:</u> Let Σ be separable and zero-symmetric, and let $Q_i = \mathrm{Rch}_i(0)$. Then, for each $i \in \mathbb{N}_0$,

(i) $Q_{i+1} = \alpha(Q_i) + \mathrm{Im}\,\beta \supseteq \alpha(Q_i)$

(ii) $Q_0 \subseteq Q_1 \subseteq \ldots \subseteq \mathrm{Rch}_\infty(0) := \bigcup_{i \in \mathbb{N}_0} Q_i \subseteq \mathrm{Reach}_\infty(0)$.

(iii) $Q_{i+1} = \alpha(Q_i) + Q_1$

(iv) If β is an epimorphism then $Q_1 = \mathrm{Rch}_\infty(0) = \mathrm{Reach}_\infty(0) = Q$.

Proof:

(i) $Q_{i+1} = \{F(0,\bar{a}) \mid |\bar{a}| = i+1\} =$
$= \{F(F(0,\tilde{a}),a) \mid \tilde{a} \in A^*,\ |\tilde{a}| = i,\ a \in A\} =$
$= \alpha(\{F(0,\tilde{a}) \mid \tilde{a} \in A^*,\ |\tilde{a}| = i\}) + \{\beta(a) \mid a \in A\} =$
$= \alpha(Q_i) + \operatorname{Im}\beta \supseteq \alpha(Q_i)$

(ii) We show $Q_i \subseteq Q_{i+1}$ by induction on i. $Q_o = \{0\}$, $Q_1 = \{\alpha(0) + \beta(a) \mid a \in A\} = \operatorname{Im}\beta \leq (Q,+)$, since β is a homomorphism and Σ is zero-symmetric. So $Q_0 \subseteq Q_1$. Now let $Q_{i-1} \subseteq Q_i$. From (i) we get $Q_{i+1} = \alpha(Q_i) + \operatorname{Im}\beta \supseteq \alpha(Q_{i-1}) + \operatorname{Im}\beta = Q_i$.

(iii) follows from (i) and (ii).

(iv) is a consequence of (iii) and of $Q_1 = \operatorname{Im}\beta$. □

Separable feedbacks do not change the reachability behaviour:

<u>Theorem 4:</u> Let Σ be a system, and Φ a feedback on Σ. Suppose that Σ and Φ are both separable and zero-symmetric with ψ being an epimorphism. Then $Rch_i(0)$, and hence $Rch_\infty(0)$ and $Reach_\infty(0)$ coincide in Σ and in Σ^Φ.

Proof: Let $Rch_i(0) =: Q_i$ in Σ and $=: Q_i^\Phi$ in Σ^Φ. We proceed again by induction. For $i = 0$ we get $Q_0 = \{0\} = Q_0^\Phi$. Let $Q_i = Q_i^\Phi$. From Proposition 8 and Proposition 9 (i) we get $Q_{i+1}^\Phi = (\alpha + \beta\phi)Q_i^\Phi + \operatorname{Im}\beta\psi = (\alpha + \beta\phi)Q_i + \beta\psi(A) = \alpha(Q_i) + \beta\phi(Q_i) + \beta(A) =$ $= \alpha(Q_i) + \operatorname{Im}\beta = Q_{i+1}$, since $\beta\phi(Q_i) \subseteq \beta(A)$, which is a subgroup of $(Q,+)$. Hence the result. □

The last considerations had to do with the syntactic near-rings $N(\Sigma)$ (lines after Definition 3); cf. also [15]. Now we turn to topics which involve the near-rings $N(A) = \{f_\Sigma \mid \Sigma \in \mathbf{S}(A)\}$ of the "second type".

If a function f_Σ in $N(A)$ is (left-, right-) invertible (in $N(A)$) then there is a separable system Σ' such that $f_{\Sigma + \Sigma'} = f_{\Sigma'} \circ f_\Sigma = id$ ($f_{\Sigma' + \Sigma} = f_\Sigma \circ f_{\Sigma'} = id$, respectively), where id is the identity function on the group $L(A)$ of all formal Laurent series on A. For many applications, however, this is too restrictive and one requires a weaker kind of "reversibility":

<u>Definition 8:</u> $f_\Sigma \in N(A)$ is *invertible with delay* L if there is a system Σ' such that $f_{\Sigma'} \circ f_\Sigma = z^L$, where $z^L(\Sigma a_i z^i) = \Sigma a_i z^{i+L}$.

Hence $\Sigma + \Sigma'$ is a separable system whose transfer function maps $(a_k, a_{k+1}, \ldots)$ into $(b_k = 0,\ b_{n+1} = 0, \ldots, b_{k+L-a} = 0,\ b_{k+L} = a_k,\ b_{k+L-1} = a_{k+1}, \ldots)$. Also, f_Σ is invertible with delay $L = 0$ iff it is left invertible.

Invertibility, as defined so far, only says something if Σ starts in state 0. In order to obtain a more general setting we follow [16] by defining a slightly more general form of invertibility:

Definition 9: Σ is *invertible with delay* L *from the state* $q \in Q$ if there is a function f which associates to each sequence $b_k, b_{k+1}, \ldots, b_{k+n}, \ldots, b_{k+n+L}$ a sequence $a_k, \ldots, a_{k+n}$ such that Σ, started in state q, responds $b_k, \ldots, b_{k+n+L}$ to the input sequence $a_k, \ldots, a_{k+n}, x_{k+n+1}, \ldots, x_{k+n+L}$ (for some appropriate $x_{k+n+1}, \ldots, x_{k+n+L}$).

Like in a good fairy-tale, three wishes will be fulfilled: we can restrict us to $k = 0$, $q = 0$, and $n = 0$. The thing with $k = 0$ is simply a consequence of time-invariance: it does not matter at which time k we start our engine. We shall assume $k = 0$ from now on. The remaining two wishes are less trivial and seemed at first to be particular to linear systems.

Theorem 5: Let Σ be separable with maps $\alpha,\beta,\gamma,\delta$ (in short: $\Sigma = (\alpha,\beta,\gamma,\delta)$), and $q \in Q$.

(i) Σ is invertible with delay L from q $\Longleftrightarrow$ there is a function $f_0 : B^{L+1} \to A$, $(b_0, \ldots, b_L) \to a_0$ with $f_\Sigma(a_0, x_1, x_2, \ldots) = (b_0, \ldots, b_L, \ldots)$ for some appropriate $x_1, x_2, \ldots$.

(ii) Σ is invertible with delay L from q $\Longleftrightarrow$ Σ is invertible with delay L from 0.

Proof:

(i) $\Rightarrow$ is clear (take $k = q = 0$ in Definition 9).
$\Leftarrow$: Let $b_0, \ldots, b_n, \ldots, b_{n+L}$ be given. We have to show that we can determine $a_0, \ldots, a_n$ such that we get $G(q, a_0 a_1 \ldots a_n x_{n+1} \ldots x_{n+L}) = b_0 b_1 \ldots b_{n+L}$ for some $x_{n+1}, \ldots, x_{n+L}$. We proceed by induction on n.
If $n = 0$ we choose $b_0' := \gamma(0) - \gamma(q) + b_0$. We get $a_0 = f_0(b_0', b_1, \ldots, b_L)$ such that $G(0, a_0) = b_0'$, which means $\gamma(0) + \delta(a_0) = \gamma(0) - \gamma(q) + b_0$, whence $\delta(a_0) = -\gamma(q) + b_0$. Hence $G(q, a_0) = \gamma(q) + \delta(a_0) = b_0$, as desired. For $1 \leq i \leq n$, let $b_i' := \gamma(0) - \gamma(F(q, a_0 a_1 \ldots a_{i-1})) + b_i$. With $a_i = f_0(b_i', b_{i+1}, \ldots, b_{i+L})$ we get $G(0, a_i) = b_i'$, hence $\gamma(0) + \delta(a_i) = \gamma(0) - \gamma(F(q, a_0 \ldots a_{i-1})) + b_i$, from which we get $\hat{G}(q, a_0 \ldots a_i)$, the last output in $G(q, a_0 \ldots a_i)$, as $\hat{G}(q, a_0 \ldots a_i) =$ $= \gamma(F(q, a_0 \ldots a_{i-1})) + \delta(a_i) = b_i$. For $i > n$ we can do the same, for an arbitrary choice of $b_{n+1}, \ldots, b_{n+L}$, in order to get some appropriate $x_{n+1}, \ldots, x_{n+L}$.
(ii) follows from (i). □

Now we give some necessary and sufficient conditions for the actual construction of a separable system Σ' such that $f_{\Sigma \# \Sigma'} = f_{\Sigma'} \circ f_\Sigma$ transforms

$(a_0,0,0,\ldots) = a_0 z^0$ into $(0,\ldots,0,a_0,0,\ldots) = a_0 z^L$.

In order to simplify the following expressions, we return to the Markov symbols $[\alpha,\beta;n]$ for a separable system with maps $\alpha,\beta,\gamma,\delta$ (Definition 5). The meaning of $[\alpha,\beta;n]\ (\phi_1,\ldots,\phi_{n-1})$ for appropriate maps $\phi_1,\ldots,\phi_{n-1}$ should be clear, too. Remember that we can take $k = 0$ in Definition 5, and we shall do so.

Proposition 10: Let $\Sigma = (Q,A,B,F,G) = (Q,A,B,\alpha,\beta,\gamma,\delta)$ be separable and zero-symmetric.

(i) If there exist zero-symmetric maps $\alpha':Q \to Q$, $\gamma':Q \to A$, and homomorphisms $\beta':B \to Q$, $\delta':B \to A$ such that, for some $L \in \mathbb{N}_0$,

(0) $\delta'\delta = 0$ (= zero map)

(1) $\gamma[\alpha',\beta';0]\delta + \delta'\gamma\beta = 0$

⋮

(*) (i) $\gamma'[\alpha',\beta';i-1]\ (\delta,\gamma\beta,\gamma\alpha\beta,\ldots,\gamma\alpha^{i-2}\beta) + \delta'\gamma\alpha^{i-1}\beta = 0 \qquad (1 \leq i \leq L-1)$

⋮

(L) $\gamma'[\alpha',\beta';L-1]\ (\delta,\gamma\beta,\gamma\alpha\beta,\ldots,\gamma\alpha^{L-2}\beta) + \delta'\gamma\alpha^{L-1}\beta = \mathrm{id}$

then $\Sigma' := (Q,B,A,\alpha',\beta',\gamma',\delta')$ fulfills for each $a_0 \in A$:

(**) $f_{\Sigma \# \Sigma'}\ (a_o,0,0,\ldots) = (0,\ldots,0,x_L,x_{L+1},\ldots)$ with $x_L = a_0$.

(ii) Conversely, if zero-symmetric maps $\alpha':Q \to Q$, $\gamma':Q \to A$, and homomorphisms $\beta':B \to Q$, $\delta':B \to A$ fulfill (**) then they fulfill (*).

Proof: Let Σ start at time 0 in state $q_0 = 0$. The input sequence $a_000\ldots$ gives a sequence of states $q_0 = 0$, $q_1 = \alpha(0) + \beta(a_0) = \beta(a_0)$, $q_2 = \alpha\beta(a_0) + \beta(0) = \alpha\beta(a_0),\ldots,q_i = \alpha^{i-1}\beta(a_0)$ $(i \geq 1)$ and produces the output sequence $b_0 = \gamma(0) + \delta(a_0) = \delta(a_0)$, $b_1 = \gamma(\beta(a_0)) + \delta(0) = \gamma\beta(a_0),\ldots,b_i = \gamma\alpha^{i-1}\beta(a_0)$ $(1 \leq i)$. Suppose there are maps $\alpha',\beta',\gamma',\delta'$ (as specified in the statement) then we get $f_{\Sigma \# \Sigma'}\ (a_0,0,0,\ldots) = (x_0,x_1,\ldots)$ with $x_0 = \gamma'(0) + \delta'(b_0) = \delta'\delta(a_0)$, and, for $1 \leq i$, $x_i = \gamma'[\alpha',\beta',i-1]\ (b_0,b_1,\ldots,b_{i-1}) + \delta'(b_i) = (\gamma'[\alpha',\beta',i-1]\ (\delta,\gamma\beta,\gamma\alpha\beta,\ldots,\gamma\alpha^{i-2}\beta) + \delta'\gamma\alpha^{i-1}\beta)(a_0)$. Hence (**) is fulfilled iff (*) holds. □

Remark: If $L = 0$ then (*) reduces to $\delta'\delta = \mathrm{id}$. This is solvable iff δ is a monomorphism such that $\mathrm{Im}\ \delta$ is a semidirect summand (retract) of B, i.e. if there exists a normal subgroup N of B such that $\mathrm{Im}\ \delta + N = B$, $\mathrm{Im}\ \delta \cap N = \{0\}$. In this case, every $b \in B$ can uniquely be written as $b = i + n$ with $i \in \mathrm{Im}\ \delta$, $n \in \mathbb{N}$. This holds, because if δ has this property then δ is an isomorphism from A to $\mathrm{Im}\ \delta$, having an inverse $\delta':\mathrm{Im}\ \delta \to A$. If $B = \mathrm{Im}\ \delta + N$, $\mathrm{Im}\ \delta \cap N = \{0\}$, we can extend δ' to a homomorphism from B to A by defining $\delta' = \delta'(i + n) := \delta'(i)$. Conversely, if $\delta'\delta = \mathrm{id}$ holds for some homomorphism $\delta':B \to A$ then

$B = \text{Im}\ \delta + \text{Ker}\ \delta\delta'$ via $b = \delta\delta'(b) + (-\delta\delta'(b) + b)$, and $\text{Im}\ \delta \cap \text{Ker}\ \delta\delta' = \{0\}$.

Let us now fix the following

Notation: Let Σ,Σ' be as in Proposition 10, with (*) holding. If $L \geq 1$, let $\Sigma^1 := (Q^L,B,A,\alpha^1,\beta^1,\gamma^1,\delta^1)$ with

$\alpha^1(b_1,\ldots,b_L) := (b_2,\ldots,b_L,0)$

$\beta^1(b) := (0,\ldots,0,b)$

$\gamma^1(b_1,\ldots,b_L) := \gamma'[\alpha',\beta',L-1]\ (b_1,\ldots,b_L)$

$\delta^1(b) := \delta'(b)$

If $L = 0$, take $\Sigma^1 := (\{0\},B,A,0,0,0,\delta')$. Low let, for $L \geq 1$, $\hat{\gamma} := \gamma'[\alpha',\beta',L-1]\ (\gamma,\gamma\alpha,\ldots,\gamma\alpha^{L-1}) + \delta'\gamma\alpha^L$. Then $\Sigma^2 := (Q,A,A,\alpha-\beta\hat{\gamma},\beta,-\hat{\gamma},id)$.

Proposition 11: Let Σ^1 be as above. If Σ^1 starts in state $q_0 = (0,\ldots,0)$ then it assumes states $q_1 = (0,\ldots,0,b_0),\ldots,q_L = (b_0,\ldots,b_L)$, $q_{L+1} = (b_1,\ldots,b_L,b_{L+1}),\ldots$ under the input sequence $(b_0,b_1,b_2,\ldots)$, and produces outputs $c_0 = \delta'(b_0)$, $c_1 = \gamma'[\alpha',\beta',L-1]\ (0,\ldots,0,b_0) + \delta'(b_1),\ldots,$ $c_L = \gamma'[\alpha',\beta',L-1]\ (b_0,\ldots,b_L) + \delta'(b_{L+1})$, $c_{L+1} = \gamma'[\alpha',\beta',L-1]\ (b_1,\ldots,b_{L+1}) + \delta'(b_{L+2}),\ldots$

Proof: We get $q_1 = \alpha^1(0,\ldots,0) + \beta^1(b_0) = (0,\ldots,0,b_0)$, $q_2 = \alpha^1(0,\ldots,0,b_0) + \beta^1(b_1) = (0,\ldots,b_0,b_1)$, and so on. The outputs are

$c_0 = \gamma'[\alpha',\beta',L-1]\ (0,\ldots,0) + \delta'(b_0) = \delta'(b_0)$,

$c_1 = \gamma'[\alpha',\beta',L-1]\ (0,\ldots,0,b_0) + \delta'(b_1)$, etc. □

Before the next proof it is good to have the following

Lemma: If Σ and Σ' are as above, we get for all $i \in \mathbb{N}$ and $b_0,\ldots,b_{i-1}$:
$[\alpha',\beta',L-1]\ (0,\ldots 0,b_0,\ldots,b_{i-1}) = [\alpha',\beta',i-1]\ (b_0,\ldots,b_{i-1})$

Proof: This can either be shown by induction or by recalling that $[\alpha',\beta',L-1]\ (0,\ldots,0,b_0,\ldots,b_{i-1})$ is the state of Σ' after starting at "time" $L-i+1$ and input sequence $b_0,\ldots,b_{i-1}$. Due to time-invariance, this is the same as the state of Σ' after starting at time 0 and input sequence $b_0,\ldots,b_{i-1}$. □

Theorem 6: Let Σ, Σ^1, Σ^2 be as above. Then we have for all $a_0 \in A$ and $L \in \mathbb{N}$
$f_{\Sigma +\!\!\!+\, \Sigma^1 +\!\!\!+\, \Sigma^2}\ (a_0,0,0,\ldots) = (0,\ldots,0,a_0,0,\ldots) = a_0 z^L$.

Proof: We insert $b_0 = \delta(a_0)$, $b_i = \gamma\alpha^{i-1}\beta(a_0)$ (see Proposition 10) into the formulas for c_i in Proposition 11 and get

$c_o = \delta'(b_o) = \delta'\delta(a_o) = 0$, and, for $1 \leq i \leq L$,

$$\begin{aligned}c_i &= \gamma'[\alpha',\beta';L-1]\ (0,\ldots,0,b_0,\ldots,b_{i-1}) + \delta'(b_i)\\ &= \gamma'[\alpha',\beta',i-1]\ (b_0,\ldots,b_{i-1}) + \delta'(b_i) \text{ (using the Lemma)}\\ &= \gamma'[\alpha',\beta',i-1]\ (\delta(a_0),\gamma\beta(a_0),\gamma\alpha\beta(a_0,\ldots,\gamma\alpha^{i-2}\beta(a_0)) + \delta'\gamma\alpha^{i-1}\beta(a_0)\\ &= \begin{cases} 0(a_o) = 0 \ \ldots\ldots & i < L \\ id(a_0) = a_0 \ \ldots & i = L. \end{cases}\end{aligned}$$

If, however, $i > L$, we get

$$\begin{aligned}c_i &= \gamma'[\alpha',\beta',L-1]\ (b_{i-L},\ldots,b_{i-1}) + \delta'(b_i)\\ &= \gamma'[\alpha',\beta',L-1]\ (\gamma\alpha^{i-L-1}\beta(a_0),\ldots,\gamma\alpha^{i-2}\beta(a_0)) + \delta'\gamma\alpha^{i-1}\beta(a_0)\\ &= (\gamma'[\alpha',\beta',L-1]\ (\gamma\alpha^{i-L-1}\beta,\ldots,\gamma\alpha^{i-2}\beta) + \delta'\gamma\alpha^{i-1}\beta)(a_o),\end{aligned}$$

or, with $i = L + j$ $(j > 0)$

$$c_{L+j} = (\gamma'[\alpha',\beta',L-1]\ (\gamma\alpha^{j-1},\gamma\alpha^{j},\ldots,\gamma\alpha^{L+j-2}) + \delta'\gamma\alpha^{L+j-1})(\beta(a_0)) = \hat{\gamma}\ \alpha^{j-1}\beta(a_0).$$

The states of Σ^2 are given by $q_0 = \ldots = q_L = 0$, $q_{L+j} = \alpha^{j-1}\beta(a_0)$ for $j \geq 1$. We show this last statement by induction on j. For $j = 1$ we get $q_{L+1} = (\alpha-\beta\hat{\gamma})(q_L) + \beta(c_L) = (\alpha-\beta\hat{\gamma})(0) + \beta(a_0) = \beta(a_0) = \alpha^0\beta(a_0)$. The induction step from j to j+1 is given by $q_{L+j+1} = (\alpha-\beta\hat{\gamma})(q_{L+j}) + \beta(c_{L+j}) = (\alpha-\beta\hat{\gamma})\alpha^{j-1}\beta(a_0) + \beta\hat{\gamma}\alpha^{j-1}\beta(a_0) = \alpha^j\beta(a_0)$, which was to be shown.

Finally, we can now compute the outputs $d_0,d_1,\ldots$ of $\Sigma + \Sigma^1 + \Sigma^2$. We get, of course, $d_0 = d_1 = \ldots = d_{L-1} = 0$, and then $d_L = -\hat{\gamma}(q_L) + id(c_L) = 0 + c_L = a_0$, and for $j \geq 1$, $d_{L+j} = -\hat{\gamma}(q_{L+j}) + id(c_{L+j}) = -\hat{\gamma}\alpha^{j-1}\beta(a_0) + \hat{\gamma}\alpha^{j-1}\beta(a_0) = 0$. Hence the result. □

Remark: (i) If $L = 0$ then we get already $f_{\Sigma + \Sigma^1}(a_0,0,0,\ldots) = (a_0,0,0,\ldots)$

(ii) It is easy to show that, if $L = 0$,

$$f_{\Sigma + \Sigma^1 + \Sigma^2}(a_0,a_1,a_2,\ldots) = (a_0,a_1,a_2,\ldots)$$

holds. Hence $f_{\Sigma^1 + \Sigma^2}$ is a left inverse of f_Σ.

(iii) The same holds if $L = 1$ and $\delta = id$, for instance, but not in general.

(iv) Also, it can be seen that Theorem 6 is not true in general if Σ is not zero-symmetric.

The construction in Theorem 6 of an inverse $\Sigma^1 + \Sigma^2$ with delay L of Σ can be used for decoupling procedures (see [16]) if A is the direct product of some other groups. If only one component is excited, usually all components respond. Decoupling is the procedure to plug another system $\bar{\Sigma}$ ($\Sigma^1 + \Sigma^2$, for instance) after Σ so that only one output channel responds to the activation of just one input channel. For more on this, see e.g. [16].

Invertibility has to do with regularity in the near-ring N(A). We call a system $\Sigma \in \mathbf{S}(A)$ *regular*, if there is some other system $\Psi \in \mathbf{S}(A)$ such that $\Sigma + \Psi + \Sigma$ has the same input-output behaviour (= transfer function) as Σ itself. $\mathbf{S}(A)$ is regular if every $\Sigma \in \mathbf{S}(A)$ is regular. A system $\Sigma \in \mathbf{S}(A)$ is *idempotent*, if $\Sigma \# \Sigma$ has the same behaviour as Σ. We say that N(A) is *strongly regular* if for each zero-symmetric system $\Sigma \in \mathbf{S}(A)$ there is a zero-symmetric $\Psi \in \mathbf{S}(A)$ such that $\Sigma \# \Sigma \# \Psi$ behaves like Σ. In this case, $\mathbf{S}(A)$ is regular and Ψ is invertible with delay $L = 0$.

Now we get from 9.155, 9.156, 9.158, 9.159 (c), (f) and 9.62 of [14]:

Theorem 7: If $\mathbf{S}(A)$, A finite, is regular then

(i) $\Sigma \# \Psi$ and $\Psi \# \Sigma$ (as defined above) are idempotent systems.

(ii) The set of all systems $X \# \Sigma$ ($X \in \mathbf{S}(A)$) coincides with the set $X \# E$ for an idempotent system E.

(iii) If the zero system Ω (with $f_\Omega = 0$) and the identity system I (with $f_I = id$) are the only idempotent systems in $\mathbf{S}(A)$ then every system in $\mathbf{S}(A)$ is invertible with delay $L = 0$ (and conversely).

(iv) The situation in (iii) happens iff for each $\Sigma \in \mathbf{S}(A)$ there is no $T \in \mathbf{S}(A)$ such that $\Sigma \# T = \Omega$.

(v) If $\mathbf{S}(A)$ is even strongly regular then every $\Sigma \in \mathbf{S}(A)$ is the parallel connection of a constant and finitely many systems which are invertible with zero delay.

The same statements hold, of course, for subnear-rings of $\mathbf{S}(A)$ if they happen to be regular (with the obvious changes).

Finally, we briefly mention the realization problem. For a given function $f:L(A) \to L(B)$ (A,B groups) we want to find a group Q and "separable" functions $F:Q \times A \to Q$, $G:Q \times A \to B$ such that for $\Sigma := (Q,A,B,F,G)$, $f = f_\Sigma$. This is certainly a non-trivial problem, and it is not always solvable, since we know from Example 1 that not every function $f:L(A) \to L(B)$ is of the form $f = f_\Sigma$ for some separable systems Σ.

If f is linear and A,B are abelian then the usual methods and results of linear realization theory, as developed by Kalman in [8], apply. In the other case, one might think about the Arbib-Manes categorical approach (see [2] and [13]). The obvious candidate is the category **C** consisting of groups and maps between these groups. Since every set can be made into a group (see e.g. [9]), **C** has products (= cartesian products), coproducts (= disjoint unions with some group structures), and free objects, hence also free dynamics in the sense of [2]. Hence we get a free realization (left adjoint) with state group A^* (any group structure on A^*) in the sense of [2] (p. 680), but the maps F,G are not separable in general. Also, since f(A) is in general not a group if f is not a homomorphism, we get difficulties with coequalizers and minimal realizations (cf. [13], p. 726). Also, the restrictions to affine maps F,G (i.e. sums of a homomorphism and a constant map) and abelian A does not work: the category **C**' of abelian groups and affine maps between these does not contain (co)products. So this remains an open problem.

REFERENCES

[1] Adler, I., Composition Rings, Duke Math. J. 29 (1962), 607-625.
[2] Bobrow, L.S. and Arbib, M.A., Discrete Mathematics, Saunders, Philadelphia, 1974.
[3] Casti, J.L., Nonlinear System Theory, Academic Press, New York, 1985.
[4] Grätzer, G., Universal Algebra, 2nd edition, Springer-Verlag, New York-Heidelberg-Berlin, 1979.
[5] Hofer, G. and Pilz, G., Near-rings and automata, Proc. Conf. Univ. Algebra, Klagenfurt, Austria, 1982, 153-162.
[6] Holcombe, W.M.L., The syntactic near-ring of a linear sequential machine, Proc. Edinbg. Math. Soc. 26 (1983), 15-24.
[7] Holcombe, W.M.L., A radical for linear sequential machines, Proc. Royal Irish Acad. 84A (1984), 27-35.
[8] Kalman, R.E., Falb, P.L. and Arbib, M.A., Topics in Mathematical System Theory, McGraw-Hill, New York, 1969.
[9] Kertesz, A., Einführung in die transfinite Algebra, Birkhäuser, Basel, 1975.
[10] Lausch, H. and Nöbauer, W., Algebra of Polynomials, North Holland/American Elsevier, Amsterdam, 1973.
[11] Lidl, R. and Pilz, G., Applied Abstract Algebra, Springer-Verlag (Undergraduate Texts in Mathematics), New York-Heidelberg-Berlin, 1984.
[12] Menger, K., Algebra of Analysis, Notre Dame Math. Lectures, No. 3, Notre Dame, Indiana, 1944.
[13] Padulo, L. and Arbib, M.A., System Theory, Saunders, Philadelphia, 1974.
[14] Pilz, G., Near-Rings, 2nd edition, North Holland/American Elsevier, Amsterdam, 1983.
[15] Pilz, G., Strictly connected group automata, submitted.
[16] Sain, M.K., Introduction to Algebraic System Theory, Academic Press, New York, 1981.

Near-rings and Near-fields, G. Betsch (editor)

REDUCED NEAR-RINGS

D. RAMAKOTAIAH and V. SAMBASIVARAO

Department of Mathematics, Nagarjuna University,
Nagarjunanagar-522 510, A.P. India

ABSTRACT

In this paper we introduce a partial order relation in a reduced near-ring and show that the set of all idempotents of a reduced near-ring with identity forms a Boolean algebra under this partial ordering. Further we introduce the notions hyper atom and orthogonal subsets in a reduced near-ring with identity and show that a reduced near-ring with identity is isomorphic to a direct product of near-fields if and only if it is hyper atomic and orthogonally complete.

Introduction: Abian [1] introduced a relation $\leq$ in a reduced ring R by defining $a \leq b$, $a,b \in R$ if and only if $a^2 = ab$ and showed that this relation is indeed a partial order relation in R. In this paper, we examine the possibility of extending the above mentioned partial order in a reduced ring to a reduced near-ring and obtain some of the consequences which of course are generalizations of the results which were already obtained for reduced rings.

This paper is divided into three sections. In section § 1, we show that the set of all idempotents of a reduced near-rings with identity forms a Boolean algebra and obtain some interesting consequences of this result. In section § 2, we introduce the notion of orthogonal subsets in a reduced near-ring and show that the following conditions for a reduced near-ring with identity with partial order $\leq$ are equivalent:

(i) $(N,\leq)$ satisfies the ascending chain condition

(ii) $(N,\leq)$ satisfies the descending chain condition and N is orthogonally complete

(iii) Any orthogonal subset of N is finite

(iv) N satisfies the ascending (descending) chain condition on annihilators.

In section § 3, we define hyper atoms in a reduced near-ring with identity and show that a reduced near-ring with identity is hyper atomic and orthogonally complete if and only if it is isomorphic to a direct product of near-fields.

§ 0

Preliminaries: We recall that an algebraic system N = (N,+,.,o) is a *(left) near-ring* if and only if (i) (N,+,o) is a (not necessarily abelian) group, (ii) (N,.) is a semi-group, (iii) a(b+c) = ab + ac for all a,b,c ∈ N and (iv) oa = o for all a ∈ N. A near-ring without non-zero nilpotent elements is called a *reduced* near-ring. For any x ∈ N, we denote the ideal generated by x by <x>. For other terminology the reader is referred to [2].

We note the following well known results in the theory of near-rings.

Lemma 0.1: Let N be a reduced near-ring. Then ab = o, a,b ∈ N implies ba = o. Further for any a ∈ N, A(a) = {x ∈ N | ax = o} is an ideal of N.

Lemma 0.2: Let N be a reduced near-ring. Then

(i) N has IFP, that is, a,b ∈ N, ab = o implies anb = o for all n ∈ N.

(ii) ne = ene for every idempotent e in N and n in N.

(iii) If N contains an identity, all idempotents in N are central.

A careful observation of (ii) in lemma 0.2 says that a reduced near-ring has a right identity if and only if it has an identity.

§ 1

Lemma 1.1: Let N be a reduced near-ring. If a,b ∈ N, ab = o, then a+b = b+a. Further for all idempotents e,f ∈ N, ef = o implies (e+f)e = e and (e+f)f = f.

Proof: Suppose a,b ∈ N such that ab = o. Then by lemma 0.1, ba = o. This fact and our supposition imply that a(a+b-a-b) = o and b(a+b-a-b) = o. Again by lemma 0.1, we have (a+b-a-b)a = o and (a+b-a-b)b = o. Consequently $(a+b-a-b)^2$ = o and hence a+b-a-b = o. This shows that a+b = b+a. The rest of the proof follows from lemma 0.2 (ii).

Lemma 1.2: Let N be a reduced near-ring with identity. Then for all idempotents e,f ∈ N, ef and e+f-fe are idempotents.

Proof: By lemma 0.2 (iii), all idempotents in N are central and the rest of the proof follows as in the case of ring theory.

We now introduce an order relation in an arbitrary near-ring N by setting $a \leqq b$ for $a,b \in N$ if and only if $a^2 = ab$. If a near-ring is a Boolean ring, than the relation $\leqq$ is a partial order in N. What is the class of near-rings for which the above introducted relation is a partial order? We answer this question in the following.

Theorem 1.3: Let N be near-ring and $\leqq$ a relation on N as defined above. Then N is a reduced near-ring if and only if $\leqq$ is a partial order relation on N. This is the case, $(N,.,\leqq)$ is a partially ordered semigroup.

Proof: Suppose N is a reduced near-ring. Clearly $\leqq$ is reflexive. Suppose $a,b \in N$ such that $a \leqq b$ and $b \leqq a$. Then $a^2 = ab$ and $b^2 = ba$, that is, $a(a-b) = o$ and $b(b-a) = o$. By lemma 0.1, we have $(a-b)a = o$ and $(b-a)b = o$. Consequently $(a-b)^2 = o$. Since N is reduced, $a-b = o$ and hence $a = b$. So $\leqq$ is antisymmetric. Suppose $a \leqq b$ and $b \leqq c$. Then $a^2 = ab$ and $b^2 = bc$, that is, $a(a-b) = o$ and $b(b-c) = o$. This implies $aca = acb$, $a^3c = a^3b$ and $a^2-ac = a(b-c)$. From these facts, we have $a^2(a^2-ac) = o$ and $ac(a^2-ac) = o$. By lemma 0.1, we have $(a^2-ac)a^2 = o$ and $(a^2-ac)ac = o$. Consequently $(a^2-ac)^2 = o$. Since N is reduced, we have $a^2-ac = o$ and hence $a^2 = ac$. So $\leqq$ is transitive. Suppose $x \leqq y$ and $a \leqq b$. Then $x(x-y) = o$ and $a(a-b) = o$. This implies $xa(x-y) = o$ and $ax(a-b) = o$, that is, $xax = xay$ and $axa = axb$. From these facts, we have $xa \leqq yb$. Therefore $(N,.,\leqq)$ is a partially ordered semigroup. Conversely, suppose $\leqq$ is a partial order relation on N. It is enough if we show that $a^2 = o$ implies $a = o$. It is clear that $o \leqq a$. Since $a^2 = o = oa$, we have $a \leqq o$. Hence $a = o$.

We introduce suprema and infima of any subset of a reduced near-ring under the partial order relation $\leqq$ in the usual way. For any subset S of N, sup S (inf S) denotes the supremum (infimum) of S. From now on, N stands for a reduced near-ring with identity, unless stated otherwise, and $\leqq$ always stands for the partial order defined above.

Proposition 1.4: For any subset S of idempotents of N, if sup S (inf S) exists, then it must be an idempotent.

Proof: Suppose sup S = x exists. Then $s = sx$ for all $s \in S$. This implies $sx^2 = s$ and $s \leqq x^2$. Since sup S = x, we have $x \leqq x^2$ and hence $x^2 = x^3$. Now $x(x-x^2) = o$ and $x^2(x-x^2) = o$. Consequently $(x-x^2)^2 = o$ and hence $x = x^2$. Suppose inf S = x. Then $x^2 = xs$ for all $s \in S$. Now $x^2 = xs = xs^2 = x^2s = x^3$. By an argument similar to the above, it can be shown that x is an idempotent, too.

Remark 1.5: In proposition 1.4, if x is a lower bound of S, then x is also an idempotent.

Theorem 1.6: For all idempotents $e,f \in N$, $e+f-fe = \sup\{e,f\}$ and $ef = \inf\{e,f\}$. Thus any two idempotents commute with each other additively.

This result can be verified in the usual way.

The partial order relation $\leq$ in N induces a partial order relation which we also denote by $\leq$, in B, the set of all idempotents of N. In view of theorem 1.6, we have

Theorem 1.7: $(B,\leq)$ is a lattice with o and 1.

Theorem 1.8: Define $\wedge$, $\vee$ and ' on B as follows. Let $e,f \in B$. Define $e \wedge f = ef$, $e \vee f = e+f-fe$ and $e' = 1-e$. Then $(B,\wedge,\vee,',o,1)$ is a Boolean algebra.

Lemma 1.9: Let $a \in N$. Then $aN = A(1-a)$ if and only if a is an idempotent. This is the case, eN is an ideal for each $e \in B$ and $\langle e\rangle = eN$.

Proof: By lemma 0.2 (iii), all idempotents in N are central and it is also clear that for each idempotent e, 1-e is also an idempotent in N. Suppose a is an idempotent in N. Let $x \in A(1-a)$. Then $(1-a)x = o$. By lemma 0.1, we have $x = xa = ax \in aN$. So $A(1-a) \subseteq aN$. Let $x \in aN$. Then $x = an$ for some $n \in N$. This implies $x(1-a) = o$. By lemma 0.1, $(1-a)x = o$ and hence $x \in A(1-a)$. So $A(1-a) = aN$. Conversely, suppose $a \in N$ such that $A(1-a) = aN$. Since $a \in aN$, we have $(1-a)a = o$. By lemma 0.1, $a(1-a) = o$ and hence a is an idempotent. By lemma 0.1 and by the above argument, for each $e \in B$, eN is an ideal and $\langle e\rangle = eN$.

Lemma 1.10: For any $e,f \in B$, $eN \cap fN = efN$.

Proof: By lemma 0.2 (ii), $efN \subseteq eN \cap fN$. Let $x \in eN \cap fN$. Then $x = en$ and $x = fn'$ for some $n,n' \in N$. Now $x = en = een = efn' \in efN$. So $en \cap fN = efN$.

Lemma 1.11: For any $e,f \in B$, $eN + fN = (e+f-fe)N$.

Proof: By lemma 1.2, e+f-fe is an idempotent and by lemma 1.9, eN, fN and (e+f-fe)N are ideals of N. Since $e,f \in B$, we have $e \in eN$ and $f-fe \in fN$. So $e+f-fe \in eN + fN$ and hence $(e+f-fe)N \subseteq eN + fN$. On the other hand, $e = e+fe-fe = e(e+f-fe) = (e+f-fe)e \in (e+f-fe)N$ and hence $eN \subseteq (e+f-fe)N$. Similarly we can show that $fN \subseteq (e+f-fe)N$. So $eN + fN \subseteq (e+f-fe)N$. Hence $eN + fN = (e+f-fe)N$.

Lemma 1.12: The set $S = \{eN \mid e \in B\}$ is a lattice under set inclusion.

Proof follows from lemmas 1.9, 1.10 and 1.11.

Theorem 1.13: Let $S = \{eN \mid e \in B\}$. Then the mapping $\alpha: B \to S$ defined by $\alpha(e) = eN$ is a lattice isomorphism and hence $(S, \subseteq)$ is a Boolean algebra.

Proof: By lemmas 1.10 and 1.11, α is a $\wedge$- and $\vee$- homomorphism. It is easy to verify that α is onto and for any $e,f \in B$, $e \leq f$ in B if and only if $\alpha(e) \subseteq \alpha(f)$. Suppose $e,f \in B$ such that $\alpha(e) = \alpha(f)$. Then $e = fn$ and $f = en'$ for some $n,n' \in N$. Now $f = en' = een' = ef = fnf = e$. So α is one-one. Therefore $\alpha: B \to S$ is a lattice isomorphism and hence S is a Boolean algebra.

We recall that a near-ring N is *biregular* if and only if for each $a \in N$, there exists a central idempotent $e \in N$ such that $\langle a \rangle = \langle e \rangle$.

Theorem 1.14: If N is a biregular reduced near-ring with identity and if $S = \{eN \mid e \in B\}$ and S' is the set of all principal ideals of N, then $S = S'$ and hence S' is a Boolean algebra. Further, the cardinality of the set of all idempotents in N is the same as the cardinality of the set of all principal ideals of N.

Proof: By lemma 1.9, $S \subseteq S'$. Since N is biregular, for each $a \in N$, there exists an idempotent $e \in N$ such that $\langle a \rangle = \langle e \rangle$ and hence by lemma 1.9, $\langle a \rangle = eN$. So $S' \subseteq S$. Hence $S = S'$ and S' is a Boolean algebra by theorem 1.13. Since the cardinality of B is the same as the cardinality of S, by the above argument, the cardinality of B is the same as the cardinality of S'.

§ 2

Throughout this section, N stands for a reduced near-ring with identity. For any subset X of N, A(X) denotes the set $\{n \in N \mid xn = o \text{ for all } x \in X\}$.

Proposition 2.1: Let X be any subset of N. Then the following are equivalent for any $c \in N$.

(i) $c = \sup X$

(ii) c is an upper bound of X and $A(X) \subseteq A(c)$.

In fact, $A(X) = A(c)$.

Proof: (i) implies (ii): Suppose sup X = c. Then $x \leq c$ for all $x \in X$. Let $d \in A(X)$. Then xd = o and $x(c+d) = x^2$ for all $x \in X$. This implies c+d is an upper bound of X. Since sup X = c, we have $c \leq c+d$. Consequently cd = o and hence $d \in A(c)$. So $A(X) \subseteq A(c)$. Let $d \in A(c)$. Then cd = o and $x^2d = xcd = o$ for all $x \in X$. Consequently xd = o. So $d \in A(X)$ and hence A(X) = A(c).
(ii) implies (i): Suppose c is an upper bound of X and $A(X) \subseteq A(c)$. Let d be any upper bound of X. Then for all $x \in X$, $x^2 = xc$ and $x^2 = xd$. This implies $c-d \in A(X) \subseteq A(c)$. Consequently c(c-d) = o and hence $c \leq d$. So sup X = c.

Corollary 2.2: If X is a subset of N such that sup X = b exists, then for all $a \in N$, sup aX exists and equals ab.

Proof: Let $a \in N$. Since $x \leq b$ for all $x \in X$, we have x(x-b) = o. Then axa(x-b) = o and hence $ax \leq ab$ for all $x \in X$. So ab is an upper bound of aX. Let $d \in A(aX)$. Then axd = o and hence xda = o for all $x \in X$. This implies $da \in A(X)$. Since sup X = b, by proposition 2.1, we have $da \in A(b)$. Consequently $d \in A(ab)$. So $A(aX) \subseteq A(ab)$. Hence by proposition 2.1, sup aX = ab.

Definition 2.3: A subset X of non-zero elements of N is said to be *orthogonal* if xy = o for all $x,y \in X$ with $x \neq y$.

Proposition 2.4: Let $X = \{x_i \mid i \in I\}$ and $Y = \{y_j \mid j \in J\}$ be two orthogonal subsets of N such that sup X = x and sup Y = y. Let $Z = \{x_iy_j \mid i \in I, j \in J\}$. Then Z is an orthogonal subset of N and sup Z = xy.

Proof: It is clear that Z is an orthogonal subset of N. Since $x_i \leq x$, we have $x_i(x_i-x) = o$ for each $i \in I$. This implies $x_iy_j(x_i-x) = o$ and hence $x_iy_jx_i = x_iy_jx$ for each $i \in I$ and for each $j \in J$. Consequently $x_iy_j \leq xy_j$ for each $i \in I$ and for each $j \in J$. Since $y_j \leq y$, we have $y_j(y_j-y) = o$ for each $j \in J$. Then $xy_jx(y_j-y) = o$ and hence $xy_j \leq xy$ for all $j \in J$. Hence by the above two arguments, $x_iy_j \leq xy$ for each $i \in I$ and for each $j \in J$. So xy is an upper bound of Z. Let $d \in A(Z)$. Then $x_jy_j = o$ for each $i \in I$ and $j \in J$. This implies $y_jd \in A(X)$ and hence $y_jd \in A(x)$ for each $j \in J$. Consequently $y_jdx = o$ for each $j \in J$ and $dx \in A(Y)$. Hence $dx \in A(y)$ and $d \in A(xy)$. So $A(Z) \subseteq A(xy)$. Hence by proposition 2.1, sup z = xy.

We note the following lemma.

Lemma 2.5: If for any $x,y \in N$ with xy = o, then (x+y)d = xd + yd for all $d \in N$.

Proof: Let $x,y \in N$ be such that $xy = o$. Then by lemma 0.1, $yx = o$. From these facts, we have $x((x+y)d-yd-xd) = 0$ and $y((x+y)d-yd-xd) = o$. Again by lemma 0.1, we have $((x+y)d-yd-xd)x = o$ and $((x+y)d-yd-xd)y = o$ and hence $((x+y)d-yd-xd)(x+y) = o$. Consequently $((x+y)d-yd-xd)^2 = o$. Since N is reduced, $(x+y)d-yd-xd = o$ and hence $(x+y)d = xd+yd$.

As a consequence of lemma 2.5, we can deduce the following.

Proposition 2.6: If X is an orthogonal subset of N, then $\sum_{x \in X} xN$ is a direct sum of N-subgroups of N.

The proof of this proposition is easy and will be omitted.

In order to obtain the main result of this section, we need the following.

Lemma 2.7: If $\{a_i\}$ is a strictly ascending chain or a descending chain of elements in N, then $\{a_i-a_{i+1}\}$ is an orthogonal subset of N.

Proof: Let $\{a_i\}$ be a strictly ascending chain. Suppose $i \neq j$ and $i < j$. Now $a_i(a_j-a_{j+1}) = a_ia_j - a_ia_{j+1} = a_i^2 - a_i^2 = o$ and $a_{i+1}(a_j - a_{j+1}) =$ $= a_{i+1}a_j - a_{i+1}a_{j+1} = a_{i+1}^2 - a_{i+1}^2 = o$. Consequently $(a_j - a_{j+1})a_i = o$ and $(a_j - a_{j+1})a_{i+1} = o$. So $(a_j - a_{j+1})(a_i - a_{i+1}) = o$ and hence $\{a_i - a_{i+1}\}$ is an orthogonal subset of N. By a similar argument, the other case can be disposed of.

Lemma 2.8: If $\{a_1,a_2,\ldots,a_n\}$ is an orthogonal subset of N, then $\sup \{a_1,a_2,\ldots,a_n\}$ exists and equals $a_1+a_2+\ldots+a_n$.

Proof: Write $a = a_1+a_2+\ldots+a_n$. Then $a_ia = a_i^2$ for all $1 \leqq i \leqq n$. So $a_i \leqq a$ for all $1 \leqq i \leqq n$. Suppose $a_i \leqq d$ for all $1 \leqq i \leqq n$. Then $a_i^2 = a_id$. By lemma 2.5, $ad = a_1d + a_2d + \ldots + a_nd = a_1^2 + a_2^2 + \ldots + a_n^2$. Since $\{a_1,a_2,\ldots,a_n\}$ is an orthogonal subset of N, by lemma 2.5, we have $a^2 = a_1^2 + a_2^2 + \ldots + a_n^2$. So $a^2 = ad$ and hence $a = \sup \{a_1,a_2,\ldots,a_n\}$.

Lemma 2.9: If $\{a_i\}$ is an orthogonal subset of N, then $\{\sup \{a_j \mid j \geqq i\}\}$ is a strictly descending chain (provided that the suprema exist) and $\{\sup \{a_i \mid 1 \leqq i \leqq j\}\}$ is a strictly ascending chain.

Proof: Write $c_i = \sup\{a_j \mid j \geq i\}$. Then $c_i \geq a_j$ for all $j \geq i$ and $c_{i+1} \geq a_j$ for all $j \geq i+1$. This implies $c_i \geq a_j$ for all $j \geq i+1$. Therefore c_i is an upper bound of $\{a_j \mid j \geq i+1\}$. So $c_{i+1} \leq c_i$ for each i and hence $\{c_i\}$ is a descending chain. If $c_i = c_{i+1}$ for some i, then $a_i c_i = a_i^2$. By corollary 2.2, $a_i c_{i+1} = a_i \sup\{a_j \mid j \geq i+1\} = \sup\{a_i a_j \mid j \geq i+1\} = 0$ since $\{a_i\}$ is an orthogonal subset of N. This shows that $a_i^2 = 0$ and hence $a_i = 0$, which is a contradiction. So $\{c_i\}$ is a strictly descending chain. Write $d_j = \sup\{a_i \mid 1 \leq i \leq j\}$. By lemma 2.8, $d_j = a_1+a_2+\ldots+a_j$. It is clear that $\{d_j\}$ is an ascending chain. Suppose $d_j = d_{j+1}$ for some j. Then $a_{j+1} = 0$, which is a contradiction. So $\{d_j\}$ is a strictly ascending chain.

The proofs of the following lemmas 2.10, 2.11 and 2.12 are easy and will be omitted.

Lemma 2.10: The descending chain and the ascending chain conditions on annihilators in N are equivalent.

Lemma 2.11: If $\{a_i\}$ is an orthogonal subset of N, then $\{A(a_1,a_2,\ldots,a_i)\}$ is a strictly descending chain of annihilators.

Lemma 2.12: If $\{I_i\}$ is a strictly ascending chain of annihilators in N, then $I_{i+1} \cap A(I_i) \neq (0)$ and if $0 \neq a_i \in I_{i+1} \cap A(I_i)$, then $\{a_i\}$ is an orthogonal subset of N.

Definition 2.13: A near-ring N is said to be *orthogonally complete* if every orthogonal subset of N has supremum.

We now state the main result of this section.

Theorem 2.14: The following conditions on N are equivalent:

(i) $(N,\leq)$ satisfies the ascending chain condition

(ii) $(N,\leq)$ satisfies the descending chain condition and N is orthogonally complete

(iii) Any orthogonal subset of N is finite

(iv) N satisfies the ascending (descending) chain condition on annihilators

The proof of this theorem is a consequence of the lemmas 2.7 to 2.12.

§ 3

In this section we introduce the notion of hyper-atoms and study some of the consequences. Throughout this section, N stands for a reduced near-ring with identity.

Definition 3.1: A non-zero element $a \in N$ is said to be a *hyper-atom* if and only if $x \leqq a$ implies either $x = o$ or $x = a$ and if $an \neq o$ for any $n \in N$, there exists $s \in N$ such that $ans = a$.

Example: Let N' be a near-field. Then N x N' is a reduced near-ring with identity and $\leqq$ is a partial ordering on N x N'. Now for each non-zero element $a \in N'$, (o,a) is a hyper-atom in N x N'.

Lemma 3.2: Let a be a hyper-atom in N. For any $n \in N$, if $an \neq o$, then an is a hyper-atom in N.

Proof: Since $an \neq o$ and a is a hyper-atom in N, there exists $s \in N$ such that $ans = a$. Suppose $x \leqq an$. Then $xs \leqq ans = a$. Since a is a hyper-atom, either $xs = o$ or $xs = a$. Case (i): If $xs = o$, then $xans = o$ and hence $xa = o$. Since $x \leqq an$, we have $x^2 = xan = o$ and hence $x = o$. Case (ii): Suppose $xs = a$. $x \leqq an$ implies $x(x-an) = o$. Then $xs(x-an) = o$ and hence $a(x-an) = o$ and $an(x-an) = o$. By lemma 0.1, $(x-an)x = o$ and $(x-an)an = o$. Consequently $(x-an)^2 = o$. Since N is reduced, $x-an = o$ and hence $x = an$. Suppose $ant \neq o$, $t \in N$. Then $ants = a$ for some $s \in N$ since a is a hyper-atom. Now $(ant)sn = (ants)n = an$. Hence an is a hyper-atom.

Lemma 3.3: Let a be a hyper-atom in N. Then there exists an element $s \in N$ such that $a = asa$, that is, a is regular in the sense of Von Neumann.

Proof: Let $a \in N$ be a hyper-atom. Then $a^2 \neq o$ and there exists an element $s \in N$ such that $a^2s = a$. This implies $a(a-asa) = o$ and $asa(a-asa) = o$. By lemma 0.1, $(a-asa)a = o$ and $(a-asa)asa = o$ and hence $(a-asa)^2 = o$. Since N is reduced, we have $a = asa$.

Lemma 3.4: Let a be a hyper-atom in N. Then there exists an element $s \in N$ such that as is an idempotent hyper-atom in N.

Proof: Follows from lemma 3.2 and lemma 3.3.

The proofs of the following lemmas 3.5 and 3.6 are easy and will be omitted.

Lemma 3.5: The set of all idempotent hyper-atoms in N is an orthogonal subset of N.

Lemma 3.6: Let $\{e_i\}_{i \in I}$ be the set of all idempotent hyper-atoms in N. Then for every $i \in I$, $N_i = \{ne_i \mid n \in N\}$ is a subnear-field of N and $N_i \cap N_j = (o)$ if $i \neq j$.

Definition 3.7: N is said to be *hyper-atomic* provided for each non-zero element $n \in N$, there exists a hyper-atom $a \in N$ such that $a \leqq n$.

Lemma 3.8: Let N be a hyper-atomic near-ring and let $\{e_i\}_{i \in I}$ be the set of all idempotent hyper-atoms in N. Then for each non-zero element $q \in N$, there exists an idempotent hyper-atom e_k such that $qe_k \neq o$. Moreover, for each $n \in N$, the sup $\{ne_i \mid i \in I\}$ exists and equals n.

Proof: Since N is hyper-atomic, if $q \neq o$, then there exists a hyper-atom $a \in N$ such that $a \leqq q$. Then $a^2 = aq \neq o$. By lemma 3.4, there exists $s \in N$ such that as is an idempotent hyper-atom. Now we will show that $qas \neq o$. Suppose $qas = o$. Then $aqas = o$. This implies $a^2as = 0$ and hence $as = o$, a contradiction. So $qas \neq o$. Thus there exists an idempotent hyper-atom $as = e_k$ in N such that $qe_k \neq o$. Since $(ne_i)^2 = ne_in$ for all $i \in I$, n is an upper bound of $\{ne_i \mid i \in I\}$. Suppose u is an upper bound of $\{ne_i \mid i \in I\}$. Then $nne_i = nue_i$ for each $i \in I$. If $n \nleqq u$, then $n^2 - nu \neq o$. By the above argument, there exists an idempotent hyper-atom e_i in N such that $(n^2-nu)e_i \neq o$, which is a contradiction. Hence $\sup \{ne_i \mid i \in I\} = n$.

Lemma 3.9: Let N be hyper-atomic and let $\{e_i\}_{i \in I}$ be the set of all idempotent hyper-atoms in N. Then $\alpha : n \to (ne_i)_{i \in I}$ is an isomorphism of N onto a subnear-ring of the direct product $\prod_{i \in I} N_i$ of near-fields N_i.

Proof: Clearly α is a homomorphism. Suppose $\alpha(n) = o$. Then $(ne_i)_{i \in I} = o$. This implies $\sup \{ne_i \mid i \in I\} = o$. By lemma 3.8, $n = o$. Hence α is a monomorphism.

We now state and prove the main result of this section.

Theorem 3.10: N is isomorphic to a direct product of near-fields if and only if N is hyper-atomic and orthogonally complete.

Proof: Suppose $N = \prod_{i\in I} F_i$, where each F_i is a near-field. Let $o \neq n \in N$. Then $n = (n_i)_{i\in I}$ and there exists $j \in I$ such that $n_j \neq o$. Now $x = (o,o,\ldots,o,n_j,o,\ldots,o)$ is a hyper-atom in N and $x \leq n$. Let X by any orthogonal subset of N. Let $k_i:N \to F_i$ be the canonical projection. Write $a = \{a_i\}_{i\in I}$, where $a_i = k_i(x)$ if $k_i(x) \neq o$ for some $x \in X$ and $a_i = o$ otherwise. Since X is orthogonal, it follows that $a = \sup X$. Conversely, suppose N is hyper-atomic and orthogonally complete. By lemma 3.9, it is enough if we show that the mapping $\alpha:N \to \prod_{i\in I} N_i$ of lemma 3.9 is onto. Let $(n_ie_i) \in \prod_{i\in I} N_i$. By lemma 3.5, the set $\{n_ie_i \mid i \in I\}$ is an orthogonal subset of N. Since N is orthogonally complete, $\sup \{n_ie_i \mid i \in I\} = h$ exists. Now by corollary 2.2 and lemma 3.5, for each $j \in I$, we have $he_j = e_jh = e_j \sup \{n_ie_i \mid i \in I\} = n_je_j$. Now $\alpha(h) = (he_i)_{i\in I} = (n_ie_i)_{i\in I}$. So α is onto.

REFERENCES

[1] Abian, A., Direct product decomposition of commutative semisimple rings, Proc. Amer. Math. Soc. 24 (1970).

[2] Pilz, G., Near-rings, North-Holland, Amsterdam 1977.

[3] Raphael, R. and Stephenson, W., Orthogonally complete rings, Can. Math. Bull. Vol. 20, No. 3 (1977).

Near-rings and Near-fields, G. Betsch (editor)

NON-COMMUTATIVE SPACES AND NEAR-RINGS INCLUDING PBIBD's PLANAR NEAR-RINGS AND NON-COMMUTATIVE GEOMETRY

Maic SASSO-SANT

INTRODUCTION

This work shows some relations between near-rings, non-commutative geometry and combinatorics (PBIBD's and BIBD's).
Many interesting problems for fruitful future research in this area may be found in [2] or [4] . We start with some prerequisites in non-commutative geometry.

I NON-COMMUTATIVE GEOMETRY

I.1.DEFINITION: A space is a structure $(X, \mathcal{B})$ where X is a non-empty set (elements are called points) and $\mathcal{B}$ is a subset of the power set of X.

I.2.DEFINITION: An L-space is a structure $(X, \sqcup)$ consisting of a non-empty set X and a mapping

$$\sqcup : X \times X \longrightarrow \mathcal{P}(X) \quad \text{(power set of X)}$$
$$(x , y) \longmapsto x \sqcup y$$

possessing the following properties:

(L0) $x \sqcup x = \{x\}$ for all $x \in X$

(L1) $x, y \in x \sqcup y$ for all $x,y \in X$ (incidence condition)

(L2) $z \in (x \sqcup y) \setminus \{x\} \Rightarrow x \sqcup y = x \sqcup z$ for all $x,y,z \in X$, $x \neq y$ (exchange condition)

The point x is called base point (Aufpunkt) of $x \sqcup y$ denoted by $x \prec x \sqcup y$. A point set of the form $x \sqcup y$ with $x \neq y$ is called line (Linie).

I.3.DEFINITION: An LG-space is an L-space with

(L3) $x \sqcup y = y \sqcup x = x \sqcup z \Rightarrow x \sqcup z = z \sqcup x$ $(x,y,z \in X)$

Lines with (L3) are called *straight lines* (Geraden).

I.4.DEFINITION: Let $(X, \sqcup)$ be an L-space and $\mathfrak{L} := \{ x \sqcup y / x,y \in X \}$.
An equivalence relation $\|$ on $\mathfrak{L}$ is called *parallelism* if

(P0) $x \sqcup x \; \| \; y \sqcup y$ for all $x,y \in X$

(P1) If $L \in \mathfrak{L}$ and $x \in X$ then there exists exactly one L' such that $x \in L'$ and $L' \| L$ (*Euclidean parallelism condition*)

(P2) If $x \sqcup y \; \| \; x' \sqcup y'$ then $y \sqcup x \; \| \; y' \sqcup x'$ $(x,y,x',y' \in X)$
(*symmetry condition*)

I.5.DEFINITION: An *LP-space* is a triple $(X, \sqcup, \|)$ with the properties

(a) $(X, \sqcup)$ is an L-space

(b) $\|$ is a parallelism

I.6.DEFINITION: An LP-space is called *selfadjoint* (selbstadjungiert) if

(P2) $x \sqcup y \; \| \; y \sqcup x$ for all $x,y \in X$
(*strong symmetry condition*)

An equivalence class $x \sqcup y \; \|$ is called *direction*.

I.7.DEFINITION: A subset $U \subseteq X$ of an LP-space $(X, \sqcup, \|)$ is called subspace, if:

(U1) $x,y \in U \Rightarrow x \sqcup y \subseteq U$

(U2) $x \in U,\ L \in \mathfrak{L},\ L \subseteq U \Rightarrow (x \| L) \subseteq U$

X, $\{x\}$ $(x \in X)$, $\emptyset$ are called trivial subspaces of $(X, \sqcup, \|)$.
An LP-space is called *primitive*, if only the trivial subspaces exist; otherwise it is called imprimitive.

I.8.DEFINITION: A *skewaffine space* is an LP-space $(X, \sqcup, \|)$ where the *Tamaschke-condition* holds:

Tam: If $\{x,y,z\}_{\neq}$, $x \sqcup y \; \| \; x' \sqcup y'$ $(x',y' \in X)$
then there exist $z' \in X$ such that
$x \sqcup z \; \| \; x' \sqcup z'$ and $y \sqcup z \; \| \; y' \sqcup z'$

The following closure condition is called *parallelogramm closure* condition:

PGM: If x,y,z are pairwise different points of an LP-space $(X, \sqcup, \|)$ then there exists $w \in X$ such that $x \sqcup y \; \| \; z \sqcup w$ and $x \sqcup z \; \| \; y \sqcup w$

An LP-space is called <u>Desarguesian</u> if the following Desarguesian closure condition holds:

<u>Des:</u> If u,x,y,z are pairwise different points of the set X and if $u \sqcup x = u \sqcup x'$ $(x' \in X)$ then there exist different points $y', z' \in X$ such that $u \sqcup y = u \sqcup y'$, $u \sqcup z = u \sqcup z'$, $x \sqcup y \parallel x' \sqcup y'$ $x \sqcup z \parallel x' \sqcup z'$, $y \sqcup z \parallel y' \sqcup z'$.

<u>I.9.DEFINITION:</u> A subset $F \subseteq X$ of an L-space $(X, \sqcup)$ is called <u>flat</u>, if

(F1) $x,y \in L \in \mathcal{L}$, $x,y \in F$, $x \neq y \Rightarrow L \subseteq F$

(F2) $x,y \in F$, $x \neq y$; $x \in L \in \mathcal{L}$, $y \in M \in \mathcal{L}$; $L,M \nsubseteq F \Rightarrow |L \cap M| \leq 1$

An imprimitive skewaffine space is called <u>semiaffine</u>, if every subspace is flat.

<u>II GEOMETRIC AND PLANANR NEAR-RINGS</u>

<u>II.1.DEFINITION:</u> A near-ring $(N,+,\cdot)$ with the distributive law $(a+b)c = ac + bc$ is called <u>geometric</u> [5] [6], if: (1)N is zerosymmetric, i.e. $oa = ao = o$ for all $a \in N$. (2)For all $a \in N$ there exists $1 \in N$ such that $a1 = a$. (3)If $o \neq a = bc$ then $aN = bN$ $(a,b,c \in N)$.

For the definition of a planar near-ring, see [3].
The following theorems are true. Some of them were proved independently by R.Scapellato.

<u>THEOREM 1:</u> IF N is a geometric near-ring and $Na \neq \{o\}$ $(a \in N)$, then Na is also a geometric near-ring.

<u>THEOREM 2:</u> If N is a planar near-ring and $|Na/_{\equiv}| \geq 3$ $(a \in N)$, then Na is a planar near-ring, too.

<u>THEOREM 3:</u> If N is a geometric near-ring, I a proper ideal of N, then $N/_I$ is a geometric near-ring.

<u>THEOREM 4:</u> (André): Every finite planar near-ring is geometric.

<u>THEOREM 5:</u> Every integral planar near-ring is geometric.

The last theorem gives a geometric interpretation of the B_a's in the structure theorem of a planar near-ring:

$$x \sqcup y = \overline{B}_{y-x} + x, \text{ where } \overline{B}_{y-x} = B_{y-x} \cup \{o\}$$

III PBIBD's, PLANAR NEAR-RINGS AND NON-COMMUTATIVE GEOMETRY

For the definition of a tactical configuration see [3] .

III.1.DEFINITION: Start with a tactical configuration $(V, \mathcal{B}, \varepsilon)$. Let $\mathcal{P} = \{A \subset V / |A| = 2\}$ and let $\mathcal{A} = \{A_1, A_2, \ldots\ldots, A_m\}$ be a partition of $\mathcal{P}$. Then $\mathcal{A}$ is an association scheme on V if, given $\{x,y\} \in A_h$, the number $z \in V$ such that $\{x,z\} \in A_i$ and $\{y,z\} \in A_j$ depends only on h,i and j and not on x and/or y. That is, there is a number p^h_{ij} such that for $\{x,y\} \in A_h$ there are exactly p^h_{ij} distinct elements $z \in V$, such that $\{x,z\} \in A_i$ and $\{y,z\} \in A_j$. Association schemes with $m = 1$ or $m = v(v-1)/2$, $v = |V|$ are declared "uninteresting".

III.2.DEFINITION: Suppose $(V, \mathcal{B}, \varepsilon, \mathcal{A})$ is a tactical configuration with association scheme. This structure is a PBIBD (partially balanced incomplete block design), if:

a.) for each $A_i \in \mathcal{A}$, there exists a number λ_i such that $\{x,y\} \in A_i$ implies x,y belong to exactly λ_i blocks of $\mathcal{A}$.

b.) for each $A_i \in \mathcal{A}$, there exists a number n_i, such that for each $x \in V$, there are exactly n_i distinct elements $y \in V$, such that $\{x,y\} \in A_i$.

So a PBIBD has parameters p^h_{ij}, λ_i, n_i in addition to those of a tactical configuration.

Now the question is: "Which planar near-rings yield PBIBD's ?"

III.3.LEMMA 1: Let N be a finite planar near-ring, $a \in N^*$ $(N^* = N \setminus \{0\})$ fixed, $b \in N$. Then $\{aN + b \mid b \in N\}$ gives a tactical configuration

III.4.LEMMA 2: Let N be a finite planar near-ring. Then $\{(\bigcup_{a \in A} aN) + b \mid b \in N\}$ for $A \subset N^*$, gives a tactical configuration, if at least one $a \notin C_1(N)$, where $C_1(N) := \{x \in N / xN = -xN + x\}$.

III.5.DEFINITION: A near-ring N is called symmetric, iff

$$xN = -xN \qquad \text{for all } x \in N$$

III.6.THEOREM: Suppose $(N, +, \cdot)$ is a finite, symmetric, planar near-ring and $\mathcal{B}$ is the set of all blocks $\{aN + b \mid a \in N^*, b \in N\}$. Then the following is true:

1.) $(N, \mathcal{B})$ is a PBIBD

2.) $(N, aN + b)$ is a PBIBD $(a \in N^*$ fixed)

3.) $(N, (\bigcup_{a \in A} aN) + b)$ for an $A \subset N^*$ is a PBIBD (at least one $a \notin C_1(N)$)

REMARK: If $C_1(N) = \{o\}$ in 6.1.), then the PBIBD is a BIBD and $(N, \mathcal{B})$ is a Desarguesian, skewaffine space with parallel closure condition (PGM).
If $C_1(N) \neq \{o\}$ then $(N, \mathcal{B})$ is a Desarguesian, semiaffine space with PGM.
If one considers 6.2.) then the blocks $aN + b$ are lines of one direction of this space and in 6.3.) the blocks are the union of lines with different directions, but fixed base point.
For parameters n_i, λ_i, p^h_{ij} of a PBIBD over a planar near-ring we have:

a.) $n_i = |x \sqcup y| - 1 = k - 1$ for all $i = 1, \ldots\ldots, m$

b.) 1.) $p^h_{ij} \geq 1 \Rightarrow p^j_{hi} \geq 1$ and $p^i_{hj} \geq 1$

2.) $p^i_{ii} = o$, $p^g_{gg} = 1$, where $k = 3$

3.) $p^g_{gg} = k - 2$, if $k \geq 3$

4.) $p^i_{gg} = o$, $i \neq g$

5.) $p^g_{jj} = o$, $j \neq g$, $p^g_{hj} \leq 1$ (semiaffine space)

6.) $p^j_{hh} = k - 2$, $j \neq g \neq h$, $j \neq h$, $k \geq 3$

The index g denotes a straight line direction, that means if $g = x \sqcup y$, then $x \sqcup y = y \sqcup x$.

EXAMPLES: 1.) Consider $(\mathbb{Z}_5, +)$ with the usual modulo addition and the fixed point free automorphism group $\Phi = \{id, \varphi\}$ with $\varphi(x) = -x$. This is a very interesting example, because the blocks are circles of a Möbius plane, they form a BIBD, they are lines of a Desarguesian skewaffine space with PGM, and all the lines form a PBIBD.
2.) We consider $(\mathbb{Z}_9, +)$ again with the usual modulo addition and the fixed point free automorphism group $\Phi = \{id, \varphi\}$, with $\varphi(x) = -x$. The blocks form a PBIBD and they are lines of a Desarguesian semiaffine space with PGM. See also [2] .

Starting with Lemma 1 one may ask the question, which planar near-rings lead to symmetric block designs? The answer is still unknown. The reader may find examples of special symmetric block designs, Hadamard designs and projective planes in [4] .

IV NON-COMMUTATIVE GEOMETRY OVER $(\Gamma, M(\Gamma))$

We begin with an additive, not necessarily abelian group $(\Gamma, +)$. Let $M(\Gamma)$ denote a subset of the set of all functions from Γ to Γ. We make the following definition:

$$\sqcup : \Gamma \times \Gamma \longrightarrow \mathcal{P}(\Gamma)$$
$$(x,y) \longmapsto x \sqcup y := M(\Gamma)(y-x) + x,$$

and we ask the question, under which conditions the space $(\Gamma, M(\Gamma), \sqcup)$ becomes a non-commutative space?

It's a generalization of the concept of Theobald [5] [6] .

If one considers the special functions $M(\Gamma) := \{\gamma_a \in M(\Gamma) / \gamma_a(n) = na\}$ where $(\Gamma, +, \cdot)$ is a near-ring and $n, a \in \Gamma$, then we get the definition: $x \sqcup y = (y - x)\Gamma + x$.

Among many other results the following is important:

IV.1.THEOREM: If $(N, +, \cdot)$ is a finite, zero-symmetric, quasi-integral near-ring, then $(N, +, \cdot)$ is a geometric near-ring, where quasi-integral means $a \cdot b = o \Rightarrow a = o \vee b = o$.

This section motivates the following section V.

V STRONGLY GEOMETRIC NEAR-RINGS FROM AUTOMORPHISM GROUPS

V.1.DEFINITION: A geometric near-ring N is called strongly geometric, if N contains a right unit 1 (i.e. $n \cdot 1 = n$ for all $n \in N$).

V.2.DEFINITION: Consider an automorphism group Φ on a group $(\Gamma, +)$. An orbit $\Phi(\gamma), \gamma \in \Gamma$, is called principal, if $|\Phi(\gamma)| = |\Phi|$, otherwise it is called secondary orbit. We denote with H the family of all principal orbits and with N the family of all secondary orbits. From every principal orbit we choose representatives. Now we define a product $\cdot$ on $(\Gamma, +)$ in the following way:

$$\alpha \cdot \beta := \begin{cases} 0 & , \text{ if } \beta = 0 \\ \varphi_\beta(\alpha) & , \text{ if } \beta \in \Phi(\alpha_j) \in H \\ 0 & , \text{ if } \beta \in \Phi(\alpha_i) \in H \end{cases}$$

where φ_β is the only automorphism, with $\varphi_\beta(\alpha_j) = \beta$.

<u>V.3.THEOREM:</u> With the last definition $(\Gamma, +, \cdot)$ becomes a strongly geometric near-ring, with distributive law $(a + b)\cdot c = a\cdot c + b\cdot c$.

<u>V.4.THEOREM:</u> The near-ring $(\Gamma, +, \cdot)$ is quasi-integral.

The proofs of all theorems in this section may be found in [4] .

REFERENCES

[1] André J. Zur Geometrie der Frobeniusgruppen. Mathem. Zeitschrift 154 (1977) S.159 - 168

[2] Betsch G.-Clay J.R. Block designs from Frobenius groups and planar near-rings. Proc. Conf. finite groups (Park City, Utah) Acad. Press (1976), pp. 473 - 502

[3] Pilz G. Near-Rings The Theory and its Application North Holland (1983)

[4] Sasso-Sant M. Nichtkommutative Räume und Fastringe. Diplomarbeit Universität Saarbrücken (1986)

[5] Theobald E. Nichtkommutative Geometrien über Fastringen. Diplomarbeit Universität Saarbrücken (1981)

[6] Theobald E. Near-rings and non-commutative geometry. In: Proceed. Conf. on Near-rings and Near-fields, ed. by G. Ferrero and C. Ferrero Cotti, San Benedetto del Tronto, pp. 211 - 218 (1981).

Author's address
Maic Sasso-Sant
IV. Gartenreihe 18
663 Saarlouis
West - Germany

Near-rings and Near-fields, G. Betsch (editor)

ON GEOMETRIC NEAR-RINGS

Raffaele SCAPELLATO

Dipartimento di Matematica, Università di Parma, Parma, Italy*

In this paper we give a short account of our results on geometric near-rings.

A zero-symmetric near-ring is said geometric if it satisfies the following conditions:

(G1) for every $a \in N$ there is an $e \in N$ such that $ea=a$;

(G2) for every $a, b \in N$, if $ab \neq 0$ then $Nab=Nb$.

An L-space (see [1]) is a pair (S,U), where S is a set and U is a map from $S \times S$ to $P(S)$, such that, if $U(x,y)$ is denoted by xUy, the following conditions hold for every $x, y, z \in S$:

(L0) $xUx=\{x\}$;

(L1) $x, y \in xUy$;

(L2) if $z \in xUy \setminus \{x\}$ then $xUz=xUy$.

Theobald [6] proved that, if N is a near-ring and U is defined by $xUy=N(y-x)+x$, then the pair (N,U) is a L-space iff N is geometric. This is a link between near-rings and non-commutative geometry in the sense of André (see [1]). For geometric applications of near-rings see also, e. g., [2], [3], [4].

The following results (except Th.3) have been proved in [5].

LEMMA 1. If I is a proper ideal of a geometric near-ring N, then $IN=0$ and N/I is geometric.

THEOREM 2. A finite non-constant near-ring is geometric iff it is strongly monogenic.

In the next statement we will denote by F the set of all non-zero maps $x \mapsto ax$.

THEOREM 3. A near-ring N is both geometric and strongly monogenic iff F is a group.

---

*Work supported by the italian MPI.

PROOF. Suppose that N is both geometric and strongly monogenic. Let $a \in N$ with $aN=N$, let $e \in N$ with $ea=a$. If $b \in N$ we get $b=ac$ for some c, so $eb=eac=ac=b$, and e is a left identity.

By (G2) we have $e \in Ne=Nae$, so $e=dae$ for some d. Since da is clearly a left identity, it is now easy to conclude that F is a group.

Conversely, if F is a group, let $a \in N$ such that $aN \neq 0$. There is $b \in N$ such that ab is a left identity: now $N=abN \subset aN$ and N is strongly monogenic.

Finally, let a, $b \in N$ and $ab \neq 0$. For $c \in N$, if $cb=0$ it is clear that $cb \in Nab$; otherwise the map $x \mapsto cx$ belongs to F and it is equal to the map $x \mapsto dax$ for a suitable d. We get $cb=dab$, thus $cb \in Nab$, and (G2) follows. This completes the proof.

REFERENCES

[1] André, J., Nichtkommutative Geometrie und verallgemeinerte Hughes-Ebenen, Math. Z. 177 (1981), 449-462.
[2] Ferrero, G., Stems planari e BIB-disegni, Riv. Mat. Univ. Parma, (2) 11 (1970), 79-96.
[3] Ney, H., Planar near-rings and their relations to some non-commutative spaces, in Proc. Conf. on Near-rings and Near-fields, Harrisonsburg, 1983.
[4] Sasso-Sant, M., Nichtkommutative Raume und Fastringe, to appear.
[5] Scapellato, R., On geometric near-rings, Boll. Un. Mat. It., (6) 2-A (1983), 389-393.
[6] Theobald, E., Near-rings and non-commutative geometry, in Proc. Conf. on Near-rings and Near-fields, S. Benedetto del Tronto, 1981.

Near-rings and Near-fields, G. Betsch (editor)

A TERNARY INTERPRETATION OF THE INFRA-NEAR RINGS

Mirela Stefănescu

1. INTRODUCTION

In studying some generalizations of rings as infra-near rings (see Stefănescu [11]),weak rings (see Climescu [3,4] , Cupona [5]),prerings (see Janin [6,7]),we have noted that certain facts can be better explained by means of a ternary interpretation of the additive composition law. This interpretation might be of considerable use in studying affine infra-near rings (Stefănescu [10]) as well as in studying ideals of those algebraic systems. Given the development of the theory of ternary groups and ternary rings,such an interpretation using a ternary operation instead of the binary addition may be interesting in itself.

First we recall some definitions and properties of generalizations of rings involved in our considerations.

(1.1) Definition. A left infra-near ring is a triple (I,+,.), where I is a nonempty set , + and . are binary operations on I (addition and multiplication) , such that the following conditions are fulfilled:

(i) (I,+) is a group (generally,noncommutative);

(ii) (I,.) is a semigroup;

(iii)the multiplication is left infra-distributive with respect to the addition, i.e.

$$x \cdot (y + z) = x.y - x.0 + x.z \text{ , for all } x,y,z \in I.$$

The concept of right infra-near ring is analogous.

If $x.0 = 0$ for all $x \in I$,then we obtain a left near-ring. If (I,+,.) is a left and a right infra-near ring with a commutative addition, we obtain an equivalent definition of a <u>weak ring</u>, in the general case. We mention that the weak rings were first introduced by Al.Climescu [3] in 1961 in a particular case, when $x.0 = 0.x = x$ for all $x \in I$. In 1964, Climescu [4] has noted a general possibility of obtaining weak rings starting from a given ring. The particular case was rediscovered by G.Cupona [5] in 1971, who called it "quasi-ring".

Let $(I,+,.)$ be a left infra-near ring. One can easily check that for each $x \in I$ the mapping $f_x : I \longrightarrow I$, given by $f_x(y) := x.y - x.0$, is an endomorphism of the additive group $(I,+)$. This almost obvious remark may help us to study the left infra-distributive associative multiplications over an additive group (see Stefănescu [12]), the discussion being unexpectedly interesting.

One of the first examples of right near-rings was the set of all affine transformations over a vector space or over an Abelian additive group endowed with the pointwise addition and the mapping composition. This is also a left infra-near ring and a few properties of this set come from its infra-near ring structure. It is surprising how much algebra can be obtained only by considering this weaker structure (see [10], a paper in which we have studied and generalized the infra-near ring of affine transformations from this point of view).

If a left infra-near ring satisfies the condition :

$0.x = 0$, for all $x \in I$,

then we call it a <u>left C-infra-near ring</u>, while in the case :

$0.x = x$, for all $x \in I$,

we call it a <u>left Z-infra-near ring</u>.

We give two examples of finite left C-infra-near rings:

1°. The cyclic group $(Z_3,+)$, with the multiplication : $0.x = 0$, $1.x = 2.x = x$, for all $x \in Z_3$.

2^0. The symmetric group $S_3 = \{0,a,b = 2a,x,y = x+a,z = x+b\}$, where $3a = 2x = 0$ and $x+a = b + x$, with the multiplication given by one of the following three possibilities:

(I) $0.s = 0, a.s = x.s = y.s = z.s = a$, $b.s = s$, for all $s \in S_3$;

(II) $0.s = 0, a.s = x.s = y.s = a$, $b.s = z.s = s$,for all $s \in S_3$;

(III) $0.s = 0, a.s = x.s = a$, $b.s = y.s = z.s = s$,for all $s \in S_3$.

Many other examples of left infra-near rings and a general study of such algebraic systems can be found in [11] . Let us remark that , if (I,+) is a group,then (I,+,+) is a left and a right infra-near ring.

In a left infra-near ring we do not obtain 0.0 = 0. For example,if we consider $I = Z^5$ and the binary operations :

$$x + y = (x_1+y_1, x_2+y_2, x_3+y_3, x_4+y_4, x_5+x_2.y_3+x_4.y_1+y_5),$$

$$x.y = (x_1.y_1, x_2.y_2, a_3, a_4, a_5),$$

with a_3,a_4,a_5 fixed elements in Z,then we obtain a left infra-near ring for which $0.0 = (0,0,a_3,a_4,a_5)$,while $0 = (0,0,0,0,0)$.

The real difference between the general case (see [8]) in which the multiplication is an "independent" operation on an additive group, - that means the multiplication does not fulfil any conditions with respect to the addition, - and our case (in which the multiplication is left infra-distributive over the addition) is perceptibly in the theory of congruences,hence in defining homomorphisms and ideals of such structures.

(1.2) Definition. A two-sided ideal of a left infra-near ring (I,+,.) is a nonempty subset J of I satisfying the conditions:

(i) (J,+) is a normal subgroup of the group (I,+);

(ii) $x.j - x.0 \in J$, for all $x \in I$ and $j \in J$;

(iii) $(j + x).y - x.y \in J$, for all $x,y \in I$ and $j \in J$.

If J only satisfies the conditions (i) and (ii) (respectively, (i) and (iii)) , J is called a left ideal (respectively,a right ideal) of I.

If $(I,+,.)$ is a left near-ring, hence $x.0 = 0$ for all $x \in I$, then we obtain the definitions of the two-sided, left and right ideals of a left near-ring. The similitude is amazing. But in fact the situation is more complicated, since even the two sided ideals of a left infra-near ring are not closed under the multiplication, hence generally they are not subinfra-near rings of I. For left near-rings a similar assertion is true only for the right ideals (see Pilz [9], 1.28 and 1.33).

Of course, if J is a two-sided ideal of the left infra-near ring I, then $I/J = \{x+J \ / \ x \in I\}$ is a left infra-near ring with respect to the usual operations between cosets.

(1.3) <u>Definition</u>. Let $(I,+,.)$ and $(I',+,.)$ be two left infra-near rings. A mapping $\varphi : I \rightarrow I'$ is called a <u>homomorphism of left infra-near rings</u> if the following two conditions are fulfilled :

(i) $\varphi(x + y) = \varphi(x) + \varphi(y)$,

(ii) $\varphi(x.y) = \varphi(x).\varphi(y)$, for all $x,y \in I$.

The <u>kernel of φ</u>, $\ker \varphi = \{x \in I \ / \ \varphi(x) = 0'\}$, is a two-sided ideal of I, while the <u>image of φ</u>, $\operatorname{Im} \varphi = \{\varphi(x) \ / \ x \in I\}$, is a subinfra-near ring of I'. Moreover, the two-sided ideals of I are exactly the kernels of the homomorphisms from I to another left infra-near ring .

(1.4) <u>Definition</u>. Let I be a left C-infra-near ring and $(M,+)$ be a group. If there exists a mapping $\mu : M \times I \rightarrow M$, $\mu(m,x) = m.x$, such that the following axioms are fulfilled :

(i) $m.(x + y) = m.x - m.0 + m.y$, for all $m \in M$ and $x,y \in I$;

(ii) $(m.x).y = m.(xy)$, for all $m \in M$ and $x,y \in I$;

(iii) $0.x = 0$, for all $x \in I$,

then we call $(M,+,\mu)$ a <u>right I-group</u>.

Obviously I is itself a right I-group and each right ideal of I is a right I-subgroup , while a right I-subgroup of I is not a right ideal of I, even in the case when it is a normal subgroup of (I,+).

Denote by $D = \{ d \in I \ / \ d.0 = 0 \}$ and $W = \{ w \in I \ / \ w.0 = w \}$, the subset of all left distributive elements of I and the subset of all weakly left distributive elements of I, respectively. If I is a left C-infra-near ring and $(x + y).0 = x.0 + y.0$ for all $x,y \in I$, hence 0 is a right distributive element of I, then D is a left subinfra-near ring of I which is a normal subgroup of (I,+) and W is a right I-subgroup of I. Moreover we have the semidirect decomposition $I = D + W$; indeed, for any $x \in I$, we have $x = (x - x.0) + x.0$, where $x - x.0 \in D$ and $x.0 \in W$.

2. TERNARY GROUPS

Let us note that the left infra-distributivity of the multiplication with respect to the addition reminds one of a left distributivity of the binary multiplication over a ternary composition defined by :

$(y,t,z) := y - t + z$, for all $y,t,z \in I$.

Indeed, we have then $(y,0,z) = y + z$ and $x.(y,0,z) = (x.y,x.0,x.z)$ for all $x,y,z \in I$. Moreover we have $x.(y - t + z) = x.y-x.t+x.z$, for all $x,y,t,z \in I$. But such triples were obtained by Dörnte and by Baer in the twenties in connection with some geometrical facts (the Erlangen Program on groups) ; in [1], R.Baer had nicely explained the geometrical and algebraic reasons to take as invariants for a group of transformations the expressions of the form $x - y + z$. Starting from this point, Certaine [2] was guided to consider a ternary operation on a group (G,+) defined by $(x,y,z) := x - y + z$, for all $x,y,z \in G$ (we translate into an additive notation the paper written by Certaine). By considering the abstract properties of such an operation, Certaine has defined ternary groups.

(2.1) Definition. Let T be a nonempty set endowed with a ternary composition $(\cdot,\cdot,\cdot) : T \times T \times T \longrightarrow T$ and with a fixed element $0 \in T$ such that the following conditions are fulfilled:

(1) $(x,0,(y,z,u)) = ((x,0,y),z,u)$, for all $x,y,z,u \in T$;

(2) $(x,0,0) = x$, for all $x \in T$;

(3) $(x,x,0) = 0$, for all $x \in T$.

Such an algebraic system is called a ternary group of first type (Certaine [2] , Definition 3) .

By defining a binary operation on T

(4) $x + y := (x,0,y)$, for all $x,y \in T$,

we obtain a group (T,+) with 0 as its neutral element ([2] , Theorem 4) .

When the binary and the ternary operations are related by the equality :

(5) $(x,y,z) = x - y + z$, for all $x,y,z \in T$,

then we say that the ternary group is regular .

In a binary group , the neutral element is unique . A similar statement does not hold for a ternary group , but if 0 and 0_1 both satisfy (1), (2) and (3) in Definition (2.1), then the corresponding binary groups by (4) are isomorphic as it is proved in [2] (the isomorphism is given by $x \longmapsto (x,0,0_1)$, for $x \in T$).

Moreover it is possible that distinct ternary groups correspond to the same binary group,as it is shown in [2] for the cyclic group with two elements. To illustrate this,we consider the cyclic group of order 3 , $(T = \{0,a,b\} ,+)$ with $b = 2a$ and $3a = 0$. We find three possibilities to define ternary groups of first type,namely :

(I) the regular ternary group $(x,y,z) = x - y + z$,for any $x,y,z \in T$;

(II) the special products are $(a,a,a) = (0,b,b) = b$, $(b,b,b) =$

= (0,a,a) = a , (b,a,a) = (a,b,b) = 0 , the other ternary products being regular ;

(III) the special products are (a,a,b) = (b,b,a) = 0 , (0,b,a) = (b,a,b) = a , (0,a,b) = (a,b,a) = b , the other products being regular.

We obtain another concept of ternary group, which we call of second type, type, while a ternary group of Certain's type is called of first type.

(2.2) Definition. A couple (T,(.,.,.)) ,where T is a nonempty set and (.,.,.) : $T\times T\times T \longrightarrow T$ is a ternary operation on T , is called a ternary group of second type , if it satisfies the following axioms for all $x,y,z \in T$ and a fixed element $0 \in T$:

(6) ((x,y,0),z,0) = (x,(y,z,0),0) ,

(7) (0,x,0) = x ,

(8) ((0,0,x),x,0) = 0 .

(2.3) Proposition. If (T,(.,.,.)) is a ternary group of second type , then (T,+) , where + is the binary composition defined by

(9) x + y := (x,y,0) , for all $x,y \in T$,

is a group.

Proof. The associativity comes from (6) , by using the definition of the binary operation. The other two axioms translated into the addition complete the definition of a group. We also have the equalities:

(7') (x,0,0) = x , for all $x \in T$,

(8') (x,(0,0,x),0) = 0 , for all $x \in T$.

(2.4) Definition. A ternary group of second type is called regular , if the ternary operation is connected with the binary composition (9) by the following equality:

(10) (x,y,z) = x + y - z , for all $x,y,z \in T$.

As in the case of ternary groups of first type, there are

many ternary groups of second type associated to the same group by the equality (9). For example, for the cyclic group of order 2, $(T = \{0,a\},+)$, we may define a ternary composition: $(0,0,0) = (a,0,a) = (a,a,0) = (a,a,a) = 0$, $(0,a,0) = (0,0,a) = (a,0,0) = (0,a,a) = a$, which provides T with the structure of a ternary group of second type with $(T,+)$ as the associated group. This ternary group is not regular, since $(a,a,a) = 0 \neq a = a + a - a$.

The following ternary compositions on $T = \{0,a\}$ satisfy only two of the three axioms in Definition (2.2), namely (7),(8) for (t_1), (6),(8) for (t_2) and (6),(7) for (t_3) :

$$(t_1)\begin{cases}(0,0,0) = (a,0,0) = (0,a,a) = (a,a,0) = 0 ,\\ (0,0,a) = (a,0,a) = (0,a,0) = (a,a,a) = a ;\end{cases}$$

$$(t_2)\begin{cases}(0,0,0) = (0,0,a) = (0,a,0) = (0,a,a) = 0 ,\\ (a,0,0) = (a,0,a) = (a,a,0) = (a,a,a) = a ;\end{cases}$$

$$(t_3)\begin{cases}(0,0,0) = (a,0,a) = (0,a,a) = 0 ,\\ (0,0,a) = (a,0,0) = (0,a,0) = (a,a,0) = (a,a,a) = a .\end{cases}$$

These models prove the following

(2.5) <u>Proposition</u>. The system of axioms in Definition (2.2) is independent.

(2.6) <u>Definition</u>. A ternary group $(I,(.,.,.))$ (of first or of second type) together with an associative binary multiplication which is left distributive with respect to the ternary composition, i.e. which satisfies the equality

(11) $x.(y,z,w) = (x.y,x.z,x.w)$, for all $x,y,z,w \in I$,

is called a <u>left ternary near-ring</u> (<u>of first or of second type</u>).

(2.7) <u>Proposition</u>. (i) If $(I,(.,.,.),.)$ is a left ternary near-ring of first type , then $(I,+,.)$, where + is defined by (4), is a left infra-near ring. (ii) If $(I,(.,.,.),.)$ is a regular left ternary near-ring of second type, then $(I,+,.)$, where + is defined by (9) , is a left infra-near ring.

<u>Proof</u>. (i) Since (I,+) is a group (Certaine [2], Theorem 4), and (I,.) is a semigroup by assumption, we have to check the left infra-distributivity of the multiplication with respect to the addition. We have : $x.(y + z) = x.(y,0,z) = (x.y,x.0,x.z) =$ $= ((x.y,0,0),x.0,x.z) = (x.y,0,(0,x.0,x.z)) = x.y + (0,x.0,x.z)$. But $x.0 + (0,x.0,x.z) = (x.0,0,(0,x.0,x.z)) = ((x.0,0,0),x.0,x.z)$ $= (x.0,x.0,x.z) = x.(0,0,z) = x.z$, therefore $(0,x.0,x.z) =$ $= - x.0 + x.z$, hence $x.(y + z) = x.y - x.0 + x.z$, for all $x, y,z \in I$.

(ii) Here we obtain immediately $x.(y + z) = x.(y,z,0) =$ $= (x.y,x.z,x.0) = x.y + x.z - x.0$, from the regularity of the ternary group. But taking $y = 0$ we obtain

(12) $x.z + x.0 = x.0 + x.z$,for all $x,z \in I$,

therefore $x.(y + z) = x.y - x.0 + x.z$, for all $x,y,z \in I$. The other axioms in the definition of left infra-near rings come from the hypotheses and the Proposition (2.3) .

We note that in the general case this commutativity given by (12) does not hold for left infra-near rings , as one can see from the example $I = \mathbb{Z}^5$ in Section 1.

This way we obtain as a particular case of the ternary near--rings the "prerings" defined by Janin [6,7] . Indeed, a prering is a ternary near-ring of first type or of second type which satisfies for all $x,y,z \in I$ a commutativity condition of the ternary composition , namely $(x,y,z) = (z,y,x)$ - for the first type and $(x,y,z) =(y,x,z)$ - for the second type ,which has an idempotent multiplication being left and right distributive with respect to the ternary composition.

We translate the definition of a two-sided ideal of a left infra-near ring into ternary language. In a left ternary near-ring, we say that a nonempty subset J is an ideal,if it satisfies the following conditions:

(i) $(a,b,0) \in J$, $((x,0,a),x,0) \in J$, for all $a,b \in J$ and $x \in I$;

(ii) $(x.a,x.0,0) \in J$, for all $a \in J$ and $x \in I$;

(iii) $((x,0,a).y,x.y,0) \in J$, for all $a \in J$ and $x,y \in I$.

Let I be a left ternary near-ring of first type in which $x.0 \neq 0$ but $0.x = 0$ for all $x \in I$. In addition, we assume that for any $x,y \in I$, $(x,0,y).0 = (x.0,0,y.0)$, hence the element 0 is right distributive with respect to the triples of the given form. Then the subsets D and W , as defined above, which are semigroups under multiplication , are also closed under the ternary operation. We note that $x.w \in W$, for any $x \in I$ and $w \in W$ In fact, the following proposition holds:

(2.8) <u>Proposition</u>. Let I be a left ternary near-ring of first type which satisfies the conditions :

(i) $0.x = x$, for all $x \in I$;

(ii) $(x,y,z).0 = (x.0,y.0,z.0)$, for all $x,y,z \in I$;

(iii) $(0,x.0,x) = (x,x.0,0)$, for all $x \in I$;

(iv) $(w,0,d).x = (w.x,0,d.x)$, for all $d \in D$, $w \in W$ and $x \in I$.

Then : (i) $I = (D,0,W) = \{(d,0,w) \ / \ d \in D, w \in W\}$ and for each $x \in I$,the elements $d \in D$ and $w \in W$ such that $x = (d,0,w)$ are uniquely determined;

(ii) Using ternary expressions for two arbitrary elements of I , the multiplication is given by the formula:

(13) $(d,0,w).(d',0,w') = (d.d',0,(w,0,d.w'))$.

<u>Proof</u>. (i) For each $x \in I$, we determine $d = (x,x.0,0)$ which belongs to D and $w = x.0 \in W$. But $x=(d,0,w)$, with d and w just determined above (one can verify this relation by straightforward calculations). If $(d,0,w) = (d',0,w')$, with $d,d' \in D$ and $w,w' \in W$, we have $(((0,d',0),0,(d,0,w)),0,(0,w,0)) =$ $= (((0,d',0),0,(d',0,w)),0,(0,w,0))$ and after some skillful calculations we obtain $(0,d',d) = (w',w,0) \in W \cap D$, hence $(0,d',d) =$ $= (w',w,0) = 0$. But then $d' = (d',0,0) = (d',0,(0,d',d)) =$

$= ((d',0,0),d',d) = (d',d',d) = (d',d',(0,0,d)) =$

$=((d',d',0),0,d) = (0,0,d) = d$. In the same manner, we prove that $w' = w$.

(ii) The formula (13) is obtained by using the condition (iv) taking into account the equality $w.x = w$ for all $x \in I$ and $w \in W$.

It might be interesting to study ternary near-rings in an independent context . Here we have used the ternary interpretation for showing the relation between infra-near rings and pre-rings and for explaining the left infra-distributivity as a reflection of a left distributivity of the multiplication with respect to a ternary composition.

REFERENCES

[1]. Baer,R. - Zur Einführung des Scharbegriffs,J.reine angew. Math.,160 (1929) , 199-207.

[2]. Certaine,J.- The ternary operation (abc) = $ab^{-1}c$ of a group, Bull.Amer.Math.Soc.,49 (1943) , 869-877.

[3]. Climescu,Al.- Anneaux faibles , Bul.Inst.Polit.Iaşi , 7 (11) (1961) , 1-6 .

[4]. Climescu,Al. - A new class of weak rings , (Romanian) , ibidem, 10 (14)(1964), 1-4.

[5]. Cupona,G. - On quasirings , (Macedonian) , Bull.Soc.Math. Phys,Macédoine , 20 (1969) ,19-22 (1971) ; MR 44 # 1703 .

[6]. Janin,P.- Une généralisation de la notion d'anneau.Pré-anneaux.C.R.Acad.Sci.Paris,Sec.A , 269 (1969) , 62-64.

[7]. Janin,P. - Une généralisation de la notion d'algèbre. Préalgèbres. C.R.Acad.Sci.Paris , Sec.A , 269 (1969) , 120-122.

[8]. Murdoch,D.C.,Ore,O. - On generalized rings, Amer.J.Math.,

63 (1941) ,73-78 .

[9]. Pilz,G.- Near-rings.The theory and its applications. North-Holland Mathematics Studies 23 , North-Holland Publ.Comp.,Amsterdam,1977.

[10]. Stefănescu,Mirela - Infra-near rings of affine type, An.St.Univ.Al.I.Cuza Iași,24 (1978) ,5-14 .

[11]. Stefănescu,Mirela - A generalization of the concept of near-ring :Infra-near rings , ibidem, 25 (1979), 45 - 56 .

[12]. Stefănescu,Mirela - Multiplications infra-distributives sur un groupe,Publ.Math.Debrecen, 27 (1980), 255 - 262.

Near-rings and Near-fields, G. Betsch (editor)

ON TWO-SIDED IDEALS IN MATRIX NEAR-RINGS

ANDRIES PJ VAN DER WALT

Department of Mathematics, University of Stellenbosch,
7600 Stellenbosch, Republic of South Africa

Matrix near-rings over arbitrary near-rings were introduced in [2], where some results about the correspondence between the two-sided ideals in the base near-ring R and those in the matrix near-ring $\mathbb{M}_n(R)$ were given. The purpose of this paper is to give a fuller account of this correspondence which turns out to be quite a bit more complex than in the ring case, although the overall picture is pleasantly similar.

Let $(R,+,\cdot)$ be a right near-ring with identity 1. R^n will denote the direct sum of n copies of $(R,+)$, and similarly for subgroups of $(R,+)$. The elements of R^n are thought of as column vectors and written in transposed form with pointed brackets, eg $\langle r_1, \ldots, r_n\rangle$. The symbols ι_j and π_j will denote the jth coordinate injection and projection functions respectively. The *elementary* $n \times n$ *matrices over* R are the functions $f^r_{ij}: R^n \to R^n$, where $f^r_{ij} = \iota_i\lambda(r)\pi_j$. Here $r \in R$ and $\lambda(r): R \to R$ is the left multiplication $s \mapsto rs$, all $s \in R$. For typographical reasons f^r_{ij} is sometimes written $[r;i,j]$. The subnear-ring of $M(R^n)$ generated by the f^r_{ij} is the *near-ring* of $n \times n$ *matrices over* R, denoted $\mathbb{M}_n(R)$.

Now let I be a (two-sided) ideal of R. There are two obvious ways in which one can let an ideal in $\mathbb{M}_n(R)$ correspond to I. The first is to define

$$I^* := \{A \in \mathbb{M}_n(R) \mid A\rho \in I^n, \forall \rho \in R^n\}$$

as was done in [2], where it was shown that the mapping $I \mapsto I^*$ is an injection from the set of all ideals of R into that of $\mathbb{M}_n(R)$ (see 4.3 of [2]). The second is to define

$$I^+ := \mathrm{Id}(\{f^a_{ij} \mid a \in I,\ 1 \leqslant i,j \leqslant n\}),$$

where $\mathrm{Id}(T)$ denotes the ideal generated by the set T. Since $f^a_{ij} \in I^*$ for all $a \in I$, $1 \leqslant i,j \leqslant n$, we immediately have the following

PROPOSITION 1 $I^+ \subseteq I^*$ for any ideal I of R. □

PROPOSITION 2 The mapping $I \mapsto I^+$ is an injection.

PROOF Suppose I_1 and I_2 are ideals of R such that $I_1 \neq I_2$; say

$a \in I_1 \setminus I_2$. Then $f_{11}^a \in I_1^+$. If $f_{11}^a \in I_2^+$, then $f_{11}^a \in I_2^*$ by Proposition 1. But this is not the case because $f_{11}^a \langle 1,0, \ldots, 0\rangle = \langle a,0, \ldots, 0\rangle \notin I_2^n$. □

Next, if J is any ideal of $\mathbb{M}_n(R)$ we define

$J_* := \{x \in R | x = \pi_j A\rho$ for some $A \in J, \rho \in R^n, 1 \leqslant j \leqslant n\}$.

It was proved in [2] that J_* is an ideal of R. The following is an important result in this connection:

PROPOSITION 3 $(J_*)^+ \subseteq J \subseteq (J_*)^*$ for every ideal J of $\mathbb{M}_n(R)$.

PROOF $J \subseteq (J_*)^*$ was shown in 4.7 of [2]. By 4.4 of the same paper $f_{11}^a \in J$ iff $a \in J_*$. This shows that $(J_*)^+ \subseteq J$. □

In view of this result it is of interest to know whether the inclusion $I^+ \subseteq I^*$ of Proposition 1 can be proper. That this is indeed the case is shown in the next

EXAMPLE 4 Let $R = \mathbb{Z}_0[x]$, the zerosymmetric polynomial near-ring in one indeterminate over the integers (see Pilz [3] p 220 onwards). Consider the ideal $I := (2)[x]$. We proceed to show that in $\mathbb{M}_2(R)$ $I^+ \neq I^*$, and to this end we prove that $A = [x^2;1,1](f_{11}^x + f_{12}^x) - [x^2;1,1] - [x^2;1,2] \in I^* \setminus I^+$.

Clearly, $A\langle p,q\rangle = \langle 2pq,0\rangle \in I^2$ for all $\langle p,q\rangle \in R^2$, so $A \in I^*$. To see that $A \notin I^+$ it suffices to carry out the following steps:

(In the following discussion we let X,Y denote arbitrary matrices in $\mathbb{M}_2(R)$; B, B_1 and B_2 will be elements of I^+; $\langle p,q\rangle$, $\langle u,v\rangle$, $\langle r,s\rangle$ are arbitrary elements of R^2; a_i, b_i, c_i are appropriate integers; and $i \in \{1,2\}$.)

(a) Show that $\pi_i X\langle p,q\rangle = a_i p + b_i q + 2c_i pq + \ldots$

(b) Show that $\pi_i(X(\langle u,v\rangle + \langle r,s\rangle) - X\langle r,s\rangle) = (a_i + 2c_i s)u + (b_i + 2c_i r)v$ $2c_i uv + \ldots$

(c) Note that $\pi_i B\langle p,q\rangle = 2f_i(p,q)$ for some polynomial f_i in 2 variables. (This is because $I^+ \subseteq I^*$.)

(d) Prove that $\pi_i B\langle p,q\rangle = 2a_i p + sb_i q + 4c_i pq + \ldots$ for any $B \in I^+$. This is done recursively by showing the assertion

(i) is true for $B = [2ax^t;i,j]$;

(ii) if true for B_1, B_2, is also true for $B_1 + B_2$;

(iii) if true for B, is also true for BX (by (a) and (c));

(iv) if true for B, is also true for $X(B + Y) - XY$ (by (b) and (c)).□

For any ideal J in $\mathbb{M}_n(R)$ we therefore have the following diagram:

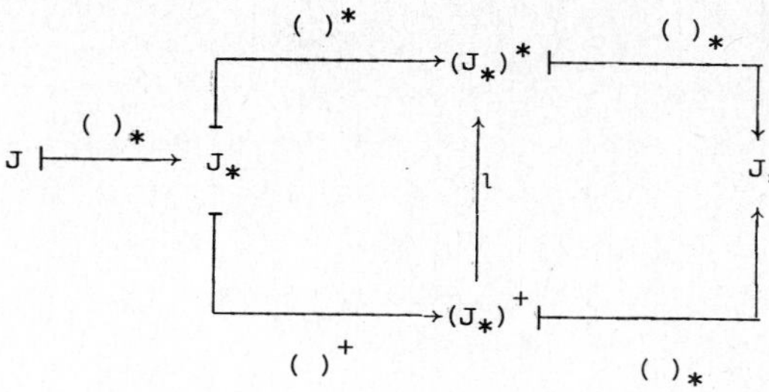

Here ι is the inclusion injection and, in general, not the identity function. This relationship reminds one of the phenomenon of *enclosing ideals* that one has in polynomial near-rings over commutative rings (see [1] or [3]). However, in that case one deals only with so-called *full ideals* of the polynomial near-ring, i.e. those ideals that are also ideals of the polynomial ring, whereas the situation we are concerned with here applies to any ideal J of $\mathbb{M}_n(R)$. The following proposition provides more information about the mappings $(\ )^*$ and $(\ )^+$.

PROPOSITION 5 For ideals A and B of R we have:

(i) $A \subseteq B$ implies $A^+ \subseteq B^+$ and $A^* \subseteq B^*$.

(ii) If $A \nsubseteq B$ then $A^+ \nsubseteq B^*$.

(iii) $(A \cap B)^* = A^* \cap B^*$. (In fact, this holds for arbitrary intersections.)

(iv) $(A \cap B)^+ \subseteq A^+ \cap B^+$

(v) $(A + B)^* \supseteq A^* + B^*$

(vi) $(A + B)^+ = A^+ + B^+$

PROOF (i) This is clear.

(ii) This was actually shown in the proof of Proposition 2.

(iii) This is a standard result about noetherian quotients. See [3], 1.44.

(iv) Obvious from (i).

(v) If $A \in A^* + B^*$, then $A\rho \in A^n + B^n = (A + B)^n$ for all $\rho \in R^n$.

(vi) It is clear that $A^+ + B^+ \subseteq (A + B)^+$. To prove the reverse inclusion, note that $(A + B)^+$ is generated by the set $\{f_{ij}^{a+b} \mid a \in A,\ b \in B,\ 1 \leqslant i,j \leqslant n\}$. Since $f_{ij}^{a+b} = f_{ij}^{a} + f_{ij}^{b} \in A^+ + B^+$ we have our result. □

Since $0^* = 0$ we conclude from (iii) above and Proposition 3.

COROLLARY 6 R is subdirectly irreducible iff $\mathbb{M}_n(R)$ is. □

The next result gives some information on the behaviour of products:

PROPOSITION 7 If $BC \subseteq A$ for ideals A, B and C of R, then $B^+C^* \subseteq A^*$.

PROOF We have to show that if $B \in \mathcal{B}^+$, $C \in \mathcal{C}^*$, then $BC\rho \in A^n$ for all $\rho \in R^n$. Since $C\rho \in \mathcal{C}^n$ for all $\rho \in R^n$ we need only show that $B\gamma \in A^n$ for all $\gamma \in \mathcal{C}^n$. This is easy enough when B is of the form f^b_{ij}, $b \in \mathcal{B}$, and this observation provides the starting point for a recursive proof based on the way in which $\mathcal{B}^+$ is generated by the set $\{f^b_{ij} | b \in \mathcal{B}, 1 \leqslant i,j \leqslant n\}$. □

COROLLARY 8 The ideal A of R is nilpotent iff A^+ is nilpotent in $\mathbb{M}_n(R)$. □

The following result takes 4.11 of [2] further:

PROPOSITION 9 P is a prime (semiprime) ideal of R iff P^* is a prime (semiprime) ideal of $\mathbb{M}_n(R)$.

PROOF In 4.11 of [2] it was proved that if P is prime (semiprime), then so is P^*. So suppose P^* is prime, and let $A, \mathcal{B} \nsubseteq P$. Then $A^+, \mathcal{B}^* \nsubseteq P^*$ by 5(ii), so $A^+\mathcal{B}^* \nsubseteq P^*$. By 7 this implies $A\mathcal{B} \nsubseteq P$, so P is prime. The semiprime case is treated similarly. □

If we denote by rad S the prime radical of the near-ring S, then we have the

COROLLARY 10 $\text{rad } \mathbb{M}_n(R) \subseteq (\text{rad } R)^*$.

PROOF By 5(iii) $(\text{rad } R)^* = \cap\{P^* | P \text{ is prime in } R\} \supseteq \text{rad } \mathbb{M}_n(R)$. □

It would be interesting to know whether the inclusion in this result can be proper, especially since it was proved in [4] that the J_2-radical of $\mathbb{M}_n(R)$ is of the form $(J_2(R))^*$.

We close with a result about A^*, where A is an ideal in a distributively generated near-ring R. First we establish a purely group theoretic result which is probably well known, and so we provide only a sketch of the proof:

LEMMA 11 Let $(G,+)$ be a group with $a, a_1, \ldots, a_m, b, b_1, \ldots, b_m \in G$ such that $a = a_1 + \ldots + a_m$, $b = b_1 + \ldots + b_m$. Then $c := a_1 + b_1 + a_2 + b_2 + \ldots + a_m + b_m = a + b + d$, where $d \in \delta_1(G)$, the commutator subgroup of G.

PROOF We have $c = a - a + c - b + b = a - (a_2 + \ldots + a_m) + b_1 + (a_2 + \ldots + a_m) - (a_3 + \ldots + a_m) + b_2 + \ldots + (a_{m-1} + a_m) - a_m + b_{m-1} + a_m - b_{m-1} - \ldots - b_1 + b$ from which the result follows. □

PROPOSITION 12 Let R be a dg near-ring with identity, and suppose $A \supseteq \delta_1(R)$, where A is an ideal of R and $\delta_1(R)$ is the commutator sugbroup of $(R,+)$. Then A^* in $\mathbb{M}_2(R)$ consists of all and only those matrices of the form $[a_1;1,1] + [b_1;1,2] + \ldots + [a_m;1,1] + [b_m;1,2] + [c_1;2,1] + [d_1;2,2] + \ldots + [c_q;2,1] + [d_q;2,2]$, where Σa_i, Σb_i, Σc_i, $\Sigma d_i \in A$.

PROOF In [2] it was noted that if R is dg, then so is $\mathbb{M}_n(R)$. By applying matrices in A^* to the vectors $<1,0>$ and $<0,1>$ it is easy to see that all matrices in A^* are of the given form. On the other hand, it follows from Lemma 11 that every matrix of the given form is in A^*. □

It will be obvious that an extension of this result to $\mathbb{M}_n(R)$, $n > 2$, is possible. However, the formulation of the extension does not provide additional insight, so we shall forego stating it.

ACKNOWLEDGEMENT

Work on this paper was financially supported by the Council for Scientific and Industrial Research.

REFERENCES

[1] Lausch, H and W Nöbauer, Algebra of polynomials, North-Holland, 1973.
[2] Meldrum, JDP and APJ van der Walt, Matrix near-rings, Archiv der Math (Basel). To appear.
[3] Pilz, G, Near-rings, Revised Edition, North-Holland, 1983.
[4] Van der Walt, APJ, Primitivity in matrix near-rings, Quaest. Math. To appear.

Near-rings and Near-fields, G. Betsch (editor)

SOME PATHOLOGY FOR RADICALS IN NON-ASSOCIATIVE NEAR-RINGS

Stefan Veldsman

Department of Mathematics, University of Port Elizabeth, Port Elizabeth
Republic of South Africa

ABSTRACT. It is well-known that every semisimple class in the class of all associative rings or alternative rings is hereditary, this being an easy consequence of the Anderson-Divinsky-Suliński property (cf. [1]). Dropping the associativity, one get degenerate radicals in the sense that a radical class has a hereditary semisimple class if and only if it is an A-radical, i.e. the radical only depends on the structure of the underlying abelian groups (cf. [4] and [5]). In the class of all near-rings (or abelian near-rings) the situation is not as bad, here we have both hereditary and non-hereditary semisimple classes, see, for example [3]. Dropping the associativity in this case, the result is even worse than in that of non-associative rings. We show that in the class of all abelian, not-necessatily associative zero-symmetric near-rings the only radicals with hereditary semisimple classes are the two trivial radical classes.

1. PRELIMINARIES

Let $\mathcal{W}$ be the class of all abelian not-necessarily associative zero-symmetric near-rings. Near-rings will be right near-rings and we use the notation and notions of Pilz [6]. $\mathcal{W}$, being a universal class of Ω-groups is hence suitable for the development of a Kurosh-Amitsur radical theory, basics of which we refer to [7].

A class $\mathcal{R} \subseteq \mathcal{W}$ is a *radical class* (in the sense of Kurosh-Amitsur) if

(i) $\mathcal{R}$ is homomorphically closed

(ii) For every $N \in \mathcal{W}$,

$$\mathcal{R}(N) := \sum(I \mid I \text{ an ideal in } N \text{ with } I \in \mathcal{R}) \in \mathcal{R}$$

(iii) $\mathcal{R}(N/\mathcal{R}(N)) = 0$.

As usual, $S\mathcal{R} = \{N \in \mathcal{W} \mid \mathcal{R}(N) = 0\}$ will denote the *semisimple class* of $\mathcal{R}$. $S\mathcal{R}$ is *hereditary* if from $N \in S\mathcal{R}$ and I an ideal in N, $I \in S\mathcal{R}$ follows. Using standard techniques, it can be proved, for any radical class $\mathcal{R}$:

(i) If K is an ideal of N and $N/K \in S\mathcal{R}$, then $\mathcal{R}(N) \subseteq \mathcal{R}(K)$.

(ii) $S\mathcal{R}$ is hereditary if and only if $\mathcal{R}(I) \subseteq \mathcal{R}(N)$ for every ideal I of N.

(iii) $\mathcal{R}(N_1 \oplus N_2) = \mathcal{R}(N_1) \oplus \mathcal{R}(N_2)$.

For $N \in \mathcal{W}$, N^+ will denote the underlying additive group of N and N^o will be the zero near-ring built on N^+ by the multiplication $ab = 0$ for all $a,b \in N$. The class of all zero near-rings in $\mathcal{W}$ will be denoted by Z. Obviously, for any $N \in \mathcal{W}$, $N^o \in \mathcal{W}$ holds.

If I is an ideal of N, it will be denoted by $I \vartriangle N$.

In the sequel, all near-rings considered, will be from $\mathcal{W}$.

2. RESULTS

Betsch and Kaarli [2] have proved the next theorem in the class of all near-rings. As it makes no use of the associativity, it remains valid in $\mathcal{W}$, and we have

2.1 <u>Theorem</u>

Let $\mathcal{R}$ be a radical class with hereditary semisimple class $\mathcal{SR}$ and assume $\mathcal{SR}$ contains a zero near-ring N (= N^o) with $N \neq 0$. Then $\mathcal{Z} \subseteq \mathcal{SR}$.

Our main result depends on the next two constructions:

For $N \in \mathcal{W}$, we definie two near-rings $\Gamma(N)$ and $\Lambda(N)$ by:

$$\Gamma(N)^+ = N^+ \oplus N^+ \oplus N^+ = \Lambda(N)^+$$

with multiplication defined by

$$(a,b,c)(x,y,z) = \begin{cases} (ax\,,\,az\,,\,0) & \text{in } \Gamma(A) \\ (ax+cy\,,\,az\,,\,cz) & \text{in } \Lambda(A). \end{cases}$$

Both $\Gamma(N)$ and $\Lambda(N)$ are abelian, right distributive, zero-symmetric, not-necessarily associative (even if N is associative) and hence in $\mathcal{W}$.

Let $B = \{(a,b,0) \mid a,b \in N\}$. Then B is an ideal in both $\Gamma(N)$ and $\Lambda(N)$ and $B \cong N \oplus N^o$. Furthermore,

$$\frac{\Gamma(N)}{B} \cong N^o \quad \text{and} \quad \frac{\Lambda(N)}{B} \cong N.$$

2.2 <u>Lemma</u>

Let $\mathcal{R}$ be a radical class with hereditary semisimple class $\mathcal{SR}$. If $N \in \mathcal{R}$ and $N^o \in \mathcal{SR}$, then $N = 0$.

Proof

Because $\mathcal{SR}$ is hereditary, we have

$$N = \mathcal{R}(N \oplus N^o) = \mathcal{R}(B) \subseteq \mathcal{R}(\ (N))\ \Delta\ \Gamma(N).$$

On the other hand,

$$\mathcal{R}\left(\frac{\Gamma(N)}{N \oplus N^o}\right) = \mathcal{R}(N^o) = 0\ , \text{ hence}$$

$$\mathcal{R}(\Gamma(N)) \subseteq \mathcal{R}(N \oplus N^o) = N.$$

By the above, we have $N = \mathcal{R}(\Gamma(N))\ \Delta\ \Gamma(N)$.

Let $a,b \in N$. Because $(a,0,0) \in (N,0,0) \cong N\ \Delta\ \Gamma(N)$,

$$(0,ab,0) = (a,0,0)(0,0,b) \in N \cong (N,0,0) \text{ follows.}$$

Hence $ab = 0$ for all $a,b \in N$. Thus $N = N^o \in \mathcal{R} \cap \mathcal{SR} = \mathcal{O}$, i.e. $N = 0$.

2.3 <u>Lemma</u>

Let $\mathcal{R}$ be a radical class with hereditary semisimple class $\mathcal{SR}$. If $N \in \mathcal{SR}$ and $N^o \in \mathcal{R}$, then $N = 0$.

Proof

As in the previous lemma, it can be shown that

$$N^o = \mathcal{R}(\Lambda(N)) \,\Delta\, \Lambda(N).$$

For $a,b \in N$, $(0,b,0) \in (0,N,0) \cong N^o \,\Delta\, \Lambda(N)$. Hence

$$(ab,0,0) = (0,0,a)[(a,0,0) + (0,b,0)] - (0,0,a)(a,0,0) \in N^o \cong (0,N,0).$$

Thus $ab = 0$ for all $a,b \in N$.

Hence $N = N^o \in \mathcal{R} \cap \mathcal{SR} = \mathcal{O}$, i.e. $N = 0$.

We can now prove our main result:

2.4 Theorem

Let $\mathcal{R}$ be a radical class in $\mathcal{W}$. Then $\mathcal{SR}$ is hereditary if and only if $\mathcal{R} = \mathcal{O}$ or $\mathcal{R} = \mathcal{W}$.

Proof

Obviously only the sufficiency needs verification. If $\mathcal{R} \neq \mathcal{O}$, then $Z \subseteq \mathcal{R}$. Indeed, if $Z \nsubseteq \mathcal{R}$, let $N = N^o \in Z$ with $N \notin \mathcal{R}$. Hence $0 \neq N/\mathcal{R}(N) \in \mathcal{SR}$ and because $N/\mathcal{R}(N)$ is a zero near-ring, $Z \subseteq \mathcal{SR}$ follows from 2.1. Because $\mathcal{R} \neq \mathcal{O}$, let $0 \neq B \in \mathcal{R}$. Then $B^o \in Z \subseteq \mathcal{SR}$ and by 2.2 the contradiction $B = 0$ follows. Hence $Z \subseteq \mathcal{R}$. For any $N \in \mathcal{W}$, we have $N/\mathcal{R}(N) \in \mathcal{SR}$ and $(N/\mathcal{R}(N))^o \in Z \subseteq \mathcal{R}$. By 2.3, $N/\mathcal{R}(N) = 0$, thus $N = \mathcal{R}(N) \in \mathcal{R}$.

REFERENCES

[1] Anderson, T., Divinsky, A. and Suliński, A., Hereditary radical in associative and alternative rings, Canad. J. Math., 17, 1965, 594-603.

[2] Betsch, G. and Kaarli, K., Supernilpotent radicals and hereditariness of semisimple classes of near-rings, Coll. Math. Soc. J. Bolyai, 38, Radical Theory, North Holland, 1985, 47-58.

[3] Betsch, G. and Wiegandt, R., Non-hereditary semisimple classes of near-rings, Studia Sci. Math. Hungar., 17, 1982, 69-75.

[4] Gardner, B.J., Some degeneracy and pathology in non-associative radical theory, Annales Univ. Sci. Budapest, 22-23, 1979/80, 65-74.

[5] Gardner, B.J., Some degeneracy and pathology in non-associative radical theory II, Bull. Austral. Math. Soc., 23, 1981, 423-428.

[6] Pilz, G., Near-rings, (North-Holland, 1977).

[7] Wiegandt, R., Radical and semisimple classes of rings, Queen's papers in pure and applied mathematics, No.37, Kingston, Ontario, 1974.

Near-rings and Near-fields, G. Betsch (editor)

PARTIALLY AND FULLY ORDERED SEMINEARRINGS AND NEARRINGS

Hanns Joachim WEINERT

1. INTRODUCTION

According to [11] and [12], a (in this paper always right distributive) *seminearring* is defined to be an algebra $S = (S,+,\cdot)$ such that $(S,+)$ and $(S,\cdot)$ are semigroups and $(a+b)c = ac+bc$ holds for all $a,b,c \in S$. Seminearrings occur very naturally as transformations of semigroups $(\Gamma,+)$ (cf. Thm. 3.1). They are also the common generalization of nearrings and semirings, and we refer to [13] and [14] for corresponding investigations of statements shared by both, nearrings and semirings. In a similar way we deal in this paper with seminearrings established with a partial order on S. These considerations will also provide new results on partially ordered (p. o.) nearrings (cf. [5], [8] - [11]) and p. o. semirings (cf. [2], [15]), in particular in the fully ordered (f. o.) case.

In the papers just cited, a nearring $(S,+,\cdot)$ which is likewise a p. o. group $(S,+,\leqq)$ is called a p. o. nearring iff the set $P(S)$ of all positive elements of $(S,+,\leqq)$ is multiplicatively closed (hence a subseminearring). The latter turns out to be equivalent with $P(S) \subseteq M_r(S)$ (called III_r in the following), where $M_r(S)$ $[M_\ell(S)]$ denotes the set of all right [left] monotone elements of $(S,\cdot,\leqq)$ (cf. (2.1)). By results of [5], there are (proper) f. o. nearfields S satisfying also $P(S) \subseteq M_\ell(S)$ (called III_ℓ), as well as those for which III_ℓ fails to be true. Taking also the sets $N(S)$ for all negative elements of $(S,+,\leqq)$ and $W_r(S)$ $[W_\ell(S)]$ of all right [left] antitone elements of $(S,\cdot,\leqq)$ into consideration (cf. (2.2)), we introduce IV_r by $N(S) \subseteq W_r(S)$ and IV_ℓ by $N(S) \subseteq W_\ell(S)$. For each ring $(S,+,\cdot)$ which is also a p. o. group $(S,+,\leqq)$, we shall see that any of these four properties III_r - IV_ℓ implies the other three (Remark 2.5). In the case of nearrings, merely $III_\ell \Leftrightarrow IV_\ell$ holds in general, $III_\ell \Rightarrow III_r$ and $III_\ell \Rightarrow IV_r$ for zero-symmetric ones, and $III_r \Leftrightarrow IV_r$ if S satisfies $a(-b) = -(ab)$ (cf. Thm. 2.4).

More general, let $(S,+,\cdot)$ be a seminearring which is also a p. o. semigroup $(S,+,\leqq)$. We call such a structure $(S,+,\cdot,\leqq)$ a *weakly partially ordered (w. p. o.) seminearring* and introduce the sets $P(S),\ldots$ and the properties $III_r,\ldots$ already used above for w. p. o. seminearrings in § 2. The correlations between these four properties, the main subject of this section, turn out to stagger from "three do not imply the forth" to "pairwise equivalent", depending on further assumptions on S. Some of these results are mentioned above. They prove it meaningful to define - generalizing the concept of a p. o. nearring - a *p. o. seminearring* as a w. p. o. seminearring satisfying III_r, which is now independent of and mostly stronger than $P(S)\cdot P(S) \subseteq P(S)$. Those p. o. seminearrings occur naturally as transformations of p. o. semigroups $(\Gamma,+,\leqq)$, say $S = (\Gamma^\Gamma,+,\cdot,\leqq)$ (cf. Thm. 3.2). Here III_r is implied by $M_r(S) = S$, whereas each of IV_r, III_ℓ and IV_ℓ is only true for a special choice of $(\Gamma,+,\leqq)$ (Cor. 3.3). Conversely, each p. o. seminearring satisfying $M_r(S) = S$ can be represented in this way (Thm. 3.4).

For statements on f. o. nearrings S, the constant subnearring of S and the zero-symmetric one, say S_1 and S_2, are of some importance (cf. [11]). We introduce in § 4 corresponding subseminearrings S_1 and S_2 for seminearrings S with a zero. If $(S,+,\cdot,\leqq)$ is a w. p. o. seminearring, the same holds for $(S_1,+,\cdot,\leqq_1)$ and $(S_2,+,\cdot,\leqq_2)$ with the induced partial orders $\leqq_i$ (cf. Thm. 2.10). For a f. o. seminearring S, the order of S_1+S_2 is uniquely determined lexicographically by $(S_1,\leqq_1)$ and $(S_2,\leqq_2)$, and, apart from some extreme cases, a f. o. seminearring S such that $S_1 \mp \{o\} \mp S_2$ holds is non-archimedean ordered (Thm. 4.2). The applications to f. o. nearrings in Thm. 4.3 provide well known as well as new results, in particular that Thm. 9.146 of [11] is only true for zero-symmetric nearrings. Since we need various examples to prove results in § 2 and § 4 and to show that they are fairly strong, we have collected them in § 5.

There is one more point central to the disposition of this paper. To our own surprise, all results on w. p. o. seminearrings $(S,+,\cdot,\leqq)$ do not depend on the associativity of $(S,+)$, which also does not simplify their proofs. Since one-sided distributive algebras $(S,+,\cdot)$ with not neccessarily associative addition are of interest in other contexts (cf. [4], [6]), we feel that we should express this generality in our presentation. So, similarly as in [13], we generalize the concept of a seminearring as follows:

<u>DEFINITION 1.1.</u> An algebra $(S,+,\cdot)$ is called a *seminearring* iff $(S,+)$ is any groupoid, $(S,\cdot)$ a semigroup and $(a+b)c = ac+bc$ holds for all $a,b,c \in S$. To avoid confusion, we use the term nearring as usual, although a seminearring for which $(S,+)$ is a quasigroup would now be the adequate concept.

Various statements on those seminearrings have been given in [13] and [14]. However, anyone not interested in this generalization, can read this paper without taking notice of Def. 1.1. In particular, all examples of seminearrings used to prove statements are of course additively associative.

Finally, we need a few conventions. For subsets A,B of a seminearring $(S,+,\cdot)$ we introduce $A+B=\{a+b \mid a \in A,\ b \in B\}$ and $AB=\{ab \mid a \in A,\ b \in B\}$, and use e. g. aB for $\{a\}B$. If a groupoid or a seminearring S has an additive neutral, this unique element is called the *zero* of S and denoted by o. In the latter case, o is said to be *left* [*right*] *absorbing* iff $oS=\{o\}$ $[So=\{o\}]$ holds. A seminearring with a (two-sided) absorbing zero is also called *zero-symmetric*.

2. BASIC STATEMENTS ON (WEAKLY) PARTIALLY ORDERED SEMINEARRINGS

According to § 1 we do not assume additions to be associative (if not explicitely stated) and define the more general concept of a *weakly partially ordered (w. p. o.) seminearring* $S=(S,+,\cdot,\leqq)$ by

I $(S,+,\cdot)$ is a seminearring, $(S,\leqq)$ a p. o. set, and

II $a<b$ implies $a+c \leqq b+c$ and $c+a \leqq c+b$ for all $a,b,c \in S$.

In particular, S is called a *weakly fully ordered (w. f. o.) seminearring* iff $(S,\leqq)$ is a f. o. set, and trivially p. o. iff the partial order equals equality. Note that a w. p. o. seminearring S is just a seminearring $(S,+,\cdot)$ and a p. o. groupoid $(S,+,\leqq)$ (cf. [1], Chap. X). By II, $a<b$ implies $a+c<b+c$ if c is right cancellable in $(S,+)$. So if $(S,+)$ is cancellative, the monotony law II yields its strict version

IIst $a<b$ implies $a+c<b+c$ and $c+a<c+b$ for all $a,b,c \in S$,

whereas IIst implies cancellativity if $(S,+,\leqq)$ is a f. o. groupoid.

By $P_r=P_r(S)=\{p \in S \mid a \leqq a+p \text{ for all } a \in S\}$ we introduce the set of all *right positive* elements of a w. p. o. seminearring $S=(S,+,\cdot,\leqq)$, which are in fact the right positive elements of the p. o. groupoid $(S,+,\leqq)$. Due to II, $p<s$ for $p \in P_r$, $s \in S$ implies $s \in P_r$, hence $P_r(S)$ is an upper set of $(S,\leqq)$ and thus a subgroupoid of $(S,+)$ if $P_r \neq \emptyset$. The same holds for the sets $P_\ell(S)$ and $P(S)=P_r(S) \cap P_\ell(S)$ of all *left positive* or *positive* elements of S, respectively. If S has a zero o, one has $P_r=P_\ell=P=\{p \in S \mid o \leqq p\}$. Dually with respect to the relation $\leqq$, we have the corresponding definitions and state-

ments for the sets $N_r(S)$, $N_\ell(S)$ and $N(S)$ of all [*right, left*] *negative* elements of S. Note that $n \leq p$ holds for all $n \in N_\ell$ and $p \in P_r$, briefly $N_\ell \leq P_r$.

An element m of a w. p. o. seminearring $S = (S,+,\cdot,\leq)$ such that

(2.1) $a < b$ implies $am \leq bm$ for all $a,b \in S$

is called *right monotone*, and *strictly right monotone* iff (2.1) holds with $am < bm$. In the context of this paper, these concepts clearly refer to the multiplication, and we need only the first one. The set of all right monotone elements of S is denoted by $M_r(S)$. Using $ma \leq mb$ in (2.1), we introduce the set $M_\ell(S)$ of all *left monotone* elements of S, and define $M(S) = M_\ell(S) \cap M_r(S)$. All these sets, if not empty, are subsemigroups of $(S,\cdot)$, and $M_\ell(S)$ is also additively closed. An element $w \in S$ such that

(2.2) $a < b$ implies $aw \geq bw$ for all $a,b \in S$

is called *right antitone*. We denote the set of all these elements by $W_r(S)$. Likewise we introduce the set $W_\ell(S)$ of all *left antitone* elements of S and $W(S) = W_\ell(S) \cap W_r(S)$. Rules like $W_rW_r \subseteq M_r$ or $W_rM_r \subseteq W_r$ are obvious. Note in this context that $(S,+,\cdot,\leq)$ remains a w. p. o. seminearring replacing $\leq$ by its converse relation $\leq_d$. But whereas $P(S)$ and $N(S)$ change their roles under this procedure, $M_r(S)$, $M_\ell(S)$, $W_r(S)$, and $W_\ell(S)$ remain invariant.

DEFINITION 2.1. Let S be a w. p. o. seminearring. We denote, in the following order of succession, the properties

(2.3) $P(S) \subseteq M_r(S)$, $N(S) \subseteq W_r(S)$, $P(S) \subseteq M_\ell(S)$, $N(S) \subseteq W_\ell(S)$

by III_r, IV_r, III_ℓ and IV_ℓ, and use a sequence as e. g. $\langle 1,1,0,1 \rangle$ to indicate by "1" which of these properties holds for S and by "0" which not. The corresponding sequence is called the *type* of S, and S a *partially ordered (p. o.) seminearring* iff III_r holds. Henceforth we write e. g. $\langle 1101 \rangle = \langle 13 \rangle$ for types in an obvious and convenient interpretation. As shown by examples collected in § 5 (all with associative addition) we state at first:

REMARK 2.2. The properties (2.3) are independent, and there are even f. o. nearrings and f. o. semirings of type $\langle 1011 \rangle = \langle 11 \rangle$, f. o. semirings of type $\langle 1101 \rangle = \langle 13 \rangle$ and of type $\langle 1110 \rangle = \langle 14 \rangle$, and w. f. o. semirings of type $\langle 0111 \rangle = \langle 7 \rangle$ (Expls. 5.3 and 5.9). Moreover, there are w. p. o. seminearrings $S = (S,+,\cdot,\leq)$ such that all possible types from $\langle 0 \rangle$ to $\langle 15 \rangle$ really occur.

On the other hand, additional assumptions on S cause implications between the properties (2.3). For instance, there are no f. o. near-rings of type ⟨13⟩ or ⟨14⟩, in fact not even p. o. ones, and also no w. f. o. nearrings of type ⟨7⟩ (cf. Thm. 2.4). For the corresponding investigations we need one more notion and call a w. p. o. seminear-ring $(S,+,\cdot,\leq)$ with a zero *partially ordered by* $X = P(S)$ *from the left* iff the following holds for all $a,b \in S$:

(2.4) $\quad a \leq b \;\Leftrightarrow\;$ there is some $x \in X$ satisfying $x+a = b$.

Likewise one defines S to be *p. o. by* $X = P(S)$ *from the right* with respect to $a+x = b$, and both concepts apply in fact to $(S,+,\leq)$.[1]

Turning now to implications between the properties (2.3) for particular w. p. o. seminearrings $S = (S,+,\cdot,\leq)$, of course in the meaning of Def. 1.1, we use the following abbreviations for additional assumptions on S:

(ℓ) S has a <u>l</u>eft absorbing zero o.
(r) S has a <u>r</u>ight absorbing zero o.
(p) S is p. o. by its set P(S) of <u>p</u>ositive elements from one side.
(n) S is a <u>n</u>earring (i. e. (S,+) a group), which yields (ℓ) and (p).
(s) S is a nearring satisfying also the <u>s</u>ign-rule $a(-b) = -(ab)$.

<u>LEMMA 2.3.</u> Let $S = (S,+,\cdot,\leq)$ be a w. p. o. seminearring and write $P(S) = P$ etc. Then we have the following implications depending on the assumption on S fixed at each arrow:

$$(2.5)\qquad \begin{array}{ccccccc} III_r & \overset{p}{\Longleftarrow}\!\underset{\ell}{\Longrightarrow} & PP \subseteq P & \underset{n}{\Longleftrightarrow} & NP \subseteq N & \underset{\ell}{\Longleftarrow} & III_r \\ & & \Updownarrow s & & \Updownarrow s & & \\ IV_r & \overset{p}{\Longleftarrow}\!\underset{\ell}{\Longrightarrow} & PN \subseteq N & \underset{n}{\Longleftrightarrow} & NN \subseteq P & \underset{\ell}{\Longleftarrow} & IV_r \end{array}$$

$$(2.6)\qquad \begin{array}{ccccccc} III_\ell & \underset{r}{\Longrightarrow} & PP \subseteq P & \underset{s}{\Longleftrightarrow} & PN \subseteq N & \underset{r}{\Longleftarrow} & III_\ell \\ & & \Updownarrow n & & \Updownarrow n & & \\ IV_\ell & \underset{r}{\Longrightarrow} & NP \subseteq N & \underset{s}{\Longleftrightarrow} & NN \subseteq P & \underset{r}{\Longleftarrow} & IV_\ell \end{array}$$

<u>Proof.</u> From $n \leq o \leq p$ for each $n \in N$, $p \in P$ and (ℓ) we obtain $NM_r \subseteq N$ and $PM_r \subseteq P$, hence III_r implies $NP \subseteq N$ and $PP \subseteq P$. All other implications indicated by (ℓ) or (r) follow similarly. Now we assume $PP \subseteq P$ $[PN \subseteq N]$ and (p). Then (2.4) yields III_r $[IV_r]$ by right distributivity and the definition of P [N]. The implications for which

(n) or (s) is assumed are clear.

The next theorem yields in particular statements on w. p. o. nearrings S ((n)), zero-symmetric ones ((n) and (r)), and subnearrings of a nearfield ((s) and (r)), which we do not formulate explicitely. In this context one also checks the equivalences

(2.7) $III_\ell \underset{n}{\Longleftrightarrow}$ $p \in P$ and $b-a \in P$ implies $pb-pa \in P$ and

(2.8) $IV_\ell \underset{n}{\Longleftrightarrow}$ $n \in N$ and $b-a \in P$ implies $nb-na \in N$,

clearly for all $a,b,n,p \in S$. The right side of (2.7) corresponds to the supplementary axiom A4 in [5] for f. o. nearfields.

THEOREM 2.4. Let $S = (S,+,\cdot,\leqq)$ be a w. p. o. seminearring. Then, using the above conventions, we have the following implications:

(2.9) $IV_\ell \underset{n}{\Longleftrightarrow} III_\ell \underset{r,p}{\Longrightarrow} III_r$ and IV_r

(2.10) $IV_r \underset{s}{\Longleftrightarrow} III_r$.

Moreover, a w. f. o. nearring satisfying IV_r is zero-symmetric.

Proof. The second implication of (2.9) and the equivalence (2.10) follow from (2.5) and (2.6). To show $IV_\ell \Leftrightarrow III_\ell$ for a w. p. o. nearring S, at first we assume IV_ℓ, $p \in P$ and $a < b$, i. e. $b-a \in P$. Then $(-p) \in N$ and (2.8) imply $(-p)b - (-p)a = -(pb) + pa \in N$, hence $pa = pb+n$ for some $n \in N$, which yields $pa \leqq pb$ by the definition of N. This shows $IV_\ell \Rightarrow III_\ell$, and similarly we get $III_\ell \Rightarrow IV_\ell$ using (2.7). For the last statement we refer to Lemma 4.1.

From the above we obtain with obvious proofs:

REMARK 2.5. For a w. p. o. ring S, from (2.9) and its left-right dual version it follows that each of the properties (2.3) implies the other three and characterizes S as a partially ordered ring.

REMARK 2.6. The concept of a p. o. nearring S introduced by III_r in Def. 2.1 is, by (2.5), equivalent with the usual one which defines a partial order on $(S,+,\cdot)$ according to (2.4) by a subseminearring $X = P$, assumed to satisfy the properties given in footnote 1).

REMARK 2.7. Each p. o. nearring S has one of the types $\langle 1111 \rangle = \langle 15 \rangle$, $\langle 1100 \rangle = \langle 12 \rangle$, $\langle 1011 \rangle = \langle 11 \rangle$ and $\langle 1000 \rangle = \langle 8 \rangle$. Only $\langle 15 \rangle$ and $\langle 12 \rangle$ are possible for p. o. nearfields, and there are even f. o. nearfields of this kind as shown in [5]. Moreover, f. o. nearrings of the types $\langle 15 \rangle$ and $\langle 12 \rangle$ are zero-symmetric, those of type $\langle 11 \rangle$ not (by (2.9)), whereas both cases occur for f. o. nearrings of type $\langle 8 \rangle$ (cf. Expls. 5.1 and 5.5).

REMARK 2.8. The next class are w. p. o. nearrings S which do not satisfy III_r, hence with $\langle 0111\rangle = \langle 7\rangle$, $\langle 0100\rangle = \langle 4\rangle$, $\langle 0011\rangle = \langle 3\rangle$ and $\langle 0000\rangle = \langle 0\rangle$ as possible types. The latter three occur even as w. f. o. nearrings (Expls. 5.2 and 5.7). By the last statement of Thm. 2.4 and (2.9), there are no w. f. o. nearrings of type $\langle 7\rangle$. We do not know whether there are w. p. o. ones (but cf. Expl. 5.3).

REMARK 2.9. There are w. f. o. semirings of the types $\langle 1101\rangle = \langle 13\rangle$, $\langle 0111\rangle = \langle 7\rangle$, $\langle 1110\rangle = \langle 14\rangle$ and $\langle 1011\rangle = \langle 11\rangle$ (cf. Expls. 5.3 and 5.9). This shows that III_r and III_ℓ as well as IV_r and IV_ℓ are independent also for w. p. o. semirings. However, since statements on semirings are multiplicatively left-right dual, one should speak about a p. o. semiring S only if III_r and III_ℓ hold. This stronger concept was introduced in [15] by $P(S) \subseteq M(S) = M_r(S) \cap M_\ell(S)$.

We close this section dealing with questions on substructures:

THEOREM 2.10. a) Let $U = (U,+,\cdot)$ be a subseminearring of a w. p. o. seminearring $S = (S,+,\cdot,\leq)$. Then $(U,+,\cdot,\leq)$ is a w. p. o. seminearring with respect to the induced partial order, and one has

(2.11) $\quad \mathfrak{C}(S) \cap U \subseteq \mathfrak{C}(U) \quad$ for $\mathfrak{C} = P, N, M_r, W_r, M_\ell$ and W_ℓ.

b) However, even if S is fully ordered, associative and commutative with respect to both operations and satisfies III_r, IV_r, III_ℓ and IV_ℓ, none of these properties need to be true for U.

c) Sufficient conditions such that III_r transfers from S to U are $M_r(S) = S$ (cf. Thm. 3.2) or $P(U) \subseteq P(S)$, which is $P(U) = P(S) \cap U$ by (2.11). The latter is also sufficient to transfer III_ℓ. In the same way work $W_r(S) = S$ for IV_r and $N(U) \subseteq N(S)$ for IV_r and IV_ℓ.

d) Assume that U and S have a common zero, which is always the case if both are nearrings. Then $P(U) = P(S) \cap U$ and $N(U) = N(S) \cap U$ hold, and if S satisfies one of the properties III_r, IV_r, III_ℓ or IV_ℓ, the same property holds for U.

Proof. All statements of a) are checked straightforward, and c) follows from (2.11) and implies d). For b) we give the following

Example 2.11. a) A semiring $(S,+,\cdot,\leq)$ satisfying all assumptions listed in Thm. 2.10. b) is given by

$S = \{o,a,b\}$
$o < a < b$

+	o	a	b
o	o	a	b
a	a	a	b
b	b	b	b

·	o	a	b
o	o	o	o
a	o	a	b
b	o	b	b

.

One also checks $P(S) = M(S) = S$ and $N(S) = W(S) = \{o\}$. For the subsemiring $U = \{a,b\}$ we have $P(U) = M(U) = U$, however, $N(U) = \{a\}$ and $W(U) = \{b\}$. So U satisfies III_r and III_ℓ, but not IV_r and IV_ℓ.

b) Another semiring $(S,+,\cdot,\leq)$ of this kind is given by

$S = \{a,b,c,d\}$
$a < b < c < d$

+	a	b	c	d
a	a	b	b	b
b	b	b	b	b
c	b	b	c	c
d	b	b	c	c

·	a	b	c	d
a	b	b	b	b
b	b	b	b	b
c	b	b	a	a
d	b	b	a	a

Here one obtains $P(S) = \emptyset$, $M(S) = \{a,b\}$, $N(S) = \{a\}$ and $W(S) = S$. The subsemiring $U = \{a,b,c\}$ satisfies $P(U) = \{c\}$, $M(U) = \{a,b\}$, $N(U) = \{a\}$ and $W(U) = U$, hence IV_r and IV_ℓ, but not III_r and III_ℓ.

3. TRANSFORMATIONS ON PARTIALLY ORDERED GROUPOIDS

The following characterization may be considered as known by its specializations to additively associative seminearrings (cf. [3], [12]) and nearrings.

THEOREM 3.1. Let $(\Gamma,+)$ be a groupoid and $S = \Gamma^\Gamma$ the set of all transformations $f\colon \Gamma \to \Gamma$. For $f,g \in S$, define

(3.1) $(f+g)(\xi) = f(\xi)+g(\xi)$ and $(f\cdot g)(\xi) = f(g(\xi))$ for all $\xi \in \Gamma$.

Then $S = (S,+,\cdot)$ is a seminearring such that $(S,+)$ is associative, commutative, left [right] cancellative or a quasi-group, respectively, iff $(\Gamma,+)$ has this property. Iff $(\Gamma,+)$ has a zero ω, S has a zero o given by $o(\xi) = \omega$, and $of = o$ holds for all $f \in S$. In this case, the sets S_1 of all constant transformations of Γ and $S_2 = \{f \in S \mid f(\omega) = \omega\}$ are subseminearrings of S (cf. § 4).

Conversely, each seminearring $(S,+,\cdot)$ can be represented in this way. Take any groupoid $(\Gamma,+)$ containing $(S,+)$ properly as a subgroupoid, and define for all $a \in S$

(3.2) $f_a \in \Gamma^\Gamma$ by $f_a(\xi) = \begin{cases} a\xi & \text{if } \xi = x \in S \\ a & \text{if } \xi \in \Gamma \setminus S. \end{cases}$

Then $a \to f_a$ provides a monomorphism of $(S,+,\cdot)$ into $(\Gamma^\Gamma,+,\cdot)$.

Note in this context that there always exists such a groupoid $(\Gamma,+)$ which has the same properties as $(S,+)$ concerning associativity, commutativity, left [right] cancellativity, or being a quasi-group. Apart from the fact that a quasi-group need not be embeddable into a loop, $(\Gamma,+)$ may be assumed to have a zero in all cases under consideration. On the other hand, $(\Gamma,+) = (S,+)$ will do the job iff $ax = bx$ for all $x \in S$ always implies $a = b$, i. e. iff $(S,\cdot)$ is right reductive.

<u>THEOREM 3.2.</u> Now we apply Thm. 3.1 to a p. o. groupoid $(\Gamma,+,\leqq)$ assumed to be non-trivially p. o., and define for $f,g \in S = \Gamma^\Gamma$

(3.3) $f \leqq g \Leftrightarrow f(\xi) \leqq g(\xi)$ for all $\xi \in \Gamma$.

Then we obtain a p. o. seminearring $S = (S,+,\cdot,\leqq)$ which always satisfies $M_r(S) = S$ and $W_r(S) = \emptyset$. If $(\Gamma,+,\leqq)$ satisfies IIst or is p. o. by $P(\Gamma)$ from one side, the same holds for $(S,+,\leqq)$ and $P(S)$. But $(S,\leqq)$ is never f. o., even if $(\Gamma,\leqq)$ is. We further state

(3.4) $P(S) = \{p \in S \mid p(\Gamma) \subseteq P(\Gamma)\}$, $N(S) = \{n \in S \mid n(\Gamma) \subseteq N(\Gamma)\}$,

(3.5) $M_\ell(S) = \{h \in S \mid \alpha < \beta \Rightarrow h(\alpha) \leqq h(\beta)$ for all $\alpha,\beta \in \Gamma\}$, and

(3.6) $W_\ell(S) = \{h \in S \mid \alpha < \beta \Rightarrow h(\alpha) \geqq h(\beta)$ for all $\alpha,\beta \in \Gamma\}$.

If Γ has a zero and hence also S, the zero-symmetric subseminearring $S_2 = (S_2,+,\cdot,\leqq)$ is a p. o. one which satisfies $M_r(S_2) = S_2$ and $W_r(S_2) = \{o\}$, and (3.4) - (3.6) remain valid for S_2.

<u>Proof.</u> The general statements on S, also $M_r(S) = S$ (which yields III_r) and (3.4) are checked straightforward. They imply the corresponding ones on S_2 by Thm. 2.10. For $W_r(S) = \emptyset$, assume $w \in W_r(S)$ by way of contradiction, and $w(\eta) = \zeta$ for some fixed $\eta \in \Gamma$. Define $f,g \in S$ by $f(\zeta) = \alpha < \beta = g(\zeta)$ for any pair $\alpha < \beta$ in Γ, and $f(\xi) = \xi$, $g(\xi) = \xi$ for all $\xi \neq \zeta$. Then $f < g$ and $(fw)(\eta) < (gw)(\eta)$ contradict $fw \geqq gw$ as claimed by $w \in W_r(S)$. In the same way one obtains $W_r(S_2) = \{o\}$, since for $o \neq w \in S_2$ there is some $\eta \in \Gamma$ such that $w(\eta) = \zeta \neq \omega$ holds. To show (3.5) for S, assume at first $h \in M_\ell(S)$. For any pair $\alpha < \beta$ of Γ, define $f,g \in S$ by $f(\xi) = \alpha$ and $g(\xi) = \beta$ for all $\xi \in \Gamma$. Then $f < g$ implies $hf \leqq hg$, hence $h(\alpha) \leqq h(\beta)$ by (3.3) and (3.1). So $M_\ell(S)$ is contained in the set of all isotone transformations of $(\Gamma,+,\leqq)$ defined on the right side of (3.5), and the other inclusion is clear. The proof of (3.5) for S_2 is similar, but it needs some care for the pairs $\omega < \beta$ and $\alpha < \omega$ in Γ, if there are those. Likewise one shows that $W_\ell(S)$ $[W_\ell(S_2)]$ coincides with the set of all antitone transformations contained in S [in S_2], i. e. (3.6).

<u>COROLLARY 3.3.</u> Concerning IV_r, III_ℓ and IV_ℓ for the p. o. seminearrings S and S_2 considered in Thm. 3.2 we state:

(3.7) $N(S) \subseteq W_r(S) \Leftrightarrow N(\Gamma) = \emptyset$

(3.8) $P(S) \subseteq M_\ell(S) \Leftrightarrow |P(\Gamma)| \leqq 1$

(3.9) $N(S) \subseteq W_\ell(S) \Leftrightarrow |N(\Gamma)| \leqq 1$

(3.10) $N(S_2) \subseteq W_r(S_2) \Leftrightarrow N(\Gamma) = \{\omega\}$.

(3.11) $P(S_2) \subseteq M_\ell(S_2) \Leftrightarrow P(\Gamma) = \{\omega\}$ or $P(\Gamma) = \{\omega,\pi\}$, the latter only in the case that there are no different elements comparable in $(\Gamma,+,\leqq)$ except $\omega < \pi$.

(3.12) $N(S_2) \subseteq W_\ell(S_2) \Leftrightarrow N(\Gamma) = \{\omega\}$.

Proof. (3.7) and (3.10) are obvious. For (3.8) we state that there is some $h \in P(S) \setminus M_\ell(S)$ iff a pair $\xi < \eta$ in $P(\Gamma)$ exists. The latter is equivalent with $|P(S)| \geqq 2$, since $\pi \neq \pi'$ in $P(\Gamma)$ implies $\pi \leqq \pi+\pi'$ and $\pi' \leqq \pi+\pi'$, hence $\pi < \pi+\pi'$ or $\pi' < \pi+\pi'$ for $\pi+\pi' \in P(\Gamma)$. In the same way one obtains (3.9). For (3.11), $P(S_2) \subseteq M_\ell(S_2)$ fails to be true iff there is some $h \in S_2$ satisfying $h(\Gamma) \subseteq P(\Gamma)$ and $h(\alpha) > h(\beta)$ for some $\alpha < \beta$ in Γ. The latter is impossible for $P(\Gamma) = \{\omega\}$, and clearly the case if $|P(\Gamma)| \geqq 3$. For $P(\Gamma) = \{\omega,\pi\}$ such a mapping h has to satisfy $h(\alpha) = \pi > \omega = h(\beta)$ for some $\alpha < \beta$, and $h \in S_2$ forces $\alpha \neq \omega$. Conversely, there is such an $h \in S_2$ if $\omega \neq \alpha < \beta$ exists in Γ. Exactly this is excluded iff $\omega < \pi$ is the only non-trivial relation. Following the same pattern, one shows (3.12) where the case $|N(\Gamma)| = 2$ makes no difficulties.

As a consequence of Cor. 3.3, $IV_r \Rightarrow IV_\ell$ holds for S and even $IV_r \Leftrightarrow IV_\ell$ for S_2. So there remain the following possible types $\langle 1111\rangle = \langle 15\rangle$, $\langle 1101\rangle = \langle 13\rangle$, $\langle 1010\rangle = \langle 10\rangle$ and $\langle 1000\rangle = \langle 8\rangle$ for S_2, and the same list including $\langle 1011\rangle = \langle 11\rangle$ and $\langle 1001\rangle = \langle 9\rangle$ for S. All ten cases really occur for additively associative seminearrings, since there are p. o. semigroups $(\Gamma,+,\leqq)$ satisfying the corresponding properties on the right sides of (3.10) - (3.12), as well as those with a zero ω selected by (3.7) - (3.9). E. g., for the f. o. semigroup $\Gamma = (\mathbb{N}_0,+,\leqq)$ of "positive" integers including 0 one obtains an S of type $\langle 9\rangle$, but an S_2 of type $\langle 13\rangle$ whereas for each p. o. group $(\Gamma,+,\leqq)$ the p. o. nearrings S and S_2 have type $\langle 8\rangle$. We close with the following representation theorem:

THEOREM 3.4. For each p. o. seminearring [nearring] $(S,+,\cdot,\leqq)$ which satisfies $M_r(S) = S$ there is an order-isomorphism into a p. o. seminearring [nearring] $(\Gamma^\Gamma,+,\cdot,\leqq)$ of all transformations of a p. o. groupoid [group] $(\Gamma,+,\leqq)$.

Proof. Clearly, each p. o. groupoid [semigroup, group] $(S,+,\leqq)$ is a proper p. o. subgroupoid of a p. o. groupoid [semigroup, group] $(\Gamma,+,\leqq)$. So we can apply the converse part of Thm. 3.1 and it remains to show that the monomorphism $a \to f_a$ of $(S,+,\cdot)$ into $(\Gamma^\Gamma,+,\cdot)$ satisfies $a \leqq b \Leftrightarrow f_a \leqq f_b$ for all $a,b \in S$. If $a \leqq b$ holds, we obtain by (3.2)

$$f_a(\xi) = a\xi \leqq b\xi = f_b(\xi) \quad \text{for all } \xi \in S = M_r(S) \text{ and}$$
$$f_a(\xi) = a \leqq b = f_b(\xi) \quad \text{for all } \xi \in \Gamma \setminus S,$$

thus $f_a \leqq f_b$ by (3.3). The converse follows from the last formula.

4. CONSTANT AND ZERO-SYMMETRIC SUBSEMINEARRINGS

For any seminearring $S = (S,+,\cdot)$ with a zero o we define

(4.1) $\quad S_1 = \{s_1 \in S \mid s_1 o = s_1\}$ and $S_2 = \{s_2 \in S \mid s_2 o = o\}$.

It is easily checked that each of these subsets is either empty or a subseminearring of S. One always has $SS_1 \subseteq S_1$ and

(4.2) $\quad S_2 \neq \emptyset \Leftrightarrow o^2 = o \Leftrightarrow S_1 \cap S_2 \neq \emptyset \Leftrightarrow S_1 \cap S_2 = \{o\}$,

since $s_2 o = o \Rightarrow o = (s_2+o)o = s_2 o + o^2 = o^2$ shows the only non-trivial implication. Recall that each nearring S is additively a semidirect sum of its subnearrings S_1 and S_2, equivalently, each $s \in S$ has a unique decomposition $s = s_1+s_2 \in S_1+S_2$. Seminearrings, however, are far away from this nice situation. Nevertheless, each seminearring S with a zero satisfies

(4.3) $\quad S_1 = \{o\} \Leftrightarrow S_2 = S \Leftrightarrow So = \{o\}$ and

(4.4) $\quad o^2 = o$ and $s_1 s = s_1$ for all $s_1 \in S_1,\ s \in S \Leftrightarrow oS = \{o\}$.

The latter follows from $os = o \Rightarrow s_1 s = (s_1 o)s = s_1(os) = s_1$, and (4.3) from $SS_1 \subseteq S_1$. For a seminearring S with a left absorbing zero, we call S_1 the *constant* and S_2 the *zero-symmetric subseminearring* of S (their occurrence in Thm. 3.1 may justify the first term). So we call a seminearring S_1 a *constant* one iff it has a zero and $s_1 t_1 = s_1$ holds for all $s_1, t_1 \in S_1$. Note that each (p. o.) groupoid $(S,+)$ with a zero provides such a (w. p. o.) constant seminearring.

Now let $(S,+,\cdot,\leqq)$ be a w. p. o. seminearring with a left absorbing zero. Then, by Thm. 2.10, $(S_1,+,\cdot,\leqq_1)$ and $(S_2,+,\cdot,\leqq_2)$ are w. p. o. seminearrings with respect to the induced relations $\leqq_i$ such that $P(S_i) = P(S) \cap S_i$ and $N(S_i) = N(S) \cap S_i$ hold and each of the properties III_r, IV_r, III_ℓ and IV_ℓ transfers from S to S_1 and to S_2. In this context we obviously have:

<u>LEMMA 4.1.</u> Each w. p. o. constant seminearring $(S_1,+,\cdot,\leqq_1)$ satisfies $M_r(S_1) = M_\ell(S_1) = W_\ell(S_1) = S_1$, whereas $W_r(S_1) = \emptyset$ holds iff $(S_1,\leqq_1)$ is not trivially p. o. In the latter case, the type of S_1 is always $\langle 1011\rangle = \langle 11\rangle$ since $o \in N(S_1)$. Hence, if a w. p. o. seminearring $(S,+,\cdot,\leqq)$ with a left absorbing zero satisfies IV_r, its subseminear-

ring S_1 satisfies $S_1 = \{o\}$ (cf. (4.3)) or is trivially p.o.

For the next theorem recall that a f.o. semigroup $(S,+,\leqq)$ with zero o is called *archimedean* iff for each $a \in S$ the subsemigroup of S generated by a, say [a], is not properly bounded from above if $o < a$ and not properly bounded from below if $a < o$ (cf. [1], Chap. XI). Obviously, this concept applies also to a f.o. groupoid $(S,+,\leqq)$ with a zero and its subgroupoids [a] generated by $a \in S$.

THEOREM 4.2. Let $(S,+,\cdot,\leqq)$ be a f.o. seminearring with a left absorbing zero such that $S_1 \neq \{o\} \neq S_2$ holds for the subseminearrings introduced above. Then the induced order on the subset S_1+S_2 of S is uniquely determined by $(S_1,\leqq_1)$ and $(S_2,\leqq_2)$ according to

(4.5) $\quad s_1+s_2 < t_1+t_2 \;\Leftrightarrow\; s_1 <_1 t_1 \;\text{ or }\; s_1 = t_1,\; s_2 <_2 t_2$

for all $s_1+s_2 \neq t_1+t_2$ of S_1+S_2. This yields in particular

(4.6) $\quad n_1 < S_2 < p_1$ for all $n_1 \in N(S_1) \setminus \{o\}$ and $p_1 \in P(S_1) \setminus \{o\}$.

Moreover, apart from the extreme cases that $S_1 \leqq o \leqq S_2$ or $S_2 \leqq o \leqq S_1$ hold, the order of $(S,+,\cdot,\leqq)$ is non-archimedean.

Proof. Multiplying $s_1+s_2 < t_1+t_2$ by $o \in P(S) \subseteq M_r(S)$ from the right yields $s_1 \leqq t_1$. If $s_1 = t_1$ holds, $s_2 \geqq t_2$ implies the contradiction $s_1+s_2 \geqq t_1+t_2$ by II, hence we obtain $s_2 < t_2$. For the converse implication of (4.5), assume at first $s_1 < t_1$. Then we have $s_1+s_2 < t_1+t_2$ for all $s_2, t_2 \in S_2$, since $s_1+s_2 \geqq t_1+t_2$ yields as above $s_1 \geqq t_1$. If $s_1 = t_1$ and $s_2 < t_2$ are assumed, again by II we get $s_1+s_2 \leqq t_1+t_2$. This completes the proof of (4.5) which is only claimed for $s_1+s_2 \neq t_1+t_2$. Now (4.6) follows by $n_1+o < o+s_2 < p_1+o$ for all $s_2 \in S_2$. If the extreme cases are excluded, there are $t_1 \in S_1$ and $a_2 \in S_2$ satisfying $o < t_1$ and $o < a_2$, or $t_1 < o$ and $a_2 < o$. The first implies $S_2 < t_1$ by (4.6) and hence $[a_2] < t_1$, the latter likewise $t_1 < [a_2]$. So $(S,+,\leqq)$ is not archimedean ordered. In the extreme cases, which really occur (cf. Expl. 5.6), the order of S may be archimedean or not.

Note that (4.5) corresponds to the usual lexicographic order of $S_1 \times S_2$ iff $s_1+s_2 = s_1+t_2$ always implies $s_2 = t_2$. Moreover, there are various seminearrings S satisfying $S = S_1+S_2$, in particular those for which $(S,+)$ is associative and cancellative. The latter are semidirect sums of $(S_1,+)$ and $(S_2,+)$, and a construction method starting from given nearrings S_1 and S_2 due to [10] can be generalized correspondingly, also including partial or full order relations. The most simple case (cf. Prop. 5.4) is to define the operations on $S_1 \times S_2$ by

(4.7) $(a_1,a_2)+(b_1,b_2) = (a_1+b_1,a_2+b_2)$, $(a_1,a_2)(b_1,b_2) = (a_1b_1,a_2b_2)$.

As usual for rings and nearrings (cf. [11], 1.55), we then call $S = (S_1 \times S_2,+,\cdot)$ the (external) *direct sum* of the seminearrings S_1 and S_2. If both have a multiplicatively idempotent zero (as assumed in our context), $a_1 \to (a_1,o_2)$ and $a_2 \to (o_1,a_2)$ provide monomorphisms of S_i into S. So one can change from (a_1,a_2) to the notion a_1+a_2 and to the corresponding internal direct sum. We need this for the next theorem and deal with references in its proof.

<u>THEOREM 4.3.</u> a) Let $(S,+,\cdot,\leqq)$ be a f. o. nearring and assume $S_1 \neq \{o\} \neq S_2$ for the constant and zero-symmetric subnearring S_1 and S_2 of S. Then the order of S is non-archimedean and uniquely determined by the induced orders on S_i as the lexicographic order on $S = S_1+S_2$ according to (4.5). In particular, we have (4.6).

b) There are f. o. nearrings S which are the direct sum of their subnearrings $S_1 \neq \{o\} \neq S_2$ such that both have no zero divisors.

c) Let $(S,+,\cdot,\leqq)$ be a f. o. nearring. Then the positive cone $P = P(S)$ determines the order of $(S,+,\cdot)$, but by no means $(S,+,\cdot,\leqq)$ itself. More precisely, a f. o. seminearring $(P,+,\cdot,\leqq)$ may be the positive cone $P(S) = P(T)$ of non-isomorphic f. o. nearrings S and T.

<u>Proof.</u> Part a) is a corollary of Thm. 4.2, since the extreme cases are impossible for a f. o. nearring. It corresponds to [11], 9.141 and generalizes Lemma 1 of [9]. Part b) is shown by Expl. 5.5. It contradicts Thm. 1 of [8] ([11], 9.146), where it is used that $S_1S_2 = S_2S_1 = \{o\}$ holds for a direct sum of nearrings S_i, which is not implied by the above definition. For c) cf. Expl. 5.1 a).

<u>THEOREM 4.4.</u> Let $(S,+,\cdot,\leqq)$ be a f. o. seminearring with a left absorbing zero. Then, for all $s_2 \in S_2$ and $p_1 \in P(S_1)$, the product s_2p_1 is additively idempotent. Hence $s_2p_1 = o$ holds if $(S_1,+)$ has no idempotents except o, in particular if $(S_1,+)$ is left [or right] cancellative.

<u>Proof.</u> At first we assume $s_2 = p_2 \in P(S_2)$. Then III_r, (2.5) and $S_2S_1 \subseteq S_1$ imply $p_2p_1 \in P(S) \cap S_1 = P(S_1)$, hence $p_2p_1 = o$ or $o < p_2p_1$. In the latter case we can apply (4.6) to get $p_2+p_2 < p_2p_1$. By III_r and $p_1p_1 = p_1$, this yields $p_2p_1 + p_2p_1 \leqq p_2p_1$, where equality holds since p_2p_1 is positive. The proof for $s_2 = n_2 \in N(S_2)$ is similar.

Again we obtain for a f. o. nearring S that $s_2p_1 = o$ holds for all $s_2 \in S_2$ and $p_1 \in P(S_1)$. Since a left identity e_ℓ of S is contained in S_2, from $p_1 = e_\ell p_1 = o$ it follows that a f.o. nearring with a left identity is zero-symmetric (cf. [11], 9.136 and 9.137).

5. EXAMPLES

The purpose of this section is to complete statements and proofs in § 2 and § 4, not to give impressive examples. In particular, we show that all types of w.p.o. seminearrings really occur and that our results concerning the properties (2.3) are fairly complete. So we present each type by a w.p.o. seminearring S chosen as strong as we could do, e.g. as a nearring, with a full order, or with a (left) absorbing zero. For the same reason, *all seminearrings considered in this section are also additively associative.*

We clearly use the f.o. nearfields of the types $\langle 1111\rangle = \langle 15\rangle$ and $\langle 1100\rangle = \langle 12\rangle$, and that each constant w.f.o. seminearring has type $\langle 1011\rangle = \langle 11\rangle$. The following examples are, as far as possible, arranged according to § 2. The inserted Prop. 5.4 prepares Expls. 5.5, 5.6 and 5.7, where the first both concern § 4. All proofs have been omitted. They are straightforward even if sometimes a little bit tedious.

Example 5.1. a) Let $(\mathbb{Z},+,\cdot,\leqq)$ be the f.o. ring of integers in the usual meaning. Define a new multiplication by $a\circ b = a\cdot|b|$. Then $S = (\mathbb{Z},+,\circ,\leqq)$ is a f.o. zero-symmetric nearring satisfying $M_r(S) = S$ and $W_r(S) = M_\ell(S) = W_\ell(S) = \{0\}$, hence of type $\langle 1000\rangle = \langle 8\rangle$. Note that the ring $\mathbb{Z}$ as well as S have the same f.o. seminearring $(\mathbb{N}_0,+,\cdot,\leqq)$ as positive cone (cf. Thm. 4.3 c)).

b) The zero-symmetric f.o. subseminearring $U = (-\mathbb{N}_0,+,\circ,\leqq)$ of S satisfies $M_r(U) = M_\ell(U) = U$ and $W_r(U) = W_\ell(U) = \{0\}$. Hence it is of type $\langle 1010\rangle = \langle 10\rangle$.

Example 5.2. Let $(S,+,\leqq)$ be the f.o. group obtained as the direct sum of two copies of $(\mathbb{Z},+,\leqq)$, ordered lexicographically: $(a_1,a_2) \leqq (b_1,b_2) \Leftrightarrow a_1 < b_1$ or $a_1 = b_1,\ a_2 \leqq b_2$. Define a product by

$$(a_1,a_2)(b_1,b_2) = \begin{cases} (0,\ -a_1) & \text{for } b_1 > 0 \\ (0,\ 0) & \text{for } b_1 \leqq 0. \end{cases}$$

Then $(S,+,\cdot,\leqq)$ is a zero-symmetric w.f.o. nearring satisfying IV_r, but not III_r. This implies type $\langle 0100\rangle = \langle 4\rangle$ by Thm. 2.4.

Example 5.3. a) A zero-symmetric w.f.o. semiring $(S,+,\cdot,\leqq)$ which satisfies $M_r(S) = W_r(S) = N(S) = \{o\}$ and $M_\ell(S) = W_\ell(S) = P(S) = S$ is given on $S = \{o,a,b,c,d\}$ by

$o < a < b < c < d$

+	o	a	b	c	d
o	o	a	b	c	d
a	a	a	b	d	d
b	b	b	b	d	d
c	c	d	d	d	d
d	d	d	d	d	d

·	o	a	b	c	d
o	o	o	o	o	o
a	o	a	a	a	a
b	o	b	b	b	b
c	o	a	a	a	a
d	o	a	a	a	a .

Hence S is of type $\langle 0111 \rangle = \langle 7 \rangle$. Recall that there are no w.f.o. nearrings of this type, and that S cannot be p.o. by P(S) from one side. By the way, $U = \langle a,b,d \rangle$ is a w.f.o. subsemiring of S with the element a as left absorbing zero, p.o. by P(U) and of type $\langle 0011 \rangle = \langle 3 \rangle$.

b) Since S is two-sided distributive, we can switch to its multiplicatively left-right dual version, say $(S',+,\cdot,\leq)$, a zero-symmetric f.o. semiring of type $\langle 1101 \rangle = \langle 13 \rangle$. Note that the positive cone of a f.o. nearfield of type $\langle 1100 \rangle$ is a f.o. seminearfield (cf. [13]) which is also of type $\langle 1101 \rangle = \langle 13 \rangle$.

<u>PROPOSITION 5.4.</u> Let $(S_1,+,\cdot,\leq_1)$ be a constant and $(S_2,+,\cdot,\leq_2)$ a zero-symmetric w.p.o. seminearring, both non-trivially p.o., and $S = S_1 \times S_2$ the direct sum of $(S_1,+)$ and $(S_2,+)$. We assume both to be cancellative semigroups. Defining also the multiplication as in (4.7), $(S,+,\cdot)$ is the direct sum of $(S_1,+,\cdot)$ and $(S_2,+,\cdot)$.

a) $(S,+,\cdot,\leq)$ is a w.p.o. seminearring with respect to the lexicographic order $(a_1,a_2) \leq (b_1,b_2)$ iff $a_1 <_1 b_1$ or $a_1 = b_1$, $a_2 \leq_2 b_2$, P(S) consists of all (p_1,s_2) such that $o_1 < p_1$ and all (o_1,p_2) such that $o_2 \leq p_2$, and N(S) is described similarly. If $P(S_1) = \{o_1\}$, III_r for S and III_r for S_2 are equivalent. If $P(S_1) \neq \{o_1\}$, III_r for S holds iff $M_r(S_2) = S_2$. IV_r for S fails to be true. III_ℓ for S holds iff $S_2S_2 = \{o_2\}$ or both, $P(S_1) = \{o_1\}$ and $P(S_2)S_2 = \{o_2\}$, are valid, and IV_ℓ iff $S_2S_2 = \{o_2\}$ or $N(S_1) = \{o_1\}$ and $N(S_2)S_2 = \{o_2\}$.

b) $(S,+,\cdot,\leq)$ is also a w.p.o. seminearring with respect to the antilexicographic order $(a_1,a_2) \leq (b_1,b_2)$ iff $a_2 <_2 b_2$ or $a_2 = b_2$, $a_1 \leq_1 b_1$. For each $(S,+,\cdot,\leq)$ obtained in this way, III_r and IV_r always fail to be true, whereas III_ℓ holds for S iff it holds for S_2, and likewise with IV_ℓ.

Together with our examples for S_1 and S_2, Prop. 5.4 provides various examples, in particular of w.f.o. seminearrings S of the types $\langle 11 \rangle$ to $\langle 8 \rangle$ and $\langle 3 \rangle$ to $\langle 0 \rangle$. Moreover, one can use other multiplications for $S = S_1 \times S_2$. A rather simple one is given by

$$(a_1,a_2)(b_1,b_2) = \begin{cases} (a_1b_1,\ a_2b_2) = (a_1,\ a_2b_2) & \text{for } b_1 = o_1 \\ (a_1b_1,\ o_2) = (a_1,\ o_2) & \text{for } b_1 \neq o_1. \end{cases}$$

But we use here only Prop. 5.4 to give some needed examples:

Example 5.5. For any f. o. constant nearring S_1 and any f. o. zero-symmetric nearring S_2 satisfying $M_r(S_2) = S_2$ and $S_2S_2 \neq \{o_2\}$, the direct sum S according to Prop. 5.4 a) is a f. o. nearring of type $\langle 1000\rangle = \langle 8\rangle$. In particular, S_1 has no zero divisors and S_2 may be chosen so (cf. Expl. 5.1). This proves Thm. 4.3 b).

Example 5.6. A f. o. seminearring S such that $S_1 \leqq o \leqq S_2$ hold (cf. Thm. 4.2) is obtained by Prop. 5.4 a) e. g. for $S_1 = (-\mathbb{N}_0,+,\leqq)$ with $s_1t_1 = s_1$ and $S_2 = (\mathbb{N}_0,+,\cdot,\leqq)$ as usual. For $S_2 \leqq o \leqq S_1$ we use similarly e. g. $S_1 = (\mathbb{N}_0,+,\leqq)$ with $s_1t_1 = s_1$ and $S_2 = (-\mathbb{N}_0,+,\circ,\leqq)$ with $s_2 \circ t_2 = -|s_2t_2|$. In both cases, S has again type $\langle 8\rangle$.

Example 5.7. From Prop. 5.4 b) we obtain w. f. o. nearrings S of the types $\langle 0011\rangle = \langle 3\rangle$ and $\langle 0000\rangle = \langle 0\rangle$ using any constant f. o. nearring S_1 and a f. o. zero-symmetric nearring S_2 either of type $\langle 1111\rangle = \langle 15\rangle$ or $\langle 1100\rangle = \langle 12\rangle$. Likewise, w. f. o. seminearrings S of the types $\langle 0010\rangle = \langle 2\rangle$ and $\langle 0001\rangle = \langle 1\rangle$ (there are no nearrings of this kind) are obtained using zero-symmetric seminearrings S_2 with types $\langle ..10\rangle$ or $\langle ..01\rangle$, e. g. those of Expls. 5.1 b) and 5.8.

Example 5.8. We define two seminearrings S and T on the same set {n,o,a,p} with the full order $n < o < a < p$ by the tables

+	n	o	a	p
n	n	n	a	p
o	n	o	a	p
a	a	a	a	p
p	p	p	p	p

·	n	o	a	p
n	o	o	o	o
o	o	o	o	o
a	o	o	o	o
p	a	o	o	o

+	n	o	a	p
n	n	n	n	n
o	n	o	a	p
a	n	a	a	p
p	n	p	p	p

·	n	o	a	p
n	o	o	o	a
o	o	o	o	o
a	o	o	o	o
p	o	o	o	o

clearly both zero-symmetric. The first one, $(S,+,\cdot,\leqq)$ is a f. o. seminearring such that $M_r(S) = W_\ell(S) = S$, but $W_r(S) = S\setminus\{n\}$ and $M_\ell(S) = S\setminus\{p\}$ hold, thus of type $\langle 1001\rangle = \langle 9\rangle$. Similarly, the other one is a w. f. o. seminearring $(T,+,\cdot,\leqq)$ of type $\langle 0110\rangle = \langle 6\rangle$.

Example 5.9. Each p. o. semigroup $(S,+,\leqq)$ provides by the trivial multiplication $st = s$ a p. o. seminearring $(S,+,\cdot,\leqq)$ such that $M_r(S) = M_\ell(S) = W_\ell(S) = S$ and $W_r(S) = \emptyset$ hold. Thus S has the type $\langle 1011\rangle = \langle 11\rangle$ iff $N(S) \neq \emptyset$ (cf. Lemma 4.1). Clearly, one obtains f. o. nearrings S in this way. On the other hand, $(S,+,\cdot,\leqq)$ is a p. o semiring if (S,+) is idempotent. Since there are various even f. o. semigroups $(S,+,\leqq)$ of this kind satisfying $N(S) \neq \emptyset$, e. g. with a zero, we get f. o. semirings S with a left absorbing zero of type ⟨11⟩. Now we can turn to the multiplicatively left-right dual f. o. semiring S' which has the type ⟨1110⟩ = ⟨14⟩.

Example 5.10. For the f. o. group $(S,+,\leqq)$ of Expl. 5.2 define a product by $(a_1,a_2)(b_1,b_2) = (a_1b_2 + a_2b_1, a_2b_2)$. Then $(S,+,\cdot,\leqq)$ is a commutative w. f. o. ring of type ⟨0000⟩, whereas its positive cone $U = P(S)$ is a zero-symmetric w. f. o. semiring of type ⟨0101⟩ = ⟨5⟩.

The latter was the last type needed to prove Remark 2.2.

NOTES AND REFERENCES

Footnote 1)

The importance of these concepts is that one can introduce partial orders on groupoids $(S,+)$ by (2.4) using suitable subsets $X \subseteq S$. For semigroups, this was extensively investigated in [15], and we note here only the following special result. Let $(S,+)$ be a semigroup with a zero o and $X \subseteq S$. Then (2.4) defines a relation on S such that $(S,+,\leqq)$ is a p. o. semigroup iff $x+y = o$ for $x,y \in X$ implies $x = y = o$ and $c+X \subseteq X+c$ holds for all $c \in S$. In this case, X coincides with the set of positive elements of $(S,+,\leqq)$.

Bibliography

[1] Fuchs, L., Partially ordered algebraic systems, Pergamon Press, Oxford 1963, Vandenhoeck & Ruprecht, Göttingen 1966.

[2] Hanumanthachari, J., K. Venu Raju and H. J. Weinert, Some results on partially ordered semirings and semigroups, Proc. First Intern. Symposium on Ordered Algebraic Structures, Marseilles, June 1984, Heldermann-Verlag, Berlin 1986.

[3] Hogewijs, H., Semi-Nearrings-Embedding, Med. Konink. Acad. Wetensch. Lett. Schone Kunst. België Kl. Wetensch. 32 (1970) 3 - 11.

[4] Karzel, H., Zusammenhänge zwischen Fastbereichen, scharf zweifach transitiven Permutationsgruppen und 2-Strukturen mit Rechtecksaxiom, Abh. Math. Sem. Univ. Hamburg 32 (1968), 191 - 206.

[5] Kerby, W., Angeordnete Fastkörper, Abh. Math. Sem. Univ. Hamburg 32 (1968), 135 - 146.

[6] Kerby. W. and H. Wefelscheid, Bemerkungen über Fastbereiche und scharf zweifach transitive Gruppen, Abh. Math. Sem. Univ. Hamburg 37 (1972), 20 - 29.

[7] Pilz, G., Geordnete Fastringe, Abh. Math. Sem. Univ. Hamburg 35 (1970), 83 - 88.

[8] Pilz, G., Direct Sums of Ordered Near-Rings, J. Algebra 18 (1971), 340 - 342.

[9] Pilz, G., Zur Charakterisierung der Ordnungen in Fastringen, Monatsh. Math. 76 (1972), 250 - 253.

[10] Pilz, G., A construction method for near-rings, Acta Math. Acad. Sci. Hungar. 24 (1973), 97 - 105.

[11] Pilz, G., Near-Rings, North-Holland, Amsterdam 1977.

[12] Weinert, H. J., Related representation theorems for rings, semirings, near-rings and semi-near-rings by partial transformations and partial endomorphisms, Proc. Edinburgh Math. Soc. 20 (1976 - 77), 307 - 315.

[13] Weinert, H. J., Seminearrings, seminearfields and their semigroup-theoretical background, Semigroup Form 24 (1982), 231 - 254.

[14] Weinert, H. J., Extensions of seminearrings by semigroups of right quotients, Lecture Notes in Mathematics 998 (1983), Proceedings Oberwolfach (1981), 412 - 486.

[15] Weinert, H. J., Partially ordered semirings and semigroups, Proc. First Intern. Symposium on Ordered Algebraic Structures, Marseilles, June 1984, Heldermann-Verlag, Berlin 1986.

Institut für Mathematik
Technische Universität Clausthal
Erzstraße 1
D-3392 Clausthal-Zellerfeld

Near-rings and Near-fields, G. Betsch (editor)
© Elsevier Science Publishers B.V. (North-Holland), 1987

ON SUBDIRECTLY IRREDUCIBLE NEAR-RINGS WHICH ARE FIELDS

Richard WIEGANDT

Mathematical Institute of the Hungarian Academy of Sciences
Budapest, Pf. 127, H-1364, Hungary

Herrn H. J. Weinert zum 60. Geburtstag am 26.01.1987 gewidmet

In this note we shall prove under certain conditions that if a left invariant subset of a subdirectly irreducible near-ring satisfies a permutation identity, then the near-ring is a field.

In the sequel a *near-ring* will always mean a right near-ring, that is, it satisfies the right distributive law

$$(x + y)z = xz + yz .$$

Let us recall that an ideal I of a near-ring A is an additive normal subgroup of A such that

$$I\,A \subseteq I$$

and

$$x(y + i) - xy \in I$$

holds for all $x, y \in A$ and $i \in I$. By Pilz [3] 1.34 (a) any ideal I of a 0-symmetric near-ring A satisfies also $A\,I \subseteq I$. A near-ring A is said to be *subdirectly irreducible*, if A contains a unique minimal ideal $H \neq 0$ which is referred to the *heart* of A. We shall call a subset L of a near-ring A a *left invariant subset*, if $A\,L \subseteq L$ holds. For a subset S of a near-ring A we shall denote by S^k the set

$$S^k = \{s = t_1 \dots t_k : t_1, \dots, t_k \in S\}$$

for any integer $k \geq 2$.

In the following k will always stand for a fixed integer ≥ 2, and σ will stand for a fixed permutation of k elements such that $\sigma(1) \neq 1$.

The next Lemma is purely semigroup theoretical, and it is a modified special case of [2] Lemma 2. For the sake of completeness we give it with proof.

LEMMA. *If a left invariant subset* L *of a near-ring* A *satisfies the permutation identity*

(P) $$t_1 \dots t_k = t_{\sigma(1)} \dots t_{\sigma(k)} \quad t_i \in L ,$$

then $$xyt_1 \dots t_k = yxt_1 \dots t_k$$

holds for all $x, y \in A$ *and* $t_1,\dots,t_k \in L$.

Proof. Let $r = yxt_1 \dots t_k$ and $\sigma(i) = 1$, $\sigma(1) = j$. Now we apply the identity (P) to the elements $xt_1, t_2, \dots, t_k$ and then multiply by y . Thus we get

$$r = yt_{\sigma(1)} \dots t_{\sigma(i-1)}xt_{\sigma(i)}t_{\sigma(i+1)} \dots t_{\sigma(k)} .$$

Now we use the identity (P) in reverse on the elements

$$yt_{\sigma(1)},t_{\sigma(2)},\dots,t_{\sigma(i-1)},xt_{\sigma(i)},t_{\sigma(i+1)},\dots,t_{\sigma(k)}$$

to obtain

$$r = xt_1 \dots t_{j-1}\, y\, t_j\, t_{j+1} \dots t_k .$$

At this point we use again (P) on the elements $t_1,\dots,t_{j-1}, yt_j, t_{j+1},\dots,t_k$ to obtain

$$r = x\, y\, t_{\sigma(1)}\, t_{\sigma(2)} \dots t_{\sigma(k)} .$$

Finally, we use (P) in reverse on $t_{\sigma(1)},\dots,t_{\sigma(k)}$ to obtain

$$r = x\, y\, t_1 \dots t_k ,$$

and the desired identity has been established.

Another proof of the Lemma has been given by H. J. Weinert (private communication). The structure of near-rings satisfying (P) has been investigated by R. Scapellato [4].

THEOREM. *Let* A *be a subdirectly irreducible near-ring with heart* H , *and* L *be a left invariant subset of* A *such that* $H L^k \neq 0$. *If* L *satisfies the permutation identity* (P), *then* A *is a commutative ring. If, in addition,* $H^2 \neq 0$, *then* A *is a field. A* 0*-symmetric, subdirectly irreducible near-ring with commutative idempotent heart is a field.*

Proof. By the Lemma we have

$$x\, y\, t_1 \dots t_k = y\, x\, t_1 \dots t_k ,$$

that is,

$$(xy - yx)\, t_1 \dots t_k = 0$$

for all $x, y \in A$ and $t_1,\dots,t_k \in L$. This means that the element $xy - yx$ is contained in the annihilator

$$(0 : L^k) = \{a \in A : aL^k\} = 0$$

of L^k . One can easily check that $(0 : L^k)$ is an ideal in A . Since A is subdirectly irreducible, either $(0 : L^k) = 0$ or $H \subseteq (0 : L^k)$. In the latter case we would get $H L^k = 0$, contradicting the assumption. Thus $(0 : L^k) = 0$ holds, implying $xy = yx$ for all $x, y \in A$. This means that in A the multiplication is commutative, and hence A is a distributive near-ring which is also called a ring with not necessarily commutative addition. As one can prove (see for instance [1], [5] or [6]), each additive commutator $-x - y + x + y$, $(x, y \in A)$, is contained in the annihilator $(0 : A)$ of A. Since $(0 : A)$ is an ideal in A and A is subdirectly irreducible, it follows $(0 : A) = 0$, otherwise namely we would get

$$0 \neq H L^k \subseteq (0 : A) A = 0 .$$

Thus we have

$$x + y = y + x$$

for all $x, y \in A$ and therefore A is a commutative ring.

The heart of a subdirectly irreducible ring is either simple or a zero-ring. If $H^2 \neq 0$, then H is a simple commutative ring, that is, a field. Hence the ideal H of A has a unity element and therefore H is a direct summand of A. Since A is subdirectly irreducible, it follows $H = A$.

The last assertion is straightforward.

For rings we can prove a stronger version of the Theorem.

PROPOSITION. *Let* A *be a subdirectly irreducible ring with idempotent heart* H. *If* $L \neq 0$ *is a left invariant subset of* A, *then* $H L^n \neq 0$ *holds for every* $n = 1,2,\dots$.

We prove the statement by induction. First, assume that $H L = 0$. Then the right annihilator $(0 : H)_r$ of H in A contains L, and hence $(0 : H)_r \neq 0$. Further, $(0 : H)_r$ is an ideal of A, so we conclude $H \subseteq (0 : H)_r$ which yields $0 \neq H = H^2 \subseteq H(0 : H)_r = 0$, a contradiction. Thus the assertion is valid for $n = 1$.

Next, assume that $H L^{n-1} = 0$ for $n \geq 2$. Suppose that $H L^n = 0$. Then by $(H L^{n-1})L = H L^n = 0$ the left annihilator $(0 : L)$ of L in A contains $(H L^{n-1})$. Hence $(0 : L)$ is a nonzero ideal of A and so we have $H \subseteq (0 : L)$. This yields $H L = 0$, contradicting $H L \neq 0$. Thus $H L^n \neq 0$ has been proved for every $n = 1,2,\dots$.

The Theorem and the Proposition yield immediately the following

COROLLARY. *Let* A *be a subdirectly irreducible ring with idempotent heart. If a left invariant subset* $L \neq 0$ *of* A *satisfies the permutation identity* (P), *then* A *is a field.*

Remark. The Proposition and hence the Corollary are valid also for distributive near-rings, that is, for rings with not necessarily commutative addition.

REFERENCES

[1] Furtwängler, Ph. and Taussky, Olga, Über Schiefringe, *Sitzungsber. Akad. der Wiss. Wien, Mathem.-naturw. Klasse*, 145 (1936), 525.

[2] Parmenter, M. M., Stewart, P. N. and Wiegandt, R., On the Groenewald - Heyman strongly prime radical, *Quaest. Math.* 7 (1984), 225-240.
[3] Pilz, G., *Near-rings*, North-Holland, 1977.
[4] Scapellato, R., Sui quasi-anelli verificanti identita semigruppale C-mobili, *Bollettino U.M.I.* (6) 4-B (1985), 789-799.
[5] Taussky, Olga, Rings with non-commutative addition, *Bull. Calcutta Math. Soc.*, 28 (1936), 245-246.
[6] Weinert, H. J., Ringe mit nichtkommutativer Addition I, *Jber. Deutsch. Math.-Verein.* 77 H. 1 (1975), 10-27.

322